Lecture Notes in Electrical Engineering

Volume 327

About this Series

"Lecture Notes in Electrical Engineering (LNEE)" is a book series which reports the latest research and developments in Electrical Engineering, namely:

- Communication, Networks, and Information Theory
- Computer Engineering
- Signal, Image, Speech and Information Processing
- Circuits and Systems
- Bioengineering

LNEE publishes authored monographs and contributed volumes which present cutting edge research information as well as new perspectives on classical fields, while maintaining Springer's high standards of academic excellence. Also considered for publication are lecture materials, proceedings, and other related materials of exceptionally high quality and interest. The subject matter should be original and timely, reporting the latest research and developments in all areas of electrical engineering.

The audience for the books in LNEE consists of advanced level students, researchers, and industry professionals working at the forefront of their fields. Much like Springer's other Lecture Notes series, LNEE will be distributed through Springer's print and electronic publishing channels.

More information about this series at http://www.springer.com/series/7818

Vivek Vijay · Sandeep Kumar Yadav
Bibhas Adhikari · Harinipriya Seshadri
Deepak Kumar Fulwani
Editors

Systems Thinking Approach for Social Problems

Proceedings of 37th National Systems Conference, December 2013

Springer

Editors
Vivek Vijay
Systems Science
Indian Institute of Technology Jodhpur
Jodhpur, Rajasthan
India

Sandeep Kumar Yadav
Information and Communication Technology
Indian Institute of Technology Jodhpur
Jodhpur, Rajasthan
India

Bibhas Adhikari
Systems Science
Indian Institute of Technology Jodhpur
Jodhpur, Rajasthan
India

Harinipriya Seshadri
Systems Science
Indian Institute of Technology Jodhpur
Jodhpur, Rajasthan
India

Deepak Kumar Fulwani
Systems Science
Indian Institute of Technology Jodhpur
Jodhpur, Rajasthan
India

ISSN 1876-1100 ISSN 1876-1119 (electronic)
Lecture Notes in Electrical Engineering
ISBN 978-81-322-3516-3 ISBN 978-81-322-2141-8 (eBook)
DOI 10.1007/978-81-322-2141-8

Springer New Delhi Heidelberg New York Dordrecht London

Softcover reprint of the hardcover 1st edition 2015

Printed on acid-free paper

Springer (India) Pvt. Ltd. is part of Springer Science+Business Media (www.springer.com)

Preface

About National Systems Conference

The National Systems Conference (NSC) is an annual event of the Systems Society of India (SSI), primarily oriented to strengthen the systems movement and its applications for the welfare of humanity. The first NSC held in 1973 served to create an awareness of Systems Engineering Methodologies amongst planners, designers, builders, and operation managers. The NSC is hosted to facilitate an enriching engagement between academicians and industrialists. By providing a creative platform for people from diverse backgrounds, the NSC seeks to address a spectrum of issues ranging from the scientific to the social needs of contemporary society.

The Theme of NSC 2013

Ever since it was conceived in 1956 by Prof. Jay Forrester (MIT), the use of systems thinking in systems dynamics has been playing a significant role in nearly every aspect of innovation and technology. The key features of systems thinking are: a paradigm shift from the part to the whole and the ability to use concepts alternatively at the systems level. At the same time, different system levels represent levels of differing complexities. At each level the observed phenomena exhibit properties that do not exist at lower levels. Thus the system properties of a particular level are called emergent properties: they emerge at that particular level. This viewpoint envisages looking at a system: be it a mechanical one like an automobile, or a living organism such as humans, not as an individual entity but as interconnected systems that work in tandem to provide a bigger picture, sometimes called the ecological viewpoint.

The revolution in communication technologies and its diverse applications in the fields of health and education in particular, have made social innovations a ubiquitous part of our everyday lives. Employing the perspectives of a systems approach to social thinking facilitates in understanding the notion that successful social innovations incorporate integrated design systems. "Systems thinking" not only facilitates the comprehension of transactional and relational fields but throws open holistic new vistas of exciting possibilities and opportunities waiting to be explored. Therefore, Systems thinking in social structures aids in community and resource building endeavors that can augment our quality of life.

The NSC 2013 hosted by IIT Jodhpur focuses on these unique approaches that can pave the way for a meaningful and sustainable future.

About the CoE in Systems Science, IIT Jodhpur

IIT Jodhpur set up the CoE in Systems Science in 2011 to promote a systems thinking approach for solving real-world problems. The CoE SS is perhaps the first to be conceived in India and has created a novel platform by evolving an interdisciplinary approach to both research and learning. A strong emphasis is laid on a holistic approach to studying and understanding systems as a whole, as opposed to the sum of its parts. The CoE facilitates the development and use of mathematical techniques in various fields including systems dynamics, multidisciplinary and integrated systems design. The CoE SS offers unique academic programs at the UG/PG and at the doctoral levels, thereby making it one of its kind in the world.

The aim of NSC 2013 was to bring together academicians and industry practitioners onto a single platform to share their expertise, experience, and diverse perspectives on Systems Thinking. We believe that this approach is the need of the hour and can help us make a significant difference to the society in which we live.

Robotics and Automation for Society

Robotics and Automation are widely used in various industrial applications such as defense research, manufacturing, production of consumer goods, etc. The future development in robotics and automation systems faces severe engineering and scientific challenges. Engineering challenges involve how to successfully integrate complex and newly developed systems due to current technical boundaries. Scientific challenges require future breakthroughs in computing efficiencies, sensor technology, etc. Original research papers are sought to investigate systems thinking approach for developing models and methods in the field of Robotics and Automation.

Systems Approach to Computational Finance

The principles of finance combined with mathematical frameworks form useful financial instruments, strategies, and models that are tested and implemented using advanced numerical and quantitative techniques. Application of computer technology has become a key throughout the process. These financial instruments, strategies, and models form an integral part of the overall financial activities. The systems approach, which uses economics (for understanding the behavior of the agents), mathematical (for formulating structure) and statistical modeling (for estimation), computer technology (for large-scale computation), and several other disciplines is therefore useful to manage these activities. Original research papers are sought on systems approach to solve challenges in computational finance.

Complex Network Modeling for Interconnected, Self-organized and Self-adaptive Systems

The literature proposes the models of self-adaptive (SA) and self-organized (SO) systems to understand and deal with fast growing and increasingly complex real systems/real networks. Engineering design of such systems is a challenging task and the success story in this field is far from satisfactory. New theories are required to accommodate, in a systematic engineering manner, traditional top-down and bottom-up approaches. The focus areas of this theme include robustness of SA/SO systems, control of emergent properties in SA/SO systems, and mathematical modeling of SA/SO systems.

Systems Approach to Healthcare Systems

Health care at its core is widely recognized to be a public good. Health care covers not merely medical care but also all aspects of pro- and preventive care. How can we develop new healthcare systems that are ideally suited for our needs? And what is "ideal?" There are four criteria that could be suggested to measure whether a healthcare system is ideal.

- First, universal access, access to an adequate level, and access without excessive burden.
- Second, fair distribution of financial costs for access and fair distribution of burden in rationing care and capacity and a constant search for improvement toward a more just system.
- Third, training providers for competence, empathy, and accountability, pursuit of quality care and cost-effective use of the results of relevant research.

- Last, special attention to vulnerable groups such as children, women, disabled, and the aged.

  Once we have taken care of the measures on which healthcare systems should be assessed, key questions arise in the design and deployment of healthcare systems that fit the measures described above:
- How can the true system level complexity of healthcare processes be modeled and measured?
- How does this system level process model and complexity measures work on a real-world healthcare process design and implementation effort?
- How does process complexity impact change and adoption in health care?

Research articles that either propose a healthcare system that is suitable on the measures, or a mathematical/system-level study of the healthcare delivery process, both at policy-level and at deployment level to minimize the delays and leaks, are invited.

Social Computing

Social interactions have always been an important part of the human experience. Social interaction research has shown results ranging from influences on our behavior from social networks, to our understanding of social belonging on health, as well as how conflicts and coordination play out in socially developed systems and knowledge platforms like Wikipedia and Quora. Additionally, in social computing, information technology facilitates organized human endeavor, e.g., crowd sourcing in which the collective action is used to tackle problems that have been computationally intractable. This theme invites research articles, discussion sessions, and tutorials that propose the use of social computing in different domains.

Additionally, in social computing, information technology facilitates organized human endeavor, e.g., crowd sourcing in which the collective action is used to tackle problems that have been computationally intractable.

This theme invites research proposals, articles, discussion sessions, and tutorials that propose the use of social computing in different domains.

Cybersecurity and Information

Cybersecurity is one of the prime challenges which include involvement of national, international, public, and private organization dimensions. Information systems and communication networks, an essential factor in economic and social development, require reliability, privacy, and security in cyberspace. Such systems are always vulnerable to sophisticated cyber crimes and thefts and thus always pose a challenge to individuals, businesses, organizations, and governments as well.

As the cyber security threats are on increase, the possible loss could well be government, business, and military secrets or individual privacy. Original contributions are sought in all aspects of cybersecurity.

Soft Computing

The objective of this theme is to provide a forum to discuss and disseminate recent and significant research efforts on the Soft Computing research area dealing with challenging applications, with the aim to facilitate cross-fertilization between methodological and applied research, and hence to give new highlights on novel applications and methods.

Systems Thinking in Social Policies, Socioeconomic Systems

The objective of this theme is to provide a forum to discuss and disseminate recent and significant research efforts on the Soft Computing research area dealing with challenging applications, with the aim to facilitate cross-fertilization between methodological and applied research, and hence to give new highlights on novel applications and methods.

Systems Thinking in Socioeconomic Processes is an upcoming area of research driven primarily by the fact that the economic and social decisions of an individual are affected by the people with whom they are connected. The root of the social influences is so deep that it pervades the decision of the big companies regarding their choice of business and conduct, the political and the policy-making decision of the parties, extent of clientelism, and others. In recent decades, the explosion and rapid growth of the Internet and mobile communication have facilitated the spread of news and information across the globe. While this has magnified the extent of connectedness of people and enhanced economic transaction trade and commerce, on the other hand it has also magnified the risk of rapid spread in epidemics and economic crisis. All these are instances of networks, incentives, and aggregate behavior of groups of people based on links that connect individuals and the ways in which each of their decisions have consequences on the outcomes of everyone else. The need of the hour is therefore to understand the behavior of the individual in Systems approach that encompasses economics, sociology, anthropology, game theory, computer science, and mathematics for a better understanding of structure and issues related to individual and group behavior and its consequences on the outcome on socioeconomic systems.

Systems Approach to Clean and Green Technologies

The projected population of the world in the year 2050 is above 9 billion and the demand for natural resources is growing enormously. The need for sustainable, clean, and green technologies to meet the requirements of the world cannot be overemphasized. In several parts of the world, such an agenda is tried to be implemented but a lot more needs to be done. Original papers pertaining to a systems thinking approach to green solutions, cleaning and waste management technologies, recycling the resources, efficient and renewable energy sources, water conservation and rainwater harvesting, etc., are invited.

Systems Dynamics

In this theme, original contributions are sought on systems dynamics, the idea founded by Prof. Jay Forrester, MIT. The methodology of systems dynamics helps us in solving systems problems having hard (physical) and soft components encompassing the integrated states space of both physical and nonphysical elements in total information space of the system, instead of the subsystems levels which fail to incorporate the internal structural dynamics of feedback loops of the whole system. Application of systems dynamics in the following areas are of concern.

- Telecommunication Systems
- Sustainable Whole Systems Design and Control
- Systems Solutions to Agriculture

The 37th National System Conference will feature contributed and invited papers, as well as tutorial sessions.

Contents

About the Editors

Dr. Bibhas Adhikari is an Assistant Professor in Centre of Systems Science at Indian Institute of Technology, Jodhpur since 2009. He obtained his doctorate from Indian Institute of Technology, Guwahati in 2009. His areas of interest are Applied Linear Algebra, Optimization Techniques, and Network Systems. Dr. Adhikari received the ISBRI fellowship and did his research work at SAM, ETH Zurich, Switzerland in 2008. He has articles in international journals and is currently handling two funded research projects by BARC Mumbai and CSIR Delhi.

Dr. Deepak Kumar Fulwani is an Assistant Professor in Centre of Information and Communication Technology at Indian Institute of Technology, Jodhpur. He obtained his doctorate from Indian Institute of Technology, Bombay in 2009. His areas of interest are Robust Control techniques, Fault diagnosis of networked control system, Control issues in power converters, and Robust control techniques with saturated actuators. Dr. Fulwani was awarded for Excellence in Ph.D. thesis work by IIT Bombay.

Dr. Harinipriya Seshadri is working as principal scientist at NFTDC, Hyderabad since June 2014. She obtained her doctorate from Indian Institute of Technology Madras in 2003. Her areas of interest are lithium ion batteries, fuel cells, electrodeposition, thermal storage systems, Monte Carlo simulations, materials synthesis, and characterization. Dr. Seshadri received the NRO fellowship for Postdoctoral Research, Tennessee Technological University, USA; European Union grant for Postdoctoral Research, University of Ulm, Germany; and The Indian National Science Academy (INSA), Young Scientist Award, 2004. She has published more than 40 papers in international journals and Conference Proceedings.

Dr. Vivek Vijay is an Assistant Professor in Centre of Systems Science at Indian Institute of Technology Jodhpur since 2010. He obtained his doctorate from Indian Institute of Technology, Bombay in 2007. His areas of specialization include

Analysis of Categorical Data and Forecasting of Financial Data. Dr. Vijay has an outstanding academic career and received the Gold Medal for the first rank at the post-graduation level. He has also published several articles in peer-reviewed international journals.

Dr. Sandeep Kumar Yadav is an Assistant Professor in Centre of Information and Communication Technology at Indian Institute of Technology, Jodhpur since 2010. He obtained his doctorate from Indian Institute of Technology, Kanpur in 2009. His areas of interest are Signal Processing, Condition Monitoring, and Fault Diagnosis Using Intelligent Techniques. Dr. Yadav received the Young Scientist Award from Systems Society of India. He has authored a book chapter on Intelligent Autonomous Systems published by Springer and also articles in international journals.

Chapter 1
Green Supply Chain: An ISM-Based Roadmap to Boundaries of Environmental Sustainability

Navin K. Dev and Ravi Shankar

Abstract Managing the green (environmentally sustainable) supply chain is an important step toward broader adoption and development issue for industry. This paper presents an approach to effectively adapt sustainable practices in a supply chain by understanding the dynamics between various enablers. Using interpretive structural modeling, the research presents a hierarchy-based model and the mutual relationships among the enablers of sustainability in a supply chain. The research shows that there exists a group of environmental sustainability boundary enablers having a high driving power and low dependence requiring maximum attention and of strategic importance while another group consists of those enablers which have high dependence and are the resultant actions. This classification provides a useful tool to supply chain managers to differentiate between independent and dependent variables and their mutual relationships which would help them to delineate those key enablers that are imperative for effective implementation of sustainability concepts in the design of a supply chain.

Keywords Green supply chain management · Interpretive structural modeling · Environmental sustainability · Reverse logistics · Closed-loop supply chain

1.1 Introduction

In a world of finite resources and disposal capacities, recovery of used products and materials has become an endemic concern in industrialized countries. As a result, many countries have started to emphasize the prevention and control of pollution

N.K. Dev (✉)
Department of Mechanical Engineering, Technical College, Dayalbagh Educational Institute, Dayalbagh, Agra, Uttar Pradesh, India
e-mail: navinkumardev@yahoo.com

R. Shankar
Department of Management Studies, Indian Institute of Technology Delhi, New Delhi, India
e-mail: r.s.research@gmail.com

V. Vijay et al. (eds.), *Systems Thinking Approach for Social Problems*,
Lecture Notes in Electrical Engineering 327,
DOI 10.1007/978-81-322-2141-8_1

caused by discarded hazardous wastes. Regulations and laws are established to restrict and regulate the procedure for the return and recycle of these hazardous wastes. The European Union has established stricter codes for the handling of products containing hazardous substances, such as Directive 2002/96/EC related to 'waste electrical and electronic equipment' (WEEE), Directive 2002/525/EC related to End of Life, and the 'Restriction of the use of certain Hazardous Substances in electrical and electronic equipment' (RoHS) regulations. To deal with this serious problem, the Indian government has also taken many steps and has come up with innovative measures. The ministry of Environment and Forests is the nodal agency for policy, planning, and coordinating the environment programmes, including electronic waste. The management of e-waste was covered under 'management, handling and trans-boundary movement' rules 2007, part of EPA 1986 and Environment and Forests Hazardous Wastes management rules 2008. In May 2012, new rules are issued by the union ministry of environment and forests (MoEF) to address the safe and environment friendly handling, transporting, storing, and recycling of e-waste. The concept of material cycles is gradually replacing a 'one-way' perception of economy. Increasingly, customers expect companies to minimize the environmental impact of their products and processes.

Take-back and recovery obligations have been enacted or are underway for a number of product categories including electronic equipment in the European Union and in Japan, cars in the European Union and in Taiwan, and packaging material in Germany. In this vein, the past two decade has witnessed an immense growth of product recovery activities. Some of the enterprises that are putting substantial efforts into remanufacturing used equipments include copy machine manufacturers Xerox and Canon. Xerox did resource recycling for collected products at 99.9 % in 2011. They also reduced new resources use by 2,272 tons. The main drivers of this achievement are the increase in both products containing reused parts and the amount of resources recycled from consumable cartridges [1]. Canon has been operating two remanufacturing factories for used copy machines in Virginia (USA) and in the UK since 1993 and is currently exploring comprehensive recycling systems for all copier parts. Toner cartridges have been collected for reuse since 1990 and have recycled around 287,000 tones of cartridges by the end of 2011, thereby saved around 430,000 of CO_2 [2]. Yet another example of product recovery concerns single-use cameras. Kodak started in 1990 to take back, reuse, and recycle its single-use cameras, which had originally been designed as disposables. Kodak is using 86 % of reused parts in manufacturing their new cameras [3]. Some companies use remanufacturing of obsolete product components as the strategy for upgrading products (e.g., HP's main frame systems [4] and Nortel's network systems) [5]. Another group of companies recover the parts and components from used products to provide remanufactured replacement parts for customer service support, called cannibalization of components (e.g., IBM's computer service parts). Lastly, a number of companies collect their used products for material recovery to provide recycled materials to support their own operations or sell to other industries, for example, the plastic components recycling programs implemented by HP for printer cartridges and by Dell for computer peripherals [6].

Currently, in India, e-waste processing is being handled in two ways: formal and informal recycling. The Indian recycling industry recycles 19,000 million tons of e-waste every year. Of which, 95 % electronic waste is recycled in the informal sector and only 5 % goes for formal recycling. There is a very well-networked informal sector in the country [7] involving key players such as vendors, scrap dealers, dismantlers, and recyclers. However, the disposal and recycling of computer-specific e-waste in the informal sector are very rudimentary. The process followed by these recyclers is product reuse, refurbish, conventional disposal in landfills, open burning, and backyard recycling [8]. Of late, formal recycling is being pursued in a big way. Some initiatives have been taken to dismantle and dispose electronic items in the most environmentally sound manner; they also comply with occupational health and safety norms of the workers. Some major e-waste recycling companies are Trishyiraya (Chennai), Infotrek (Mumbai), and E-parisaraa (Bangalore).

Supply chain management is concerned with the efficient and responsive system of production and distribution from raw material stage to final consumer. However, the contemporary supply chains are facing much wider challenges from environmental sustainability issue perspectives [9]. Sustainability remained a fragmented approach based on anecdotic best practices for the different phases of the supply chain. In such approaches, it was difficult for companies to understand that sustainable supply chains could be much more than an obligation, a cost, a constraint, or a charitable deed. Since the Brundtland's report [10], sustainable development, defined as aiming to meet "the needs of the present generation without compromising (…) future generations,' has progressively been incorporated in governmental policy and corporate strategy. Not surprisingly, as the issue of sustainability matured, theory and more complete understanding for this inchoate discipline are currently evolving. Given the ambiguities that surround the sustainability issue, hundreds of different interpretations, related tools, and methods have been proposed to operationalize sustainability. Within the supply chain field, environmental issues have been analyzed through the traditional supply chain segments: from purchasing and inbound logistics through production and manufacturing, to distribution and outbound logistics.

1.1.1 Green Supply Chain Management

According to the moving direction, the supply chain can be divided into the forward and the reverse supply chain (RSC) [11]. The forward supply chain (FSC) means acquiring the original material products from the suppliers and increasing their additional values by creating values in them through corporate managerial functions. On the contrary, the RSC process involves product return, source reduction, recycling, material substitution, item reuse, waste disposal, reprocessing, repairing, and remanufacturing are all examples of recovery options that can represent an attractive business opportunity, a positive answer to sustainable development, and a way of achieving competitive advantage [12–14]. Thus, when putting the RSC processes into a more environmental and thus sustainable 'green' context, it is called green supply chain management (GSCM). Srivastava [13] defined the GSCM as 'incorporating environmental thoughts into supply chain management, including product design, material sourcing and selection, manufacturing processes, delivery of the final product to the customer moreover, the end-of-life cycle.'

Further, supply chains consist of and span many boundaries. Typically, every boundary can be presented at many levels of analysis [15]. Sarkis [15] suggested nine forms of interrelated boundaries leading to GSCM. These boundaries include: organizational, informational, proximal, political, temporal, legal, cultural, economic, and technological.

An *organization* comprises various sub-attributes that lead to greening the supply chain. The continuum of various sub-attributes of an organization ranges from scheduling decisions planned on individual processing equipment to the relationships between different departments. All the activities at different levels of an organization intend to minimize the time and thereby reduction of energy wastage. Thus, the desired outcome for GSCM is the result of internal response practices adopted by an organization.

Lean practices are one of the ways of internal response to sustainability. Lean practices aim to minimize the waste. 'Environmental wastes, including wasted energy, can cost companies thousands of dollars a year'—addressed by Chris Reed, a member of the United States Environmental Protection Agency's Lean and Environmental Initiatives. He further highlights that as public become more aware about the issues of climate changes, organizations are beginning to incorporate environmental concerns into their lean activities. Reed [16] points out

> … A Lean event does not have to focus on environmental targets to achieve environmental benefits – often these come on the coattails of other activities. However, altering the scope of an activity to incorporate 'green' aspects can have an *exponential effect* on both Environmental and *Economic* benefits.

Therefore, with lean practices (to save time), and thereby saving of energy at a very diminutive level of scheduling decisions, we can say

> … that Environmental sustainability accrues from exponential manifestation of a '*Nano-Activity*' performed at operational level in an organization with coalition of supply chain network and environment.

The integration of FSC and RSC shows its potential benefit when timely (*temporal* boundary) communication (*information* boundary) is made between RSC and FSC. Information is not only critical for internal supply chain management operations, but can be a very effective regulatory tool that may cause organizations to re-evaluate their supply chain processes [17]. That is, environmental information flows may be used to provide certain public images of the supply chain and its members. Having this information made public can cause significant pressures from external stakeholders on the overall supply chain to improve environmental and social performance [18]. As the company reconfigure its processes based on gathered information, its market starts moving from local to other states or countries (expanding *legal*, *cultural*, and *political* boundaries).

The present research extends the knowledge under the premise of GSCM by finding the hierarchy of interactions between the boundaries stated by Sarkis [15] mentioned above. We used interpretive structure modeling (ISM) technique to find the interactions between various GSCM enablers. For developing the structural self-interaction matrix (SSIM), we consulted five renowned researchers from top academic research institutes of India. These researchers are extensively engaged in the research of GSCM. With the consensus of these experts, we reached to final SSIM.

1.2 Literature Review: Evolution of Green Supply Chain Management

As GSCM grew and gained increased attention from researchers, the early literature in the area focused on the significant impact of the approach on the well-being of the environments in which corporations work. Much of the work of early theorists focused directly on the green approach as an economic plan for survival. Porter and Van Der Linde [19, 20] discussed the rudiments behind the movement toward green practices, which were (a) increasing supply savings, (b) reducing waste, and (c) increasing productivity. According to Srivastava [13], three advancements emerged in GSCM: the reactive, the proactive, and the value-seeking. Of these, the reactive approach requires the least supply investment and usually involves updating product labeling and exploring ways to lower the impact of production on the environment. The proactive approach is a midlevel undertaking in which organizations invest modest capital in an attempt to self-regulate and focus research and design on creating greener products while taking steps to create a recycling program. The last approach is value-seeking, through which companies focus on implementing ISO design and employing a green approach to purchasing. Hervani et al. [21] argued that because of changing supply chain requirements, environmental managers need to focus on a green approach in order to handle necessary supply chain change. The product life cycle has presented an increasingly important issue, especially when appropriation of material is involved, as well as the specific impact on product supplier relations in selecting materials for product development [22]. Several alternative methods exist to address the need for reverse logistics [23]. Those methods foster guidelines, according to Mukhopadhyay and Setoputro [24], which focus on return procedures of manufacturers when ordering merchandise. Kuo [25] analyzed a supply chain model for manufacturing/remanufacturing scenario. The author analyzed different inventory management policies using discrete event simulation. Akçali and Çetinkaya [26] classified the structural framework at single echelon (manufacturing/remanufacturing) for the inventory and production planning models in closed-loop supply chain. Hong et al. [27] suggest that the firms firstly strive for responsiveness to improve environmental performance. Secondly, lean practices are the important mediator to achieve environmental performance, and thirdly, focal organization takes the lead in achieving environmental performance, and suppliers are in the supportive circle of influence.

1.3 ISM Methodology and Model Development

ISM falls into the soft operations research (OR) family of approaches. The term ISM refers to the systematic application of graph theory in such a way that theoretical, conceptual, and computational leverage is exploited to efficiently construct a directed graph, or network representation, of the complex pattern of a contextual relationship among a set of elements. In other words, it helps to identify structure within a system of related elements. It may represent this information either by a digraph (directed graph) or by a matrix. ISM model also portrays the hierarchy of the enablers. The development of a hierarchy helps in the classification and categorization of the enablers, and thereby formulates their respective strategies and policies while providing clarity of thought. A model depicting relationships among key enablers would be of great value to the top management to delineate the focus areas. ISM methodology helps to impose order and direction on the complex relationships among elements of a system [28, 29]. The ISM methodology is interpretive from the fact that the judgment of the group decides whether and how the variables are related. The process of structural modeling consists of several elements: an object system, which is typically poorly defied system to be described by the model; a representation system, which is a well-defined set of relations; and an embedding of perceptions of some relevant features of the object system into the representation system. Interpretation of the embedded object or representation system in terms of the object system results in an interpretive structural model [29].

1.3.1 Structural Self-interaction Matrix

As previously mentioned, we consulted five experts for developing SSIM. Contextual relationship of 'leads to' type is chosen which means that one enabler helps to alleviate another enabler leading to the contribution toward GSCM. Keeping in mind, the contextual relationship for each enabler, the existence of a relation between any two enablers (i and j), and the associated direction of the relation is questioned. Four symbols are used to denote the direction of relationship between the enablers (i and j). V: Enabler i will alleviate Enabler j; A: Enabler j will be alleviated by Enabler i; X: Enabler i and j will alleviate each other; and O: Enablers i and j are unrelated. Based on the opinion of experts, Table 1.1 is developed.

1.3.2 Reachability Matrix and Level Partitions

The SSIM is transformed into a binary matrix, called the initial reachability matrix by substituting V, A, X, O by 1 and 0 as per the case shown in Table 1.2. Further, various iterations are carried out to develop a final digraph of the model. For the

Table 1.1 Structural self-interaction matrix

X	O	V	A	V	V	O	V	(1)
X	X	A	O	V	O	A	(2)	
X	O	V	O	X	O	(3)		
O	A	A	A	O	(4)			
A	A	A	O	(5)				
O	V	O	(6)					
O	V	(7)						
X	(8)							
(9)								

Table 1.2 Reachability matrix

	(1)	(2)	(3)	(4)	(5)	(6)	(7)	(8)	(9)	Driving power
(1)	1	1	0	1	1	0	1	0	1	6
(2)	0	1	0	0	1	0	0	1	1	4
(3)	0	1	1	0	1	0	1	0	1	5
(4)	0	0	0	1	0	0	0	0	0	1
(5)	0	0	1	0	1	0	0	0	0	2
(6)	1	0	0	1	0	1	0	1	0	4
(7)	0	1	0	1	1	0	1	1	0	5
(8)	0	1	0	1	1	0	0	1	1	5
(9)	1	1	1	0	1	0	0	1	1	6
Dependence	3	6	3	5	7	1	3	5	5	

Table 1.3 Iteration 1

Element p_i	Reachability set $R(p_i)$	Antecedent set $A(p_i)$	Intersection $R(p_i) \cap A(p_i)$
1	1, 2, 4, 5, 7, 9	1, 9	1, 9
2	2, 5, 8, 9	2, 8, 9	2, 8, 9
3	2, 3, 5, 7, 9	3, 5, 9	3, 5, 9
4	4	4	4
5	3, 5	3, 5	3, 5
6	1, 4, 6, 8	6	6
7	2, 4, 5, 7, 8	7	7
8	2, 4, 5, 8, 9	2, 8, 9	2, 8, 9
9	1, 2, 3, 5, 8, 9	1, 2, 3, 8, 9	1, 2, 3, 8, 9

sake of brevity, we mentioned various iterations in Tables 1.3, 1.4, 1.5, 1.6 and 1.7. From the final reachability matrix, the digraph is formed. Removing the transitivity, the digraph is finally converted into the ISM model as shown in Fig. 1.1.

Table 1.4 Iteration 2

Element p_i	Reachability set $R(p_i)$	Antecedent set $A(p_i)$	Intersection $R(p_i) \cap A(p_i)$
1	1, 2, 7, 9	1, 9	1, 9
2	2, 8, 9	2, 8, 9	2, 8, 9
3	2, 3, 7, 9	3, 9	3, 9
6	1, 6, 8	6	6
7	2, 7, 8	7	7
8	2, 8, 9	2, 8, 9	2, 8, 9
9	1, 2, 3, 8, 9	1, 2, 3, 8, 9	1, 2, 3, 8, 9

Table 1.5 Iteration 3

Element p_i	Reachability set $R(p_i)$	Antecedent set $A(p_i)$	Intersection $R(p_i) \cap A(p_i)$
1	1, 7	1	1
3	3, 7	3	3
6	1, 6	6	6
7	7	7	7

Table 1.6 Iteration 4

Element p_i	Reachability set $R(p_i)$	Antecedent set $A(p_i)$	Intersection $R(p_i) \cap A(p_i)$
1	1	1	1
3	3	3	3
6	1, 6	6	6

Table 1.7 Iteration 5

Element p_i	Reachability set $R(p_i)$	Antecedent set $A(p_i)$	Intersection $R(p_i) \cap A(p_i)$
6	6	6	6

1.3.3 Matrix of Cross Impact Multiplications Applied to Classification (MICMAC) Analysis

The objective of the MICMAC analysis is to analyze the driver power and the dependence power of the enablers [30]. In a MICMAC analysis, enablers are classified into four clusters. The first cluster consists of the 'autonomous enablers' that have weak driver power and weak dependence. These enablers are relatively disconnected from the system, with which they have only few links, which may be strong. Second cluster consists of the 'dependent enablers' that have weak driver

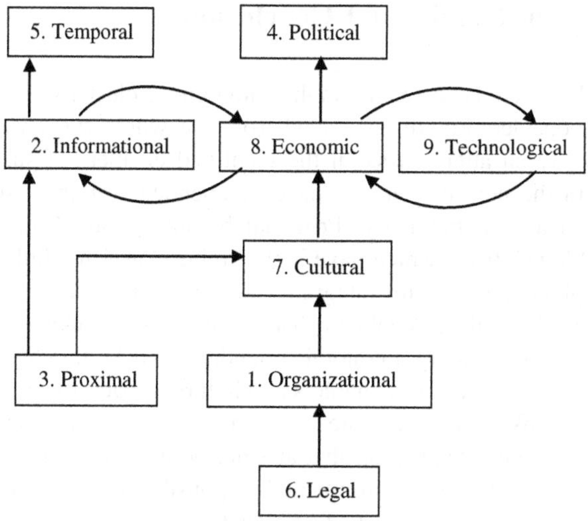

Fig. 1.1 ISM model for contribution toward GSCM

power but strong dependence. Third cluster has the 'linkage enablers' that have strong driving power and also strong dependence. These enablers are unstable in the fact that any action on these enablers will have an effect on others and also a feedback on themselves. Fourth cluster includes the 'independent enablers' having strong driving power but weak dependence. It is observed that an enabler with a very strong driving power called the key enablers, falls into the category of independent enablers. Subsequently, the driver power-dependence diagram is constructed which is shown in Fig. 1.2.

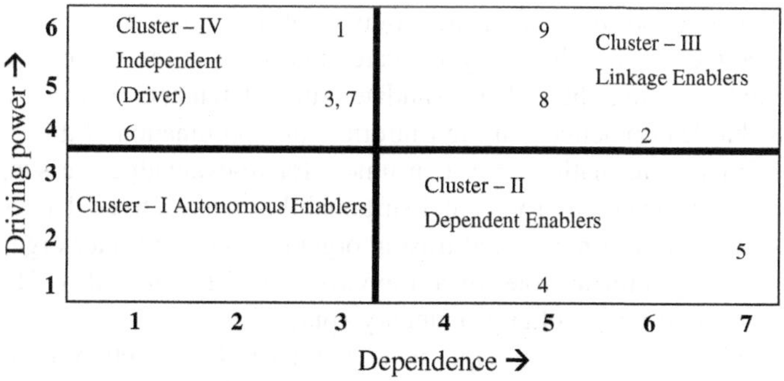

Fig. 1.2 Driver power and dependence diagram

1.4 Discussion of Results and Conclusion

As seen in Fig. 1.2, there is no enabler in the autonomous cluster which indicates no enabler can be considered as disconnected from the whole system and the management has to pay an attention to all the identified enablers of sustainability in supply chains. In the next cluster, we have independent variables such as Legal boundary, Organizational boundary, Proximal boundary, and Cultural boundary toward sustainable practices, which have high driving power but little dependence. These enablers play a key role for integrating sustainability in a supply chain. The next cluster consists of those enablers that are termed as linkage variables and include Informational boundary, Economic boundary, and Technological boundary, which is influenced by lower level enablers and in turn impacts other enablers in the model. The three linkage enablers are interconnected in a way that information boundary facilitates the organization that at what point of time it is economically infeasible to carry on with the environmental responsibilities due to end-of-pipe or incompatible technologies. The last cluster includes enablers such as Temporal boundaries and Political boundaries. In this cluster, particularly the Temporal boundary has highest dependence and form the topmost level in the ISM hierarchy. It represents the enabler that is the resultant action for effective integration of sustainability in a supply chain. Its strong dependence indicates that it requires all the other enablers to come together for effective implementation of sustainability practices. But it is important as it is finally required by the supply chain to measure the effectiveness of sustainability in the supply chain.

The ISM model developed in this paper clearly portrays that for carrying out green practices, the supply chains require that each echelon within a supply chain should morally and ethically be involved in environmental sustainability as these can be considered as the attributes of legal boundary [15]. Similarly, extraneous to proximity, may be at international level, or at individual, group, and organizational levels, the management should inculcate the culture of environmental sustainability. Further, the model portrays that it is imperative that 'Information boundary' leads to 'Temporal boundary' which clearly indicates that time is the ultimate action for environmental sustainability. This validates the statements made by various researchers that lean practices through information coordination within an organization and timely information from consumers for re-evaluating the supply chain are very important from environmental sustainability perspectives. Under political boundaries, the issues of power and trust at organizational and inter-organizational level also play a significant role toward environmental sustainability. However, it may largely depend upon cultural boundary context.

Toward this end, the present paper is an attempt in the direction to identify the key enablers which affect the environmental sustainability decision. There is scope for further research by developing a broader consensus on the order of priority of these enablers, especially in the context of reverse supply chain, and studying separately their impact on the performance of forward supply chain. For instance, a time-based performance can be analyzed for a closed-loop supply chain when different scheduling decisions are adopted at operational level under the premise of information sharing so as to mitigate system wide cycle time and thereby saving energy.

References

1. Fuji Xerox Sustainability Report (2012). http://www.fujixerox.com/eng/company/sr/booklet/2012e.pdf
2. Canon Europe Sustainability Report (2011–2012). http://www.canon.lv/Images/CANON%20EUROPE%20MAIN%20REPORT%2011-12%20INTERNET_tcm211-959831.pdf
3. David PA, Stewart RD (2010) International logistics: the management of international trade operations, 3rd edn. Cengage Learning, Mason, pp 35–36
4. Kupér A (2003) Personal communication. Hewlett & Packard, Palo Alto
5. Linton JD, Johnston DA (2000) A decision support system for planning remanufacturing at Nortel Networks. Interfaces 30:17–31
6. Guide VDR Jr, Van Wassenhove LN (2003) Business aspects of closed loop supply chain. Carnegie Mellon University Press, Pittsburg
7. Sinha S, Mahesh P (2007) Into the future: managing e-waste for protecting lives and livelihoods. www.toxicslink.org/pub-view.php?pubnum=171
8. Dixit N (2007) E-waste: a disaster in the making, CHANGE. Goorej House Mag 7:10–13
9. Seitz M, Wells P (2006) Challenging the implementation of corporate sustainability. Bus Process Manage J 12:822
10. Brundtland GH (1987) World commission on environment and development. Our common future. Oxford University Press, Oxford
11. Fleischmann M, Bloemhof-Ruwaard J, Dekker R, van der Laan E, van Nunen JAEE, Van Wassenhove LN (1997) Quantitative models for reverse logistics: a review. Eur J Oper Res 103:1–17
12. Toffel MW (2004) Strategic management of product recovery. Calif Manage Rev 46:120–141
13. Srivastava SK (2007) Green supply chain management: a state-of-the-art literature review. Int J Manage Rev 9:53–80
14. Jack EP, Powers TL, Skinner L (2010) Reverse logistics capabilities: antecedents and cost savings. Int J Phys Distrib Logist Manage 40:228–246
15. Sarkis J (2012) A boundaries and flows perspective of green supply chain management. Supply Chain Manage Int J 17:202–216
16. Reed C (2008) Finding the green in Lean Six Sigma. http://www.processexcellencenetwork.com/contributors/115-chris-reed/
17. Brown HS, de Jong M, Levy DL (2009) Building institutions based on information disclosure: lessons from GRI's sustainability reporting. J Clean Prod 17:571–580
18. Kovács G (2008) Corporate environmental responsibility in the supply chain. J Clean Prod 16:1571–1578
19. Porter M, Van Der Linde C (1995) Green and competitive: ending the stalemate. Harvard Bus Rev 73:120–134
20. Porter M, Van Der Linde C (1995) Toward a new conception of the environment—competitiveness relationship. J Econ Perspect 9:97–118
21. Hervani AA, Helms MM, Sarkis J (2005) Performance measurement for green supply chain management. Benchmarking Int J. 12:330–353
22. Stonebraker PW, Liao J (2006) Supply chain integration: exploring product and environmental contingencies. Supply Chain Manage Int J 11:34–43
23. Ravi V, Shankar R, Tiwari MK (2005) Analysing alternatives in reverse logistics for end-of-life computers: ANP and balanced scorecard approach. Comp Ind Eng 48:327–356

24. Mukhopadhyay SK, Setoputro R (2005) Optimal return policy and modular design for build-to-order products. J Oper Manage 23:496–506
25. Kuo TC (2011) The study of production and inventory policy of manufacturing/ remanufacturing environment in a closed-loop supply chain. Int J Sust Eng 4:323–329
26. Akçali E, Çetinkaya S (2011) Quantitative models for inventory and production planning in closed-loop supply chain. Int J Prod Res 49:2373–2407
27. Hong P, Roh JJ, Rawski G (2012) Benchmarking sustainability practices: evidence from manufacturing firms. Benchmarking An Int J 19:634–648
28. Warfield JW (1974) Developing interconnected matrices in structural modeling. IEEE Trans Syst Men Cyber 4:51–81
29. Sage AP (1977) Interpretive structural modeling: methodology for large-scale systems. McGraw-Hill, New York, pp 91–164
30. Mandal A, Deshmukh SG (1994) Vendor selection using interpretive structural modeling (ISM). Int J Oper Prod Manage 14:52–59

Chapter 2
An Optimum Setting of PID Controller for Boost Converter Using Bacterial Foraging Optimization Technique

P. Siva Subramanian and R. Kayalvizhi

Abstract In this paper, a maiden attempt is made to examine and highlight the effective application of bacterial foraging (BF) algorithm to optimize the PID controller parameters for boost converter and to compare its performance to establish its superiority over other methods. The proposed BF-PID controller maintains the output voltage constant irrespective of line and load disturbances than particle swarm optimization (PSO)-based PID controller and conventional PID controllers.

Keywords PID controller · Boost converter · State space modeling · Bacterial foraging algorithm

2.1 Introduction

The main target of power electronics is to convert electrical energy from one form to another. To make the electrical energy to reach the load with the highest efficiency is the target to be achieved. Power electronics also targets to reduce the size of the device which aims to reduce cost, size and high availability. In this project, the power electronic device is DC–DC boost converter. Sometimes, it is necessary to increases dc voltage. Boost converter is a DC–DC converter in which the output voltage is always greater than the input voltage which depends on switching frequency [1]. From the energy point of view. From the energy point of view, output

P. Siva Subramanian (✉)
Department of Electronics and Instrumentation Engineering, SSN Institue of Engineering and Technology, Dindigul, Tamil Nadu, India
e-mail: sivasubramanian.v.p@gmail.com

R. Kayalvizhi
Department of Electronics and Instrumentation Engineering, Annamalai University, Chidambaram, Tamil Nadu, India
e-mail: mithuvigknr@gmail.com

V. Vijay et al. (eds.), *Systems Thinking Approach for Social Problems*,
Lecture Notes in Electrical Engineering 327,
DOI 10.1007/978-81-322-2141-8_2

voltage regulation in the DC–DC converter is achieved by constantly adjusting the amount of energy absorbed from the source and that injected into the load. These two basic processes of energy absorption and injection constitute a switching cycle [2]. Some control methods have stated the issue of control through pole placement [3]. Another is the design of boost converter incorporated with PID controller that is used to control the behaviors of the system in linear. This system is a closed-loop system with feedback. A proportional integral derivative controller (PID controller) is a generic control loop feedback mechanism widely used in industrial control systems. A PID controller attempts to correct the error between a measured process variable and a desired set point by calculating and then outputting a corrective action that can adjust the process accordingly. The PID controller tuning involves the calculation of three separate parameters: the proportional, the integral, and the derivative values.

The values of these parameters to control a process depend on the process dynamics and the required response of the process. The adjustment of the controller parameters to achieve satisfactory control is called controller tuning. In many process industries, the process dynamics are poorly known. In such situation, the process model is obtained through the experimental data. The required response of the process is determined based on some performance criteria such as settling time, peak amplitude, peak time, ISE, and IAE. The PID controller is simple but cannot always effectively control systems with changing parameters and may need frequent online retuning. Most of the articles have concentrated on designing PI and PID controllers [4]. In order to obtain better response, a bacterial foraging (BF) algorithm-based PID controller is designed.

BF algorithm proposed by Passino [5] is a newcomer to the family of nature-inspired optimization algorithms. Application of group foraging strategy of a swarm of *Escherichia coli* bacteria to multioptimal function optimization is the key idea of the new algorithm. Bacteria search for nutrients in a manner to maximize energy obtained per unit time. Individual bacterium also communicates with others by sending signals. A bacterium takes foraging decisions after considering two previous factors. The process, in which a bacterium moves by taking small steps while searching for nutrients, is called chemotaxis, and key idea of BF algorithm is mimicking chemotactic movement of virtual bacteria in the problem search space. It is used as optimization method and has shown its effectiveness in various problems. BF algorithm is a powerful search tool that can reduce the time and effort involved in designing systems for which no systematic design procedure exists [6]. They can quickly find close-to-optimal solutions. They are certainly useful tools when trying to solve analytically difficult problems.

In this paper, BF optimization algorithm is developed for tuning the parameters of PID controller. The developed controller is simulated for a DC–DC boost converter and to compare the response of optimized PID controller with the conventional PID controller, particle swarm optimization (PSO)-based PID controller [7]. Simulation results indicate that BF-PID controller guarantees the good performance under various line and load disturbance conditions than others.

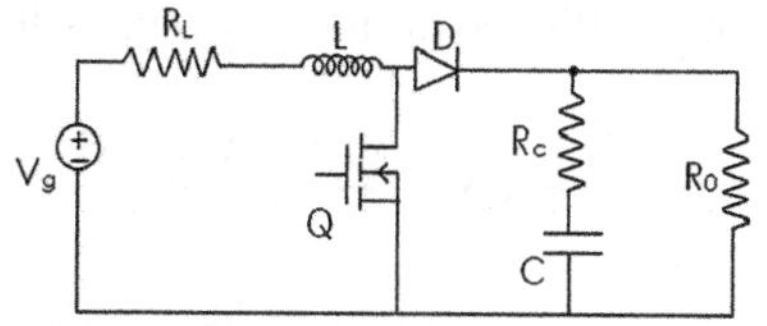

Fig. 2.1 Circuit diagram of boost converter

2.2 Boost Converter

Consider the DC–DC boost converter circuit shown in Fig. 2.1. During the interval, when switch Q is off, diode D conducts the current i_L of inductor L toward the capacitor C_0 and the load R_0. During the interval, when switch Q is on, diode D opens and the capacitor C_0 discharges through the load R. The converter transfers the energy between input and output by using the inductor.

The transfer function in Fig. 2.1 is derived using the standard state space averaging technique. In this approach, the circuits for two modes of operation (ON mode and OFF mode) for the converter are modeled as follows:

$$\begin{aligned} \dot{x} &= Ax + Bu \\ y &= Cx + Du \end{aligned} \tag{2.1}$$

where

x state variable
u input voltage (V_{in})
y output voltage (V_{o})

After modeling, the two modes are averaged over a single switching period T.

2.3 PID Controller

The PID controller shown in Fig. 2.2 is used to improve the dynamic response and to reduce the steady-state error. The derivative controller improves the transient response, and the integral controller will reduce steady-state error of the system.

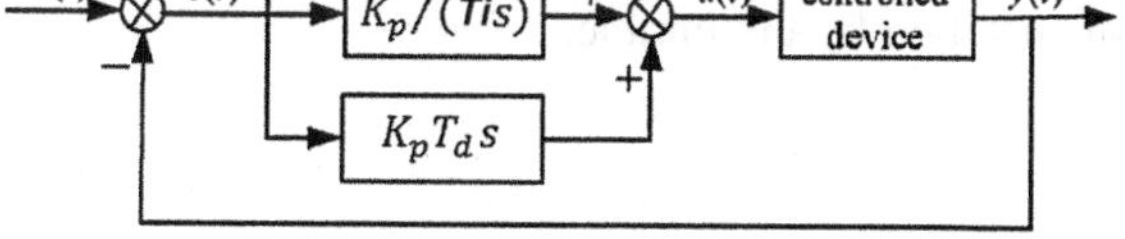

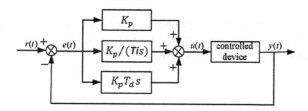

Fig. 2.2 Schematic diagram of PID controller

The transfer function of the PID controller is given as follows:

$$k_p + \frac{k_i}{s} + k_d s = \frac{k_d s^2 + k_p s + k_i}{s} \tag{2.2}$$

The PID controller works in a closed-loop system. The signal *u(t)* output of the controller is equal to the K_p times of the magnitude of the error plus K_i times the integral of the error plus K_d times the derivative of the error as follows:

$$k_p e + k_i \int e\mathrm{d}t + k_d \frac{\mathrm{d}e}{\mathrm{d}t} \tag{2.3}$$

This control signal will be then sent to the plant, and the new output *y*(*t*) will be obtained. This new output will be then sent back to the sensor again to find the new error signal *e*(*t*). The controller takes this new error as input signal and computes the gain values (K_p, K_i, K_d).

2.4 Bacterial Foraging Optimization Technique

BF algorithm is a new division of bioinspired algorithm. This technique is developed by inspiring the foraging behavior of *E. coli* bacteria. In the BF optimization process, four motile behaviors of *E. coli* bacteria are mimicked.

2.4.1 Chemotaxis

During the foraging operation, an *E. coli* bacterium moves toward the food location with the aid of swimming and tumbling via flagella. Depending upon the rotation of flagella in each bacterium, it decides whether it should move in a specified direction (swimming) or altogether in modified directions (tumbling), in the entire lifetime. To represent a tumble, a unit length random direction, say $\Phi(j)$, is generated; this will be used to define the direction of movement after a tumble. In particular,

$$\theta^i(j+1,k,l) = \theta^i(j,k,l) + C(i) * \Phi(j) \tag{2.4}$$

where $\theta^i(j,k,l)$ represents the *i*th bacterium, at *j*th chemotactic, *k*th reproductive, and *l*th elimination and dispersal step. *C*(*i*) is the size of the step taken in the random direction specified by the tumble.

2.4.2 Swarming

In this process, after finding the direction of the best food position, the bacterium will attempt to communicate with other by using an attraction signal. The signal communication between cells in *E. coli* bacteria is represented by

$$J_{cc}(\theta, D(j,k,l)) = J^{i}_{cc}(\theta, \theta^{i}(j,k,l)) = X + Y \tag{2.5}$$

where

$$X = \sum_{i=1}^{S} [-D_{\text{attract}} * \exp(-W_{\text{attract}} * \sum_{m=0}^{P} (\theta m - \theta^{i} m))^{2}]$$

$$Y = \sum_{i=1}^{S} [H_{\text{repellant}} * \exp(-W_{\text{repellant}} * \sum_{m=0}^{P} (\theta m - \theta^{i} m))^{2}]$$

where $J_{cc}(\theta, D(j,k,l))$ is the cost function value to be added to the actual cost function to be minimized to present a time-varying cost function, S is the total number of bacteria, P is the number of parameters to be optimized which are present in each bacterium, and D_{attract}, W_{attract}, $H_{\text{repellant}}$, and $W_{\text{repellent}}$ are different coefficients that should be chosen properly.

2.4.3 Reproduction

The least healthy bacteria die, while each of the healthier bacteria (those yielding lower value cost function) asexually splits into two bacteria, which are then placed in the same location. This keeps the swarm size constant.

2.4.4 Elimination and Dispersal

It is possible that in the local environment, the life of bacteria changes either gradually or suddenly due to some other influences. Events can occur such that all the bacteria in a region are killed or a group is dispersed into a new environment. They have the effect of possibly destroying the chemotactic progress, but they also have the effect of assisting in chemotaxis, since dispersal may place bacteria near attractive food sources. From a broader perspective, elimination and dispersal are part of the population-level long-distance motile behavior.

The searching procedure to develop BF-PID controller is as follows:

Step 1 *Initialization*

D Number of parameters to be optimized. In this project, it is K_P, K_I, and K_D.

S Number of bacteria to be used for searching the total region.

N_S Swimming length after which tumbling of bacteria will be done in a chemotactic step.

N_c Number of iterations to be taken in the chemotactic step.

N_{re} Maximum number of reproductions to be undertaken.

N_{ed} Maximum number of elimination and dispersal events to be imposed over bacteria.

P_{ed} Probability with which the elimination–dispersal will continue.

θ Location of the each bacterium which is specified by random numbers on [0,1].

$C(i)$ Chemotactic step size assumed to be constant for our design.

The value of D_{attract}, W_{attract}, $H_{\text{repellant}}$, and $W_{\text{repellent}}$ is to denote here that the value of D_{attract} and $H_{\text{repellant}}$ should be the same so that the penalty imposed on the cost function through "J_{cc}" will be "0" when all the bacteria will have the same value, i.e., they have converged.

Step 2 *Elimination and dispersal loop*

$l = l + 1$

Step 3 *Reproduction loop*

$k = k + 1$

Step 4 *Chemotactic loop*

$j = j + 1$

(i) For i = 1, 2, 3 … S, take a chemotactic step for each bacterium i as follows:

(ii) Compute the value of $J(i, j, k, l)$.
Let $J(i,j,k,l) = J(i,j,k,l) + J_{\text{cc}}(\theta^i(j,k,l), P(j,k,l))$ (i.e., add on the cell to cell attractant effect to the nutrient concentration).

(iii) Let $J_{\text{Last}} = J(i, j, k, l)$ to save this value since we may find a better cost via run.
End of for loop.

(iv) For i = 1, 2, 3 … S, take the tumbling/swimming decision
Tumble: Generate a random number vector R^p with each element $m(i)$, (m = 1, 2, 3 … D), a random number on the interval [−1,1].

(v) **Move**: Let

$$\theta^i(j+1,k,l) = \theta^i(j,k,l) + C(i)\frac{\Delta(i)}{\sqrt{\Delta T(i) \times \Delta(i)}}$$

This results in a step of size $C(i)$ in the direction of the tumble for bacterium i.

(vi) Compute = $J(i, j, k, l)$
Let $J(i,j,k,l) = J(i,j,k,l) + J_{\text{cc}}(\theta^i(j,k,l), P(j,k,l))$

(vii) **Swim**: Note that we use an approximation since we decide swimming behavior of each cell as if the bacteria numbered {1, 2, 3 … *i*} have moved and {*i* + 1, *i* + 2, *i* + 3 … *S*} have not; this is simpler to simulate than simultaneous decision about swimming and tumbling by all the bacteria at the same time.
Let $m = 0$ (counter for swim length)
While $m <$ Ns (if doing better), let
$J_{\text{Last}} = J(i, j+1, k, l)$ and
let $\theta^i(j+1, k, l) = \theta^i(j, k, l) + C(i)\frac{\Delta(i)}{\sqrt{\Delta T(i) \times \Delta(i)}}$
And use this $\theta^i(j+1, k, l)$ to compute the new $J(i, j+1, k, l)$ as in step (vi).
Else, let $m = Ns$, and this is the end of the while statement.

(viii) Go to next bacterium (i + 1) if '*i*' is not equal to *S* (i.e., go to (b)) to process the next bacterium.

Step 5 If $j < N_c$, go to Step 4. In this case, continue chemotaxis since the life of the bacteria is over.

Step 6 *Reproduction*

(i) For the given value of k and l, and for each i = 1, 2, 3 … *S*. Let $J^i_{\text{Health}} = \sum J(i, j, k, l)$ be the health of bacterium. Sort bacteria on chemotactic parameters $C(i)$ in order of interesting cost J_{Health} (higher cost means lower health).

(ii) The Sr = *S*/2 bacteria with the highest J_{Health} values die and other Sr bacteria with the best value split and the copies that are made are placed at the same location as their parent.

Step 7 If $k < N_{\text{re}}$, go to Step 2; in this case, we have not reached the number of specified reproduction steps, so we start the next generation in the next chemotactic step.

Step 8 *Elimination and Dispersal*: For i = 1, 2, 3 … *S*, with probability P_{ed}, eliminate and disperse each bacterium (keeps the bacterium population constant). To do this, if we eliminate a bacterium, simply disperse one into a random location in the optimization domain.

Step 9 If $l < N_{\text{ed}}$, then go to Step 1; otherwise, end.

2.5 Performance Indices for BF Algorithm

The objective function considered is based on the error criterion. The performance of a controller is best evaluated in terms of error criterion. In this work, controller's performance is evaluated in terms of integral square error (ISE).

$$\text{ISE} = \int_0^t e^2 \mathrm{d}t \tag{2.6}$$

Table 2.1 Circuit parameters of the boost converter

Parameter	Symbol	Value
Input voltage	V_{in}	12 V
Output voltage	V_o	20 V
Inductor	L	162 μH
Capacitor	C	220 μF
Internal resistors	R_L, R_C	5 mΩ
Load resistor	R	10 Ω
Duty ratio	D	0.53

The ISE weighs the error with time and hence minimizes the error values nearer to zero.

2.6 Simulation Results

The circuit parameters of the boost converter are shown in Table 2.1.

The responses of boost converter using conventional PID controller and BF-PID controllers are shown in Figs. 2.3, 2.4, 2.5 and 2.6.

The figures show that BF-PID controller will drastically reduce the overshoot and ISE and IAE values as compared to the conventional PID controller. Table 2.2

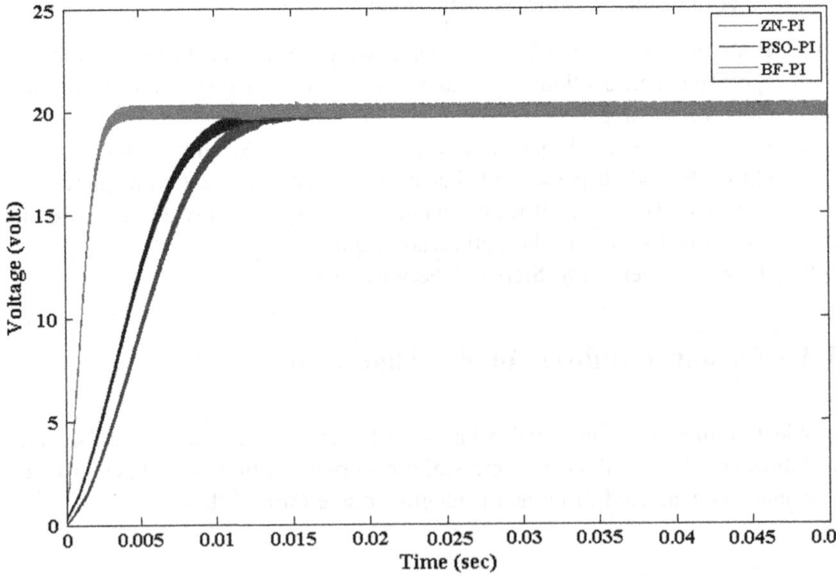

Fig. 2.3 Output voltage of conventional PID controller, PSO-based PID controller, and BF-based PID controller

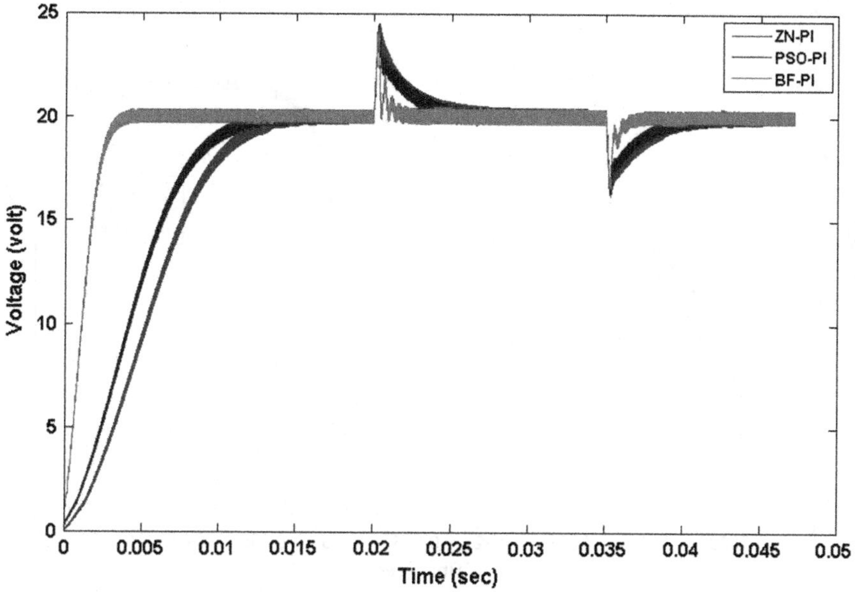

Fig. 2.4 Comparison of closed-loop response of boost converter under sudden change in line voltage of 12–15 V (25 % increase) and 12–10 V (25 % decrease)

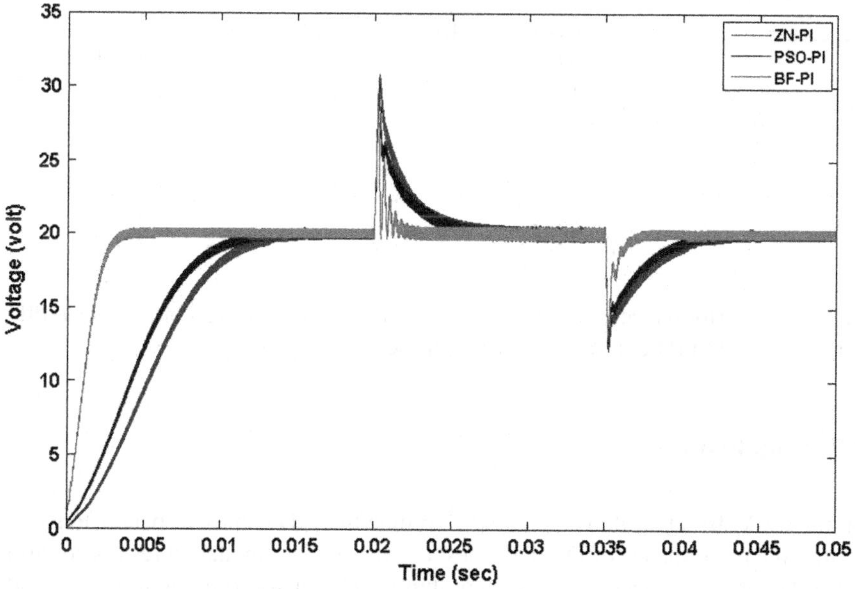

Fig. 2.5 Comparison of closed-loop response of boost converter under sudden change in load disturbance from 10 to 12 Ω (25 % increase) and 10 to 8 Ω (25 % decrease)

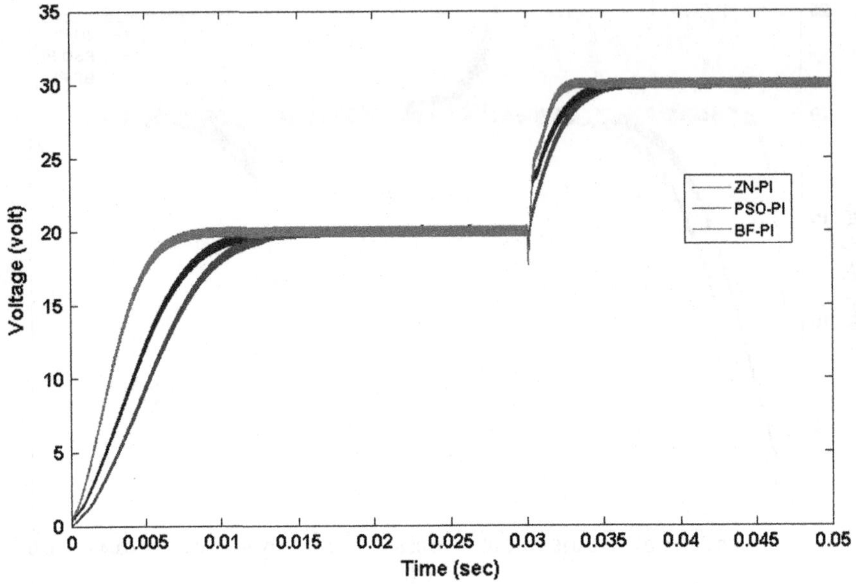

Fig. 2.6 Comparison of servo response of boost converter under sudden change in reference voltage 20–30 V

Table 2.2 Performance analysis of the boost converter

Parameters	Conventional PID controller	PSO-PID controller	BF-PID controller
Peak amplitude (V)	20	20	20
Overshoot (%)	0	0	0
Rise time (ms)	2.3	1.4	0.5
Settling time (ms)	18	13	5
ISE	1.52	1.17	0.35

shows the performance analysis of the boost converter using conventional PID controller, PSO-PID controller, and BF-PID controller.

2.7 Conclusion

In this work, BF algorithm is developed to tune the PID controller parameters that control the performance of DC–DC boost converter. The simulation results confirm that PID controller tuned with BF algorithm rejects satisfactorily both the line and load disturbances. Also, the result proved that BF-based PID controller gives the smooth response for the reference tracking and maintains the output voltage of the boost converter according to the desired voltage.

References

1. Khanchandani KB, Singh MD (2005) Power electronics. Tata McGraw Hill, New Delhi
2. Hasaneen BM, Mohammed AAE (2008) Design and simulation of DC/DC boost converter. 12th international middle-east power system conference, pp 335–340
3. Kelly A, Rinne K (2005) Control of DC–DC converters by direct pole placement and adaptive feed forward gain adjustment. In: Applied power electronics conference and exposition (APEC), vol 3, pp 1970–1975
4. Iruthayarajan MW, Baskar S (2007) Optimization of PID parameters using genetic algorithm and particle swarm optimization. In: International conference on information and communication technology in electrical sciences, pp 81–86
5. Passino KM (2002) Biomimicry of bacterial foraging for distributed optimization and control. IEEE Control syst Mag, pp 52–67
6. Mary Synthia Regis Prabha DM, Pushpa Kumar S, Glan Devdhas G (2011) An optimum setting of controller for a DC–DC converter using bacterial intelligence technique. In: IEEE PES innovative smart grid technologies, New Jersey
7. Suresh Kumar R, Suganthi J (2012) Improving the boost converter PID controller performance using particle swarm optimization. Eur J Sci Res 85(3):327–335. ISSN 1450-216X

Chapter 3
A Command Splitter for a Mini–Macro-Manipulator for Online Telemanipulation

Amar Banerji

Abstract A mini–macro-system consists of two manipulators cascaded to each other, one of them being much lighter than the other. The advantage is that the lighter manipulator which has better dynamic response can quickly respond to the command and the sluggish partner can start contributing when it is ready. The control systems of both the manipulators are good enough to execute any joint motion independently, but they need precise individual reference velocity signals for all the joints in real time so that it could move all of them appropriately even in cascaded condition. This paper presents an algorithm to compute the joint rates in real time so that both the manipulators cooperate with each other in moving the end-effector in best possible way.

Keywords Master–slave manipulator · Parallel manipulator · Dual robot

3.1 Introduction

It is obvious that a robotic arm that responds to an *online* command (*online* here is meant as interactive, such as moving the computer's curser with the mouse or arrow keys) in real time is ideal for many difficult-to-program tasks. For example, the specialized tasks, such as surgery, bomb disposal, and handling of hazardous chemicals and biological wastes, need interactive manipulation, also called 'master–slave' manipulation.

The inexpressible nature of the maneuvers rules out preprogrammed motions. These tasks may be done using an interactively operated *slave robotic arm*. For example, consider the robotic arm shown in Fig. 3.1, marked as 'main robot' (the mini-part can be ignored at this moment). It is a planar manipulator with 2

A. Banerji (✉)
Division of Remote Handling and Robotics, BARC, Mumbai 400085, India
e-mail: amar.banerji@gmail.com

V. Vijay et al. (eds.), *Systems Thinking Approach for Social Problems*,
Lecture Notes in Electrical Engineering 327,
DOI 10.1007/978-81-322-2141-8_3

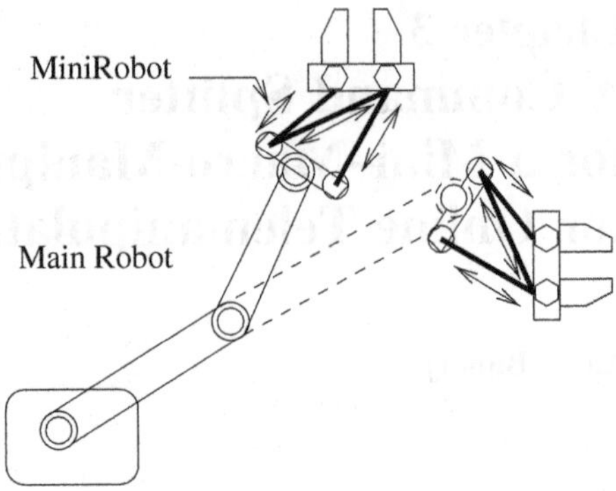

Fig. 3.1 *Left* A 3-DOF planar manipulator with mini-parallel manipulator

DOFs in position and one in rotation. It could be manipulated as a slave arm by electronically linking it with another identical manipulator (the master), but with only the encoders at its joints and no motors. The operator manually manipulates the master, and its movements are imitated by the slave. This principle can also be used in manipulating robots of much bigger sizes and with various configurations.

The master arm, however, cannot be made arbitrary large. Usually, a small 6-DOF device called Phantom™ (Fig. 3.2) is used as the master arm to feed the motion commands. This, then, requires the tracking of the end-effectors' (grippers) motions in spatial coordinate (i.e., Phantom's tip's motion in x axis will result in robot's gripper's motion in x direction) because the joint motions and postures of master and slave can no longer be identical.

The master arm is grasped by the operator, and it can be manipulated effortlessly and rapidly by the operator's hand because of its negligible mass. However, often, the experience of performing a master–slave operation has been quite unsettling and disorienting for the operator due to sluggish inertial response of the main (macro) robotic arm compared to almost effortless movement of the *Phantom* master arm (e.g., a quick maneuver required for writing letter 'M' may not get properly

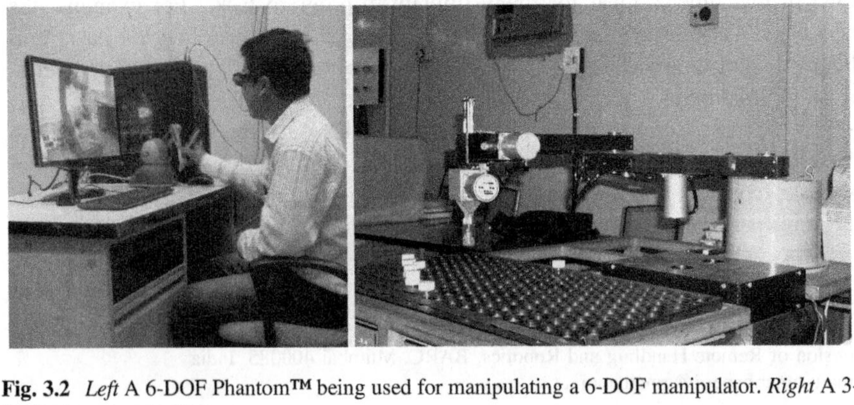

Fig. 3.2 *Left* A 6-DOF Phantom™ being used for manipulating a 6-DOF manipulator. *Right* A 3-DOF planar manipulator

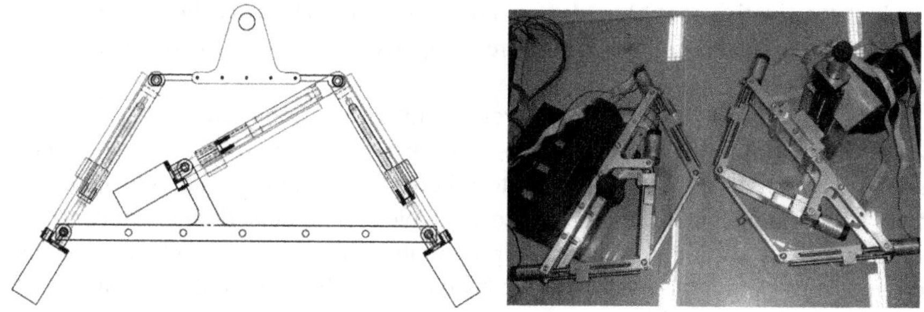

Fig. 3.3 Three-DOF planar parallel manipulators: each prismatic joint stroke range = 85 mm. Details are available in [3]

executed.). It is not due to the communication delay or control system's computation delay. The fact is that the operator maneuvers the master's stylus using his/her much agile fingers, seldom using his elbow, shoulder, or wrist. The robotic arm, however, is constrained to reproduce the same dexterity using bigger links. This is not feasible in typical robots. Basically, they are designed for executing programmed motions with well-planned trajectories suitable for the manipulator's dynamics. Thus, the desired maneuvers executed by the master arm are often poorly imitated by the slave arm in fine manipulation tasks. We investigated the possibility of creating a manipulator with two manipulators with identical DOFs in cascade to improve its dynamic response as reported in the literature. The main (macro) manipulator is connected with a small mini-manipulator at its end-effector's location. The mini-manipulator is an in-house developed parallel manipulator with two positional DOFs and one rotational DOF [3]. It has three prismatic joints actuated by three ball–screw mechanisms, but they are connected in parallel and not in series, see Fig. 3.3. It is connected to a bigger but sluggish 3-DOF planar manipulator as shown in Fig. 3.1, creating a redundant robotic system. In fact, this mini-manipulator can be conveniently retrofitted to any other existing robotic arm whose response time has to be improved, subject to its degrees of freedom. Additionally, no hardware modification of the existing robotic arm is required. The cascaded system converts a sluggish manipulator to a quick response manipulator mechanism in 3-DOF manipulator as discussed in the following sections. This attachment also neutralizes the disturbances caused by backlash and dynamics of the serial robotic arm due to its higher bandwidth and negligible backlash.

This paper describes an algorithm to generate the commands for macro- and mini-manipulator using instantaneous states of their respective joints. The approach in this algorithm is similar to that used by other researchers in computing the trajectory of flexible manipulators and mini–macro-system. But in those cases, the manipulators were not used in master–slave configuration. In most of the previous works, the joint trajectories for a given target for each manipulator were computed using dynamic models off-line and the motion was executed. However, the present algorithm does not require dynamic parameters of the manipulators. The macro-manipulator's native control hardware and program module is also not altered. Only the mini-manipulator's joint motions are computed fast enough to complement the macro-manipulators response.

The master–slave manipulation is very common in the nuclear energy industry where mechanically coupled master–slave pair is extensively used to handle radioactive substances. The transition to robotic manipulators is slowly happening there, but not much published works are available in the public domain. In contrast, other industries deploy robots as preprogrammed machines; therefore, research activities regarding mini–macro-system have been on the control of the system with a specified reference trajectory.

The mini–macro-robotic system has been extensively studied in the past. The structure of the modified manipulator is quite similar to macro–mini-manipulator investigated by Khatib [4]. His analysis reveals that in such structures, the effective inertia of a macro–mini-manipulator is bounded above by the mass/inertia of the mini-manipulator alone. Thus, the system acquires a reduced effective inertia. Yoshikawa et al. [8] attached a mini-manipulator at the tip of a flexible manipulator. The mini-manipulator was rigid, whereas macro-manipulator was flexible. A better steady-state accuracy at the end-effector of a long arm manipulator was achieved using mini–macro-manipulator, and very fine motion was generated [7]. Our emphasis, however, is on computation of instantaneous joint rates of mini- and macro-manipulators so that the controller of these two manipulators can receive independent reference commands. It may be noted that the control electronics of commercial systems are seldom amenable to modifications; hence, the new approach presented here can be used to cascade two manipulators without any risks.

3.2 Kinematics of Mini–Micro-System

As explained in the last section, the manipulator's end-effector is now replaced with the base of mini-manipulator. Thus, the base of the mini-manipulator keeps shifting as the macro-manipulator moves, and its location can be computed using the forward kinematics of the macro-manipulator. The forward kinematics is (often) a nonlinear mapping $f : \theta \mapsto x$ which maps the manipulator's joint angles to the workspace coordinates x.

It is well known that a linear map between the joint velocity and workspace velocity can be found by differentiating the forward kinematics map. This new map is called a Jacobian J. In an alternative approach, we can use a different version of Jacobian, called manipulator Jacobian, which is not the differentiation of the forward kinematic map; further details can be found in [1, 5]. Advantage of this is that we can write the Jacobian by examining the axes of rotations and translations (called *twists*), and it is symbolically denoted as ξ. A twist vector consists of linear velocity $\dot{x}$ and angular velocity ω components. In the case of a planar manipulator, all the axes of rotations are same, i.e., perpendicular to the plane. Using this mapping, the velocity (twist) of the manipulator tip $[\dot{x} \quad \omega]^T$ due to a joint at $q \in \Re^2$, rotating at speed $\dot{\theta}_{\mathrm{R}}$ is given as

$$\begin{bmatrix} \dot{x} \\ \omega \end{bmatrix} = \begin{bmatrix} q_y \\ -q_x \\ 1 \end{bmatrix} \dot{\theta}_{\mathrm{R}}, \tag{3.1}$$

where (q_x, q_y) is the point about which the rotation (right-hand coordinate system) is taking place. In the case of a prismatic joint aligned to a vector u_x, u_y, no rotation takes place and the actuation results in a translation $\dot{x}$; notice the 0 in the last row:

$$\begin{bmatrix} \dot{x} \\ 0 \end{bmatrix} = \begin{bmatrix} u_x \\ u_y \\ 0 \end{bmatrix} \dot{\theta}_{\mathrm{P}}, \tag{3.2}$$

where $\dot{\theta}_{\mathrm{P}}$ denotes the linear speed of the prismatic joint and not a revolution.

Now, since the velocity map is a linear function, we can add macro-manipulator's and mini-manipulator's individual joint velocities and have a combined Jacobian for the new structure, yielding a mapping from the joint velocities to the mini–macro-manipulator's tip velocity (which is commanded by phantom's tip velocity **U**—the manipulator tracks the master arm), in Cartesian space:

$$[\mathbf{U} \quad \omega]^T = [J_{\mathrm{M}}(\theta) \quad J_{\mathrm{m}}(\theta)][\{\dot{\theta}_{\mathrm{R}}\} \quad \{\dot{\theta}_{\mathrm{P}}\}]^T. \tag{3.3}$$

where $J_{\mathrm{M}}, J_{\mathrm{m}}$ are the Jacobian for the macro- and mini-manipulator in cascaded condition. *This matrix* $[J_{\mathrm{M}}, J_{\mathrm{m}}]$ *maps the joint-velocities to spatial coordinate system. It should not be interpreted as dynamic equations.*

We can interpret (3.3) in a very interesting way. The Jacobian has more columns than rows and has a non trivial null space. It has three rows and six columns. Any of the three columns are, therefore, a linear combination of the three other columns. It tells us that there exist a certain combinations of joint speeds which may give the same net velocity (or twist) of the tip. This actually allows us to compute a convenient set of joint velocities which will generate a net end-effector speed **U** as described below using null space motions.

Notice that in the planar workspace case, the mini–macro-robotic structure has six degrees of freedom, whereas the workspace has only three (position $x \in \Re^2$ and the orientation can be described using a rotation about z axis). This results in the possibility of a few redundant internal motions of the manipulator joints that generate the same desired movement of the end-effector. These motions correspond to the *Null* space of the Jacobian in (3.3); see [2, 5, 6] for further details. The basis vectors of the null subspace are given by the four columns of the following matrix that can be computed using MATLAB™ function `null()`:

$$\mathcal{N}_J = \{\, \psi_1 \quad \ldots \quad \psi_3 \,\}, \tag{3.4}$$

where $\psi_i \in \Re^6$; we can construct any vector in this subspace by multiplying the columns with scalar coefficients α_is and adding them:

$$[\psi_1 \quad \ldots \quad \psi_3] \begin{bmatrix} \alpha_1 \\ \vdots \\ \alpha_3 \end{bmatrix} = \begin{bmatrix} \dot{\beta}_1 \\ \vdots \\ \dot{\beta}_3 \\ \dot{\gamma}_1 \\ \vdots \\ \dot{\gamma}_3 \end{bmatrix}. \tag{3.5}$$

We denote the macro-manipulator's actual encoder rates with $\dot{\beta}_i$ and mini-manipulator's computed joint rates with $\dot{\gamma}_i$, such that $J([\theta]) = J([\beta\gamma]^T)$. The interesting part of this exercise, as mentioned before, is that if we move the joints according to these velocities, the net velocity of the end-effector of the macro–mini-manipulator will remain undisturbed. Now, all that is required is to command the macro-manipulator's joints with required velocities as usual (ignoring the existence of mini-manipulator)—and simultaneously commanding the mini-manipulator's joints with values computed in real time from the response of dual manipultors' joints. The algorithm given below ensures that all the motions happen within the null space. The step response of each manipulator along with the commanded trajectory for mini-manipulator working in tandem with macro-manipulator is shown in Fig. 3.4.

1. Store the Jacobian $J(\theta_k)$ and its null space $[\,\psi_1 \quad \ldots \quad \psi_3\,]$. A command is generated by the motion of the phantom and sent to PID controller of the macro-manipulator.
2. Read the joint speeds $\dot{\beta}_i$s of the macro-manipulator. Due to the slow dynamics of the heavy mechanism, the actual speed will be different from the commanded value.
3. Substitute $\dot{\beta}_i$s in (3.5). Calculate the value of the α_i (the three unknown scalars) with three equations having known $\dot{\beta}_i$ [current joint speeds of big (macro) manipulator] at its RHS [using first three equations of (3.5)].
4. Now, it is straightforward to compute mini-manipulator's joint rates, γ_i. Substituting α_i in (3.5), compute the joint speed requirement for the mini-manipulator.
5. Command the mini-manipulator joints by sending γ_is to its controller. As its dynamics is better (fast response), it will attain the commanded speed.
6. Refresh the elements of $J(\theta_k)$ and $\{\mathcal{N}_J\}$ with updated $\theta_{(k+1)}$ from feedback.
7. Continue till the task is completed; that is, the manipulator finishes its moves.

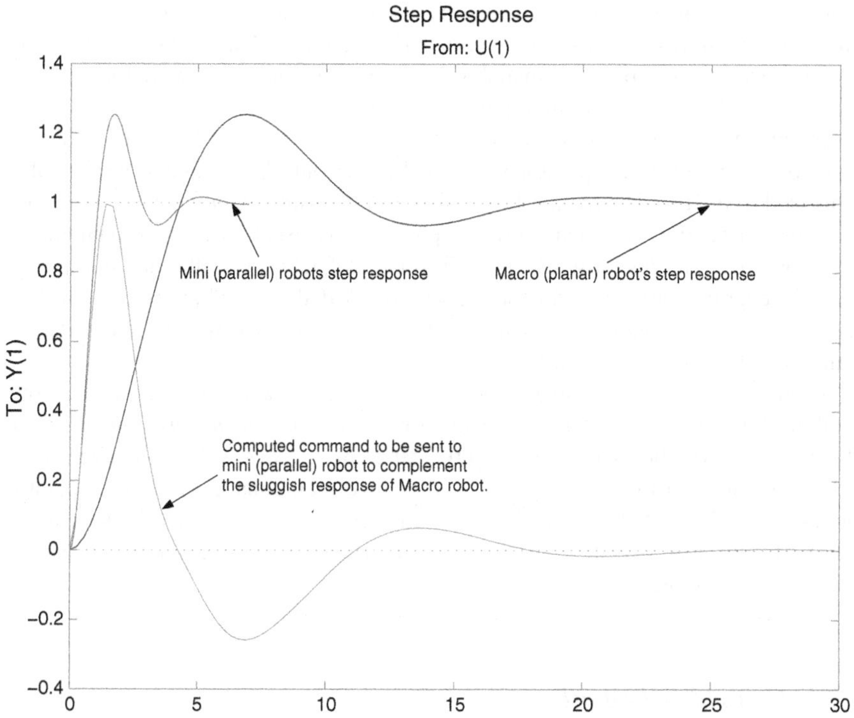

Fig. 3.4 Simulation of typical end-effector step responses and the corresponding computed command for mini-manipulator. Real-time plotting of six individual joints and its superimposition with net response is too dense for presentation

The controller of mini-manipulator is developed in-house and can be modified easily. Therefore, any transient deficiency in the performance of macro-manipulator can be comfortably compensated by the superior dynamics of the mini-manipulator. As the macro-manipulator picks up the speed, the burden on the mini-manipulator diminishes.

3.3 Implementation

A 3-link lightweight but quick acting mini-manipulator is designed based on *parallel manipulator*. What is a parallel manipulator? In comparison with a serial robotic arm where all the links of the arm are joined in series, a parallel manipulator has two or more links joined in parallel to form a tripod-like structure. This structure, naturally, can handle much larger stress. It also has quicker dynamic response due to its low mass. However, its range of motion is severely limited and a combination of both, a serial and a parallel manipulator, can work to our advantage—if a proper control structure is designed. However, it may be noted here that *serial mini-manipulator can also be used without any modification in the algorithm.* Only the value of J_{m} (the mapping from joint space to spatial space) needs to be substituted for the new manipulator in the Jacobian, be it parallel or serial.

The phantom is connected to a PC using the installed library functions in Visual C++ code. The mini-manipulator is servoed by a control card and amplifier, also connected to the PC. The main robot is linked to PC using ethernet. The main robot continues to receive command through phantom as usual as if mini-manipulator does not exist. The response of the main robot is fetched via ethernet, and this is used to generate commands for the mini-robot. A dedicated C program computes control card's input and sends it using D/A card. The control loop, thereafter, is maintained by this control card.

In our approach, we redesign the control structure appropriately so that our combined manipulator generates the motion *in response to the movement of the phantom.* In [4], the issue was slightly different—it was related to exerting a controlled force on the environment by the slave manipulator. Here, the operator moves the end-tip of the phantom with some velocity $U = |U|\hat{u}$ and the combination of mini–macro-system ensures that the gripper at the last link of the system follows it immediately.

3.3.1 Control Schematic

The control strategy, Fig. 3.5, is based on capturing the operator's command by computing the phantom's tip velocity. It is done by reading its joint encoders and substituting it in the following equation:

$$[\dot{x} \quad \omega]^T = \mathbf{U} = J_{\mathrm{Phantom}}(\theta)\dot{\theta}_{\mathrm{Phantom}}, \tag{3.6}$$

where J_{Phantom} is the instantaneous Jacobian of the phantom arm. The tip velocity $\dot{x} \in \Re^2$ is the desired linear velocity of the robotic arm, and $\omega \in \Re$ is its angular velocity. We find corresponding joint speed of the macro-manipulator using macro-manipulator's Jacobian for this velocity (since the Phantom's and manipulator's structures are kinematically different, for the identical tip velocities, their joint speed may be different):

$$\dot{\beta}_{\mathrm{M}} = J_{\mathrm{M}}(\beta_{\mathrm{act.}})^{-1}U. \tag{3.7}$$

This $\dot{\beta}_{\mathrm{M}}$ is the reference input for the macro-manipulator which is sent to its built-in PID controller; Jacobian $J_{\mathrm{M}}(\beta_{\mathrm{act.}})$ is created using actual encoder data of the macro-manipulator. Next, we compute the null space matrix of the combined Jacobian with all encoder data and desired $\mathbf{U}$ using (3.3).

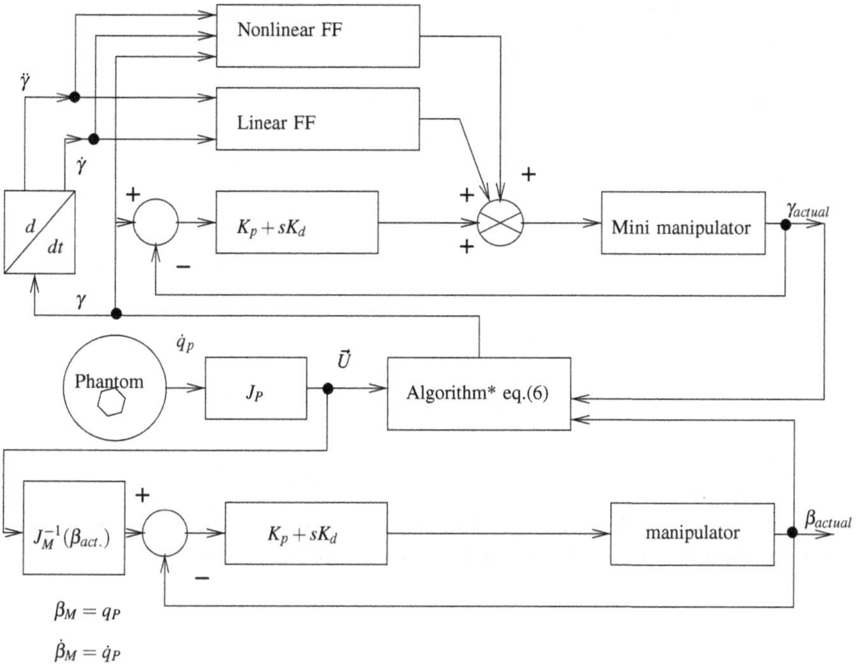

Fig. 3.5 Control schematic uses robots' existing control system and encoders (*the algorithm only needs to compute mini-manipulator's joint rates. The macro-manipulator works oblivious of the mini-manipulator's existence)

As anticipated, the joint speed of the macro-manipulator will be regulated by the slow dynamics of the system as before and they are the inputs for computing the mini-manipulator's speed. Using (3.5), we get the required speeds of mini-manipulator's joints. The mini-manipulator controller can be designed using various techniques including simple PID controller. Since it is a lightweight and small mechanism, the computed torque feedforward terms can be added to it to improve its response even more. For example, we can compute the torque requirement for the motors using the dynamic model of the mini-manipulator:

$$\Gamma = M(\gamma)\ddot{\gamma} + C(\gamma, \dot{\gamma})\dot{\gamma} + G(\gamma), \tag{3.8}$$

where $M(\gamma)$ is the inertial matrix of the system, $C(\gamma, \dot{\gamma})$ is the matrix computing coriolis and centrifugal forces [5], and G consists of load due to gravity. The overall stability is assured using a PID controller in the loop. We avoid any modification to the macro-manipulator's control system. Practically, the macro-manipulator's control circuits and amplifiers are not accessible for intervention.

The schematic functions as follows. Whenever phantom is manipulated, the macro-manipulator starts as usual, but since it has a large inertia, it is the mini-manipulator fitted to its end-effector that quickly steps in and moves the tip in desired direction. Eventually, the macro-manipulator gathers momentum and mini-

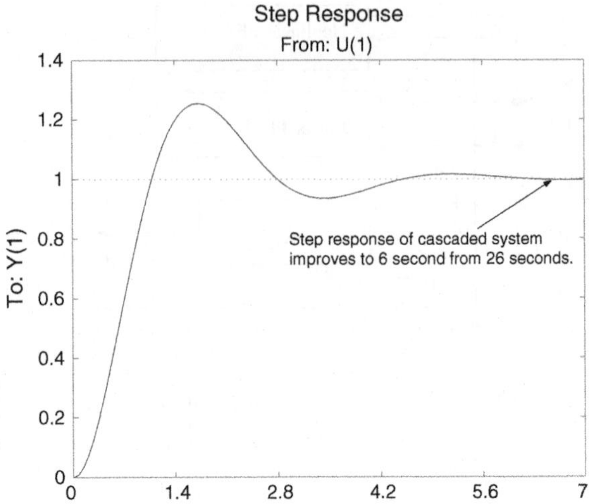

Fig. 3.6 Response of the combined robotic system

manipulator yields to it, gradually retracting its motion. The net result is that the combined manipulator's end-effector becomes much more responsive than earlier, Fig. 3.6.

3.4 Conclusion

A simple and efficient algorithm for computation of joint rates for a mini-macro-manipulator is presented. The algorithm reads the instantaneous joint rates of the two cascaded manipulators and updates the desired joint rates for the mini-manipulator so that its motions compensate the slow response of macro-manipulator. Thus, all the joints get automatically adjusted to compensate each other. The scheme is implemented at the command level only, and there is no need to change any of the manipulators' control system.

In situations where the mini-manipulator's actions are not required for slow tasks, it can be simply switched off without any change in algorithm. The algorithm does not affect macro-manipulator's function at all and merely uses its position feedback for computation of mini-manipulator's joint velocities.

References

1. Banerji A, Banavar RN, Venkatesh D (2005) A non-dexterous dual arm robot's feasible orientations along desired trajectories: analysis and synthesis. In: IEEE international conference on decision and control, pp 4391–4396
2. Banerji A, Banavar RN, Venkatesh D (2008) A task planner for a dual arm robot: a geometric formulation. In: Proceedings of the IEEE conference TENCON 2008, Hyderabad, 18–21 Nov 2008. http://ieeexplore.ieee.org

3. Dwarakanath TA, Bhutani G (2011) Novel design solution to high precision 3 axes translational parallel mechanism. In: Proceedings of the 15th national conference on machines and mechanisms (NaCoMM 2011), IIT Madras, Chennai, India
4. Khatib O (1995) Inertial properties in robotics manipulation: an object-level framework. Int J Robot Res 13(1):19–36
5. Murray RM, Li Z, Sastry SS (1994) A mathematical introduction to robotic manipulator. CRC Press, Boca Raton
6. Sciavicco L, Siciliano B (1996) Modeling and control of robot manipulators. McGraw-Hill, New York
7. Sharon A, Hardt D (1984) Enhancement of robot accuracy using endpoint feedback and a macro-micro manipulator system. In: Proceedings of the American control conference, pp 1836–1842
8. Yoshikawa T, Harada K, Matsumoto A (1996) Hybrid position/force control of flexible-macro/rigid-micro manipulator systems. IEEE Trans Robot Autom 12(4):633–640

Chapter 4
A Modified NSGA-II for Fuzzy Relational Multiobjective Optimization Problem

Garima Singh, Dhaneshwar Pandey and Antika Thapar

Abstract This study presents a multiple objective optimization problem with the solution space designed by a system of fuzzy relational equations based on max-product algebraic composition. The solution set of the fuzzy relation equation is generally characterized by a unique maximal solution and finite number of minimal solutions and is non-convex by nature. Owing to the nature of feasible space, the traditional metaheuristics cannot be applied in their original form. To overcome this situation, a modified version of NSGA-II has been presented. The original NSGA-II has set standards in the area of multiobjective optimization in terms of efficiency. But in our case, the algorithm fails to give feasible solutions at the end. For this, the algorithm is modified to adapt the algorithm in our problem domain. The whole procedure is illustrated by some test problems.

Keywords Fuzzy relation equation · Multiobjective optimization · Hybridized genetic algorithm

4.1 Introduction

Fuzzy relational models occur at various levels and stages in real-life usage systems whether materialistic or non-materialistic. Also, they are extremely important for their applications to fuzzy systems. Fuzzy systems have been widely applied in areas such as image processing, approximation reasoning, decision-making support systems, control systems, and data analysis [4, 5, 14]. Fuzzy relational models occur frequently in many problems in engineering design and soft sciences (e.g., psychology, economics, and social sciences). Basically, fuzzy systems are knowledge-based or rule-based systems that are implemented originally using the fuzzy

G. Singh (✉) · D. Pandey · A. Thapar
Department of Mathematics, Faculty of Science, Dayalbagh Educational Institute (Deemed University), Dayalbagh, Agra, India
e-mail: singhgarima.dei@gamil.com

V. Vijay et al. (eds.), *Systems Thinking Approach for Social Problems*, Lecture Notes in Electrical Engineering 327,
DOI 10.1007/978-81-322-2141-8_4

relations. The notion of fuzzy relation equations (FRE) and fuzzy relational calculus lies in the center of the fuzzy set theory and its applications, particularly in the area of fuzzy modeling and diagnostic and fuzzy control, etc. [4, 5]. Many problems in these areas end up in solving the FRE.

In real-life optimization problems, occurrence of several objectives is natural. The problem is addressed as the multiobjective optimization problem (MOOP). In case when the objectives are not conflicting, single optimal solution is accepted, but the situation becomes complex when the objectives are conflicting. In this case, in place of a single optimal solution, a set of good solutions called Pareto-optimal solutions or efficient solutions is obtained with the property that they are superior from the other existing solutions in the search space but incomparable to each other in terms of objective values. Their corresponding objective vectors in objective space are referred to as the Pareto front or non-dominated set. The rest of the solutions are known as dominated solutions.

In recent times, the increasing focus of researches is on the application of metaheuristics to solve multiobjective optimization problems. In class of multiobjective optimization metaheuristics, genetic algorithm reserves special attention. Genetic algorithm was first conceived by Holland [9]. Although genetic algorithms were not well known at the beginning, after the publication of Goldberg's book [6], they have been established as an effective and powerful global optimization algorithm providing robust search in multimodal and nonlinear complex search spaces. Their ability to simultaneously search different regions of a solution space makes it possible to find a diverse set of solutions for difficult problems with non-convex, discontinuous, and multimodal search spaces. Therefore, genetic algorithms have been the most popular approach to multiobjective design and optimization problems. Jones et al. [12] reported that 90 % of the approaches to multiobjective optimization aimed to approximate the true Pareto front for the underlying problem. A majority of these used a metaheuristic technique, and 70 % of all metaheuristic approaches were based on evolutionary approaches. Some best-known genetic algorithm approaches used to deal with the multiobjective optimization problems include plain aggregating approach, the population-based non-Pareto approach, the Pareto-based approach, etc. Schaffer [18] presented a Pareto-based approach to deal with MOOP using genetic algorithms. A comprehensive survey of optimization methods for MOOP is presented in [1]. Marler and Arora [16] presented an analysis of methods for multiobjective optimization for engineering design problems.

Fuzzy relational multiobjective optimization problem (FRMOOP) presents a class of mathematical programming problems with classical objectives constrained by a system of fuzzy relational constraints based on certain algebraic composition. A great deal of literature has been devoted to the area of multiobjective optimization. In the area of fuzzy relational optimization, it is still in the budding stage. Firstly, Wang [21] studied the problem of multiobjective mathematical programming for medical applications with multiple linear objective functions subjected to constraints defined by max–min composite fuzzy relation equation. But the work required the knowledge of all minimal solutions of system of fuzzy relational equations, which is not trivial at all.

Loetamonphong et al. [15] studied MOOP with multiple objective functions subjected to a set of max–min fuzzy relational equations. Since the feasible domain of such a problem is non-convex, taking advantage of the special structure of the solution set, they developed a reduction procedure to simplify the problem. They proposed a genetic-based algorithm to find the Pareto-optimal solutions.

Khorram and Zarei [13] considered a multiple objective optimization model subjected to a system of fuzzy relational equations with max-average composition and presented a reduction procedure in order to reduce problem dimension and then used a modified genetic algorithm to solve the problem.

Jiménez et al. [10] considered multiobjective linear programming problems. By using the idea of fuzzy goals for each of the objective functions, they showed that in the case that one of our goals is fully achieved, a fuzzy-efficient solution may not be Pareto optimal, and therefore, they proposed a general procedure to obtain a non-dominated solution, which is also fuzzy efficient. Further, Jiménez and Bilbao [11] proposed that in fuzzy optimization, it is desirable that all fuzzy solutions under consideration are attainable, so that the decision maker will be able to make a posteriori decisions according to current decision environments. A case study was analyzed, and the proposed solutions from the evolutionary algorithm considered were given.

Thapar et al. [20] considered a multiobjective optimization problem subjected to a system of fuzzy relational equations based upon the max-product composition. A well-structured non-dominated sorting genetic algorithm was applied to solve the problem.

Recently, Singh et al. [19] discussed the application of some decidable utility function for a multiobjective fuzzy relational optimization problem with max-product composition.

From the past decade, the non-dominated sorting in genetic algorithms (NSGA) [2] have established as a popular and efficient algorithm for solving MOOP. The algorithm has set certain landmarks in the area and successfully been applied to a variety of engineering design problems that involve multiple objectives. Despite of its effectiveness, it is criticized due to the high computational complexity of non-dominated sorting, lack of elitism. Also, the algorithm generally faces difficulty in achieving diversity in solutions. To overcome the demerits of NSGA, an improved version of NSGA known as NSGA-II [3] was introduced that alleviates all the demerits of NSGA with some additional features and results in more diversified solution set with lesser computational complexity.

Here, we present a modified version of the standard NSGA-II [3] to solve the FRMOOP. Depending on the problem, the modifications are introduced to adapt the original algorithm to our problem environment.

The work is organized as follows: Section 4.1 presents the basic idea and motivation behind the problem. Section 4.2 discusses the main problem with the description of the solution space. Section 4.3 describes the method used to solve the problem. Section 4.4 presents the illustration of the whole procedure by some test problems. At the end, the concluding remark and the references are presented.

4.2 Problem Description

We are interested in the following multiobjective optimization model with max-product fuzzy relation equations as constraints:

$$\begin{aligned} &\text{Min}\{f_1(x), f_2(x), \ldots, f_k(x)\} \\ &\text{s.t.} \max_{i \in I} (x_i \odot a_{ij}) = b_j, \quad \forall j \in J \\ &0 \leq x_i \leq 1, \quad \forall i \in I \end{aligned} \tag{4.1}$$

where $f_k(x)$ is a linear or nonlinear objective function, $k \in K = \{1, 2, \ldots, s\}$ and $A = [a_{ij}]$, $0 \leq a_{ij} \leq 1$, be a $m \times n$ dimensional fuzzy matrix and $b = [b_1, b_2, \ldots, b_n]$, $0 \leq b_j \leq 1$, be a n-dimensional vector, and then, the following system of FRE is defined by A and b:

$$x \circ A = b \tag{4.2}$$

where $\circ$ denotes max-$\odot$ composition of x and A, $\odot$ denotes a compositional operator from the Goguen algebra over the residuated lattice $L = \langle [0, 1], \wedge, \vee, \odot, \rightarrow, 0, 1 \rangle$, and $I = \{1, 2, \ldots, m\}$ and $J = \{1, 2, \ldots, n\}$ be the index sets.

Let $X(A, b) = \{x \in [0, 1]^m | x \circ A = b\}$ be the solution set of FRE (4.2). For any $x^1, x^2 \in X$, we say $x^1 \leq x^2$ if and only if $x_i^1 \leq x_i^2$, $\forall\, i \in I$. Therefore, $\leq$ forms a partial ordering relation on X and $(X, \leq)$ becomes a lattice. Equations in (4.2) form a system of latticized polynomial equations. $\widehat{x} \in X(A, b)$ is the maximum solution, if $x \leq \widehat{x}$, $\forall\, x \in X(A, b)$. Similarly, $\breve{x} \in X(A, b)$ is a minimal solution, if $x \leq \breve{x}$ implies $x = \breve{x}$, $\forall\, x \in X(A, b)$. According to [4, 5, 7, 8], if $X(A, b) \neq \phi$, then it is, in general, a non-convex set which can be completely determined by unique maximum solution $\widehat{x}$ and several minimal solutions $\breve{x}$.

The maximum solution can be computed explicitly by the residual implicator (pseudo-complement) by assigning

$$\widehat{x} = A \rightarrow b = \left[\min_{j \in J} (a_{ij} \rightarrow b_j) \right]_{i \in I} \tag{4.3}$$

where

$$a_{ij} \rightarrow b_j = \begin{cases} 1 & \text{if } a_{ij} \leq b_j \\ b_j / a_{ij} & \text{if } a_{ij} > b_j \end{cases}$$

If $\breve{X}(A, b)$ denotes the set of all minimal solutions, then the complete solution set of FRE (4.2) can be formed as follows:

$$X(A, b) = \bigcup_{\breve{x} \in \breve{X}(A,b)} \{x \in [0, 1]^m | \breve{x} \leq x \leq \hat{x}\} \tag{4.4}$$

The max–min composition is commonly used when a system requires conservative solutions in the sense that the goodness of one value cannot compensate the badness of another value [15]. In reality, there are situations that allow compensation among the values of a solution vector. In such cases, the min operator is not the best choice for the intersection of fuzzy sets, but max-product composition is preferred since it can yield better or at least equivalent result [7].

4.3 A Modified NSGA-II for FRMOOP

In recent years, metaheuristics has been proved as a potent technique to solve MOOP, as they offer more competencies in comparison with the traditional methods used to solve them.

In our case, the feasible domain of the considered optimization problem given by Eq. (4.4) has been extensively investigated by numerous researchers [8, 14, 17]. It has been well established that the solution space in this case is non-convex in general, so general metaheuristics used to solve multiobjective optimization problems cannot be applied in their original form as they might result in infeasible solutions.

This idea is adapted to our problem domain. As we have mentioned that any of the metaheuristics cannot be applied to the FRMOOP in its original form due to the nature of solution space. So the case is with NSGA-II. For this, we adopt NSGA-II with some modifications that are required to be made in the algorithm and not in the operations. The modified version of NSGA-II that we have used presents the combination of elements of the original NSGA-II with the feasible recombination operators designed so as to keep the newly generated solutions feasible. For this, the algorithm is modified by introducing a repair algorithm that keeps repairing the individuals resulted from the genetic operators as used in the original NSGA-II. The modified NSGA-II has some distinct characteristics that are described as follows.

4.3.1 Initialization

In general, in NSGA-II, the population is initialized randomly. This works well when dealing with unconstrained optimization problems. However, for our case, randomly generated solutions may not be feasible. Since GA intends to keep the solutions (chromosomes) feasible, we present an initialization module to initialize a population by randomly generating the individuals inside the feasible domain.

The feasible domain for the considered problem has a special structure. As the feasible region is designed by a system of fuzzy relational equations, it might be the case that some of the variables assume some specific values. The values of such variables needed to be fixed for the sake of solvability of the system. To identify such variables, we define sets:

$$I_j = \{i \in I | \widehat{x}_i \cdot a_{ij} = b_j\}, \quad \forall j \in J.$$

Definition If I_j is singleton, say $I_j = \{i'\}_{i' \in I}$ for some $j \in J$, then $x^*_{i'} = \widehat{x}_{i'}$, $i \in I$, and then, the variable $x_{i'}$ is fixed as it is the only variable to satisfy the *j*th equation. Such variables assume fixed values in all solutions.

Once fixed variables have been detected, their value is fixed as $x_i = \widehat{x}_i = \breve{x}_i$. Then, an initial population of fixed size is created with the fixed variables assuming the value $\widehat{x}_i$ and the variables that are not fixed assuming a random value in the range$(0, \widehat{x}_i)$ in the solutions. Now, the feasibility of generated solutions is examined. The following algorithm has been used to maintain the feasibility of the solutions:

Algorithm 1: For maintaining feasibility of solutions

1. Choose a violated constraint j. Let $D_j = \{i \in I | a_{ij} \geq b_j\}$
2. Randomly choose an element $k \in D_j$. For $a_{kj} > b_j$ or $\widehat{x}_k = b_j / a_{kj}$, set $x_k = b_j / a_{kj}$. Otherwise, assign a random number between $[b_j / a_{kj}, \widehat{x}_k]$ to x_k.
3. Check the feasibility of the new solution. If the solution is still infeasible, then go to Step 1 and repeat the process. Otherwise, stop.

To begin the algorithm, the initialization described above is applied, which avoids the unnecessary exploration of the search space. Once the population of finite size has been generated, the algorithm preserves its original elements except the initialization and recombination operators for the sake of feasibility of solutions, as the general real-coded genetic algorithm operators do not produce the feasible individuals at the end. The specially designed genetic operators *crossover* and *mutation* are applied on the population. Then, the infeasible individuals are repaired with the Algorithm 1 described above. The introduction of the new operators does not affect the efficiency of the algorithm much, as the structure of the algorithm remains the same. Basically, the introduced operators are the key elements that adopt the algorithm to our domain or application.

4.3.2 Crossover

Owing to the nature of solution space of the problem, the conventional real-coded crossover techniques are not feasible. So, a specific crossover scheme is designed, which generates feasible individuals at the end. The proposed algorithm used for crossover can be described in the following steps:

Algorithm 2: Crossover

Get the matrices A and b and find the maximum solution $\widehat{x}$ by (4.4) and set parameters $0 \leq \alpha \leq 1,\ \beta \geq 1,\ 0 \leq \zeta \leq 1,\ 0 \leq \delta \leq 1$
Randomly select two individuals x_1 and x_2 from the selected population.
For i = 1, 2
Generate a random number $\varepsilon \in [0, 1]$
If $(\varepsilon \geq \zeta)$

$$x_i = \beta x_i - (\beta - 1)\widehat{x}$$

Else

$$x_i = \alpha x_i + (1 - \alpha)\widehat{x}$$

Generate a random number $\varepsilon_2 \in [0, 1]$
If $(\varepsilon_2 \geq \delta)$
Go to evaluation procedure
Else

$$x_1^{\text{next}} \leftarrow x_1$$
$$x_1 = \alpha x_1 + (1 - \alpha)x_2$$

If $x_1 \circ A = b$
Go to evaluation procedure
Else

$$x_1 \leftarrow x_1^{\text{next}}$$
$$x_1 = \beta x_1 - (\beta - 1)x_2$$

If $x_1 \circ A \neq b$

$$x_1 \leftarrow x_1^{\text{next}}$$

If $x_1 \circ A \neq b$
Make x_1 feasible using Algorithm 1
Go to evaluation procedure.
End

The repeated linear combinations of individuals draw the generated individuals inside the feasible space. Here, α and β are small numbers close to 1, respectively, and are generally kept small. For our problem, we are taking $\alpha = 0.99$, $\beta = 1.0085$, $\zeta = 0.012$, $\delta = 0.99$.

4.3.3 *Mutation*

Mutation randomly perturbs a candidate solution by exploiting the search space with a hope to create a better solution in the problem domain. We adopt the following mutation procedure to solve our problem:

Algorithm 3: Mutation

1. Get the matrices A and b and find the maximum solution $\widehat{x}$ by (4.4) and set the mutation probability $\theta = 0.1$.
2. Generate $r_i \in [0, 1]$ for each bit of every individual in the crossed population.
3. For $\forall i \in I$ if $r_i \leq \theta$, randomly assign x_i a number from $[0, \widehat{x}_i]$.
4. For the modified $x = (x_1, x_2, \ldots, x_m)$ check feasibility $x \circ A = b$.
5. If $x \circ A = b$, go to the evaluation procedure, else make the solution feasible via Algorithm 1.

The mutated individuals further undergo the original process of NSGA-II. The whole designed algorithm of modified NSGA-II is shown in Fig. 4.1.

4.4 Results and Discussion

In this section, results of the proposed genetic algorithm for multiple linear and nonlinear optimization problems are discussed. We consider few examples of multiobjective linear and nonlinear optimization problems and systems of fuzzy relational equations with product t-norm-based compositions to investigate the nature of the solutions obtained using the proposed procedure.

Example 1 Consider a four-dimensional problem with randomly generated fuzzy matrices A and b as follows:

$$A = \begin{bmatrix} 0.5042 & 0.0569 & 0.3641 & 0.2527 \\ 0.9398 & 0.6578 & 0.0359 & 0.1663 \\ 0.4979 & 0.3937 & 0.5715 & 0.9849 \\ 0.7182 & 0.0330 & 0.9476 & 0.1271 \end{bmatrix},$$

$$b = [0.6120 \quad 0.4284 \quad 0.8075 \quad 0.1083]$$

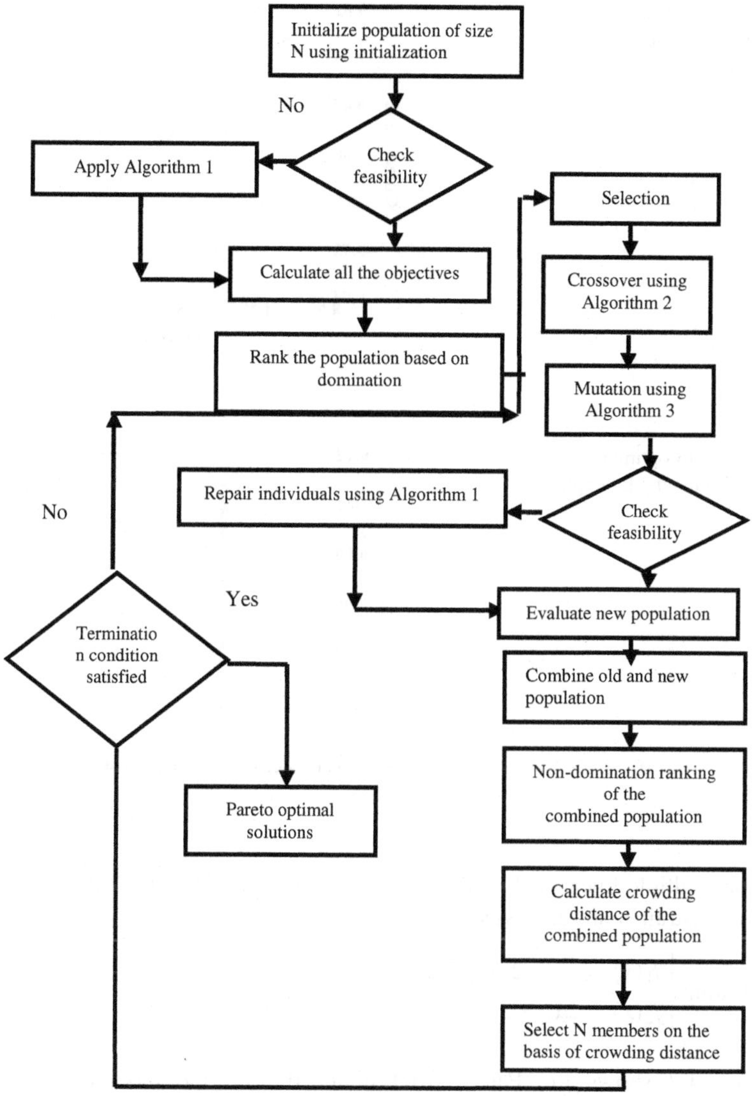

Fig. 4.1 Flowchart of modified NSGA-II algorithm

The maximum solution obtained is $[0.4286 \quad 0.6512 \quad 0.1100 \quad 0.8521]$. For this particular problem, the values of x_2 and x_4 of all solution vectors have to be fixed at 0.6512 and 0.8521, respectively. Therefore, we can focus on values of x_1 and x_3 only, and the problem dimension reduces to two. Since the problem is reduced as a two-dimensional problem, the result can be presented graphically. The test results

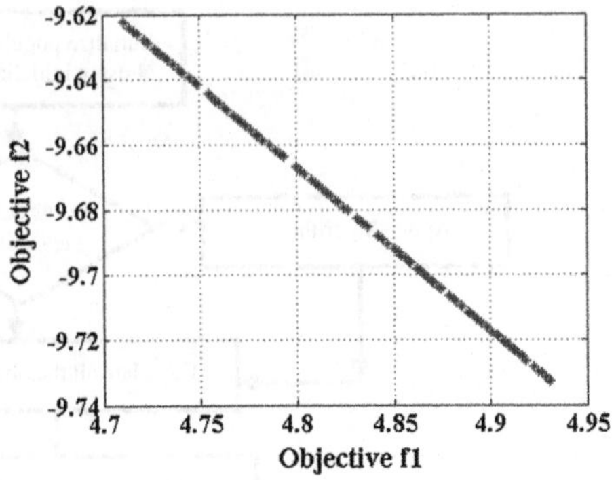

Fig. 4.2 Pareto front obtained—Example 1—Case 1

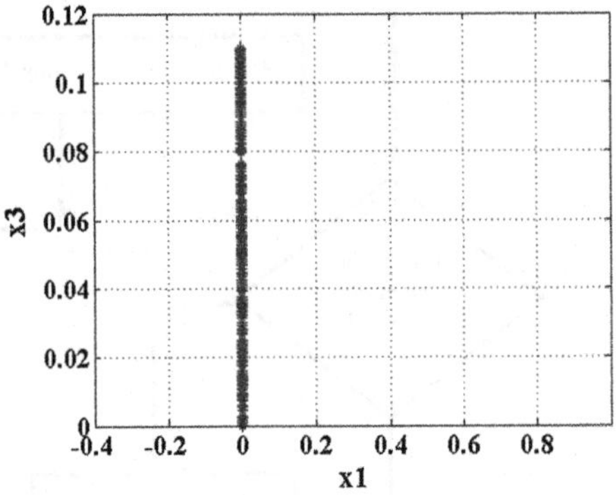

Fig. 4.3 Pareto-optimal solutions obtained—Example 1—Case 1

for some multiple linear and nonlinear optimization problems with this system of FRE as constraints are discussed below.

Case 1 min $\begin{bmatrix} f_1(x) = x_1 + 2x_2 + 2x_3 + 4x_4, \\ f_2(x) = 2x_1 - 3x_2 - x_3 - 9x_4. \end{bmatrix}$

Figure 4.2 shows the Pareto front obtained with the modified NSGA-II. Figure 4.3 shows Pareto-optimal solutions obtained in this case.

Case 2 min $\begin{bmatrix} f_1(x) = 10(x_1 - 0.45)^2 + 10(x_3 - 0.35)^2, \\ f_2(x) = -6(x_1 - 0.7)^2 + 10(x_3 - 0.45)^2. \end{bmatrix}$

Figure 4.4 shows the Pareto front obtained with the modified NSGA-II. Figure 4.5 shows Pareto-optimal solutions obtained in this case.

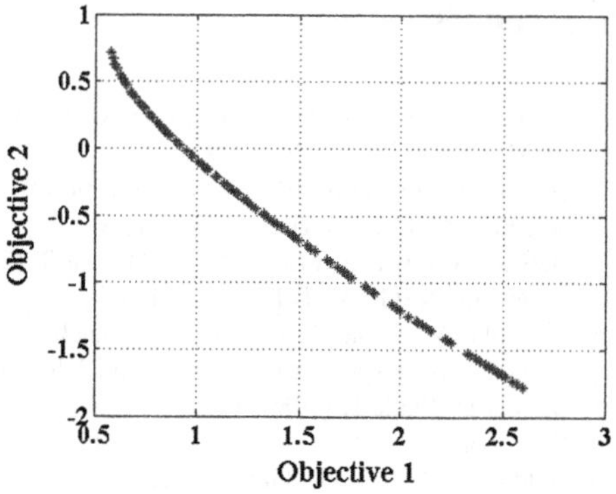

Fig. 4.4 Pareto front obtained—Example 1—Case 2

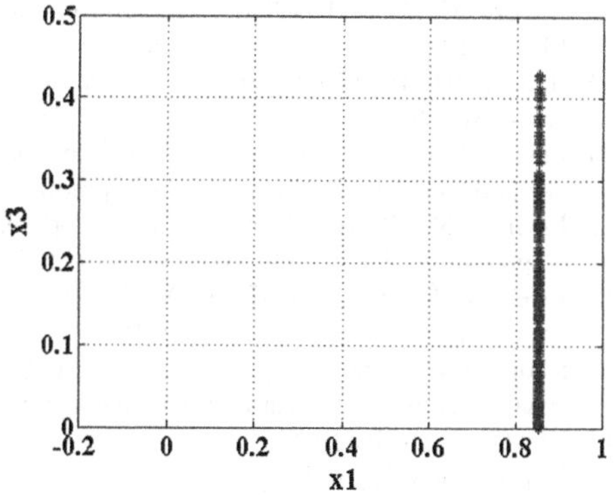

Fig. 4.5 Pareto-optimal solutions obtained—Example 1—Case 2

4.5 Conclusion

This paper considers a FRMOOP. With a careful observation of the feasible domain, a modified NSGA-II has been proposed. The original NSGA-II has established as a standard algorithm for multiobjective optimization. But this may not work in our problem due to the non-convex nature of the solution space. The modified NSGA-II possesses the original accelerating elements of the NSGA-II except the feasible generation of the individuals and continuous repairing of the infeasible individuals at the end of genetic operations mutation and crossover. The original algorithm has a special cadre in class of methods to solve MOOP. Following the same idea, the proposed algorithm performs well, offering much less computational complexity and CPU time.

Acknowledgments Authors are thankful to referees for their valuable suggestions.

References

1. Coello CAC (2000) An updated survey of GA-based multiobjective optimization techniques. ACM Comput Surv 32(2):109–143
2. Deb K (2001) Multi-objective optimization using evolutionary algorithms for solving multi-objective problems. Wiley, New Jersey
3. Deb K, Agrawal S, Pratap A, Meyarivan T (2002) A fast and elitist multiobjective genetic algorithm: NSGA-II. IEEE Trans Evol Comput 6(2):182–197
4. Di Nola A, Sessa S, Pedrycz W, Sanchez E (1989) Fuzzy relation equations and their applications to knowledge engineering. Kluwer, Boston
5. Dubois D, Prade H (1980) Fuzzy sets and systems: theory and applications. Academic Press, New York
6. Goldberg DE (1989) Genetic algorithms in search, optimization, and machine learning. Addison-Wesley, Reading
7. Hassanzadeh R, Khorram E, Mahdavi I, Mahdavi-Amiri N (2011) A genetic algorithm for optimization problems with fuzzy relation constraints, using max-product composition. Appl Soft Comput 11:551–560
8. Higashi M, Klir GJ (1984) Resolution of finite fuzzy relation equations. Fuzzy Sets Syst 13:65–82
9. Holland JH (1975) Adaptation in natural and artificial systems. The University of Michigan Press, Ann Arbor
10. Jiménez F, Cadenas JM, Sánchez G, Gómez-Skarmeta AF, Verdegay JL (2006) Multiobjective evolutionary computation and fuzzy optimization. Int J Approximate Reasoning 43:59–75
11. Jiménez M, Bilbao A (2009) Pareto-optimal solutions in fuzzy multiobjective linear programming. Fuzzy Sets Syst 160:2714–2721
12. Jones DF, Mirrazavi SK, Tamiz M (2002) Multiobjective meta-heuristics: an overview of the current state-of-the-art. Eur J Oper Res 137:1–9
13. Khorram E, Zarei H (2009) Multi-objective optimization problems with fuzzy relation equation constraints regarding max-average composition. Math Comput Model 49:856–867
14. Klir GJ, Yuan B (1995) Fuzzy sets and fuzzy logic: theory and applications. Prentice Hall, Upper Saddle River
15. Loetamonphong J, Fang SC, Young RE (2002) Multi-objective optimization problems with fuzzy relation equation constraints. Fuzzy Sets Syst 127:141–164
16. Marler RT, Arora JS (2004) Survey of multi-objective optimization methods for engineering. Struct Multidisc Optim 26:369–395
17. Sanchez E (1976) Resolution of composite fuzzy relation equations. Inf Control 30:38–48
18. Schaffer JD (1985) Multiple objective optimization with vector evaluated genetic algorithms. In: Proceedings of 1st international conference on genetic algorithms, pp 141–153
19. Singh G, Thapar A, Pandey D (2013) Decidable utility functions restricted to a system of fuzzy relational equations. In: Recent advancements in system modeling applications. Lecture notes in electrical engineering, vol 188. Springer, New Delhi, pp 89–102
20. Thapar A, Pandey D, Gaur SK (2011) Satisficing solutions of multi-objective fuzzy optimization problems using genetic algorithm. Appl Soft Comput 12:2178–2187
21. Wang WF (1995) A multiobjective mathematical programming problem with fuzzy relation equation constraints. J Multi-criteria Dec Anal 4:23–35

Chapter 5
Ground-Based Measurement for Solar Power Variability Forecasting Modeling Using Generalized Neural Network

Vikas Pratap Singh, Vivek Vijay, B. Ravindra, S. Jothi Basu and D.K. Chaturvedi

Abstract The primary aim of this paper is to analyze solar power variability. Ground-based measurements of solar photovoltaic power are used for the forecasting of 43-kW A-Si SPV system. In this study, we describe the variability in the power production of solar photovoltaic plant at IIT, Jodhpur. Solar PV generation forecasting is playing a key role in accurate solar power dispatchability as well as scheduling of PV power for hybrid power generation systems. The actual power produced by a PV power system varies according to variation in meteorological parameters and efficiency of PV system components. For the purpose of forecasting as per the schedule in the Indian power sector, a time slot of 15 min is considered for each forecasting. The proposed generalized neural network technique will be appropriated for modeling of solar power variability forecasting. In this paper, we used generalized neural network for forecasting the PV power variability.

Keywords Solar power · Power forecasting · Solar power variability

5.1 Introduction

In the twenty-first century, photovoltaic solar power is a source of clean renewable energy in a large quantity. It is highly reliable and safe to install and to maintain. In the last years, solar photovoltaic power has experienced a wonderful and fast growth due to the Jawaharlal Nehru National (JNNS) Solar Mission in our Indian

V.P. Singh (✉) · V. Vijay · B. Ravindra
Indian Institute of Technology Rajasthan, Jodhpur, Rajasthan, India
e-mail: vikasforsmile@iitj.ac.in

S. Jothi Basu
Central Power Research Institute, Bangalore, Karnataka, India

D.K. Chaturvedi
Dayalbagh Educational Institute, Agra, Uttar Pradesh, India

V. Vijay et al. (eds.), *Systems Thinking Approach for Social Problems*,
Lecture Notes in Electrical Engineering 327,
DOI 10.1007/978-81-322-2141-8_5

power sector. Solar photovoltaic systems help in hybrid power generation system and conventional electric grid to add value as systems of power generation. The electric power produced by the PV system can then be consumed by the connected load, and no power is taken from the main grid unless the load connected to the system is less than the capacity of PV systems [1]. Roof and facades of existing buildings represent a huge potential area for PV system installation, allowing the possibility to combine energy production with other functions of the building or non-building structure. Bi-PV systems seem to provide the most cost-effective and energy-effective application of grid-tied PV systems [2–4]. As a developing country, India needs a better energy management and environmental protection. It is the greatest challenge for any developing country.

Solar insolation variation is the main issue of photovoltaic power production from a PV system. Compared to conventional power, solar power is very difficult to dispatch due to uncertainty. So there is a need for the study of solar power variability and analysis. In this experimental work, data collected from a 43-kW grid-connected amorphous-silicon (A-Si) solar PV system installed at IIT Jodhpur are used.

Jodhpur city, which is called as *Sun City*, has 320 days of sun availability in a year, i.e., there is huge potential in the field of solar systems which can be harnessed to solve numerous local problems such urban and rural electricity, cooling, and water-related problems.

The above installed grid-connected PV system is going to function as a test bed to the future proliferation of solar systems. So, it becomes almost necessary that data regarding their performance must be studied in local climatic conditions and a comparative conclusion may be drawn. This is because the power generated by a grid-tied photovoltaic system depends on meteorological parameters such as incident solar radiation, ambient temperature, and wind velocity. In this study, we measured ambient temperature, module temperature, DC voltage, DC current, inverter output energy, and solar irradiance. Because of variations in solar irradiation and weather parameter, solar power is variable. PV power variability forecasting helps in the following power system areas such as [4]:

(a) Unit commitment
(b) Scheduling of PV power
(c) Hybrid power generation
(d) Risk analysis

These forecasting approaches have traditionally been used to calculate the performance of renewable energy systems. These approaches usually require extensive resources and lengthy computational time for forecast the PV power variability [5]. The climate of our Indian continental is variable according to seasons. Weather and solar power relationship are related to each other. Therefore, in forecasting, it is sufficient to use past solar power generation data. In solar power variability forecasting, the following data are considered as inputs.

- Numerical weather prediction (NWP) data
- Satellite-based measurement data
- Ground-based measurement data

In general, NWP forecasts are not yet precise enough for solar power variability forecasting applications. There are a number of existing short-term solar variability forecasting methods that make use of cloud images as input. But ground-based measurement data are very valuable because it is simple to measure, accurate, and inexpensive compared to NWP and satellite-based measurement data.

Conventional methods used for forecasting are given below [6]:

- Regression methods
- Time series methods
- Knowledge-based approach.

In this paper, ground-based measurement data and neural network methodology are used for modeling solar power variability forecasting.

5.2 Ground-Based Measurement Data

This paper uses ground-based measurement of solar radiation and weather parameter to forecast the photovoltaic power variability of solar power plant. In this process, we need power generation data and weather data, and then, there is a standard procedure for collection of weather parameters and system parameters. The data are described in detail below.

5.2.1 Collection and Analysis of Data

Plant generation monitored at the inverter side by a data acquisition system is controlled by a Sunny sensor web box. Each measurement is recorded every 15 min. The Sunny sensor web box is mounted on PV mounting frame shown in Fig. 5.1.

The location of the plant is the main building of IIT Jodhpur shown in Figs. 5.2 and 5.3.

Data recorded in each 15-min interval for AC power output on August 2, 2011. The meteorological parameter used in this paper is also recorded by sunny sensor box. The observations had 15-min basis values for temperature, wind velocity, humidity, and solar radiation.

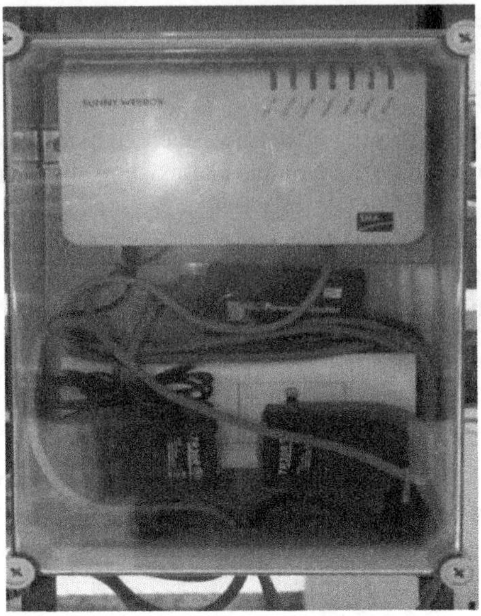

Fig. 5.1 Sunny sensor box

Fig. 5.2 Solar PV system at block-1 rooftop

5.2.2 Input Variables for Modeling

Figures 5.4, 5.5 and 5.6 show the variation of input variable with respect to time. The input variables considered in these models are as follows:

- Solar radiation (SR)
- Ambient temperature (AT)
- Module temperature (MT)
  Figure 5.5 shows the module temperature variation with ambient temperature.
- Wind velocity (WV)

Fig. 5.3 43-kW A-Si-based solar PV system at rooftop

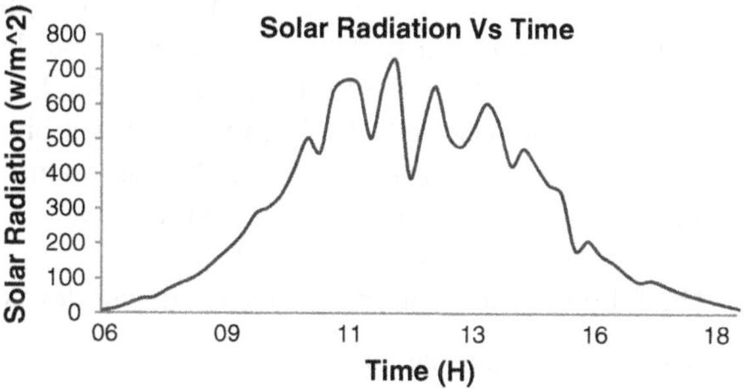

Fig. 5.4 Variation of solar radiation versus hours

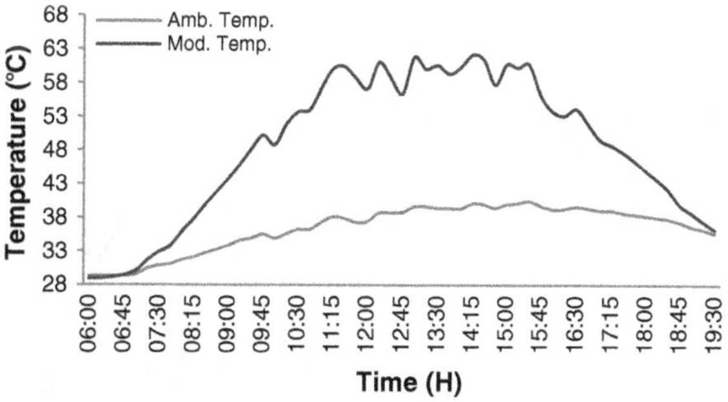

Fig. 5.5 Variations in ambient temperature versus hours

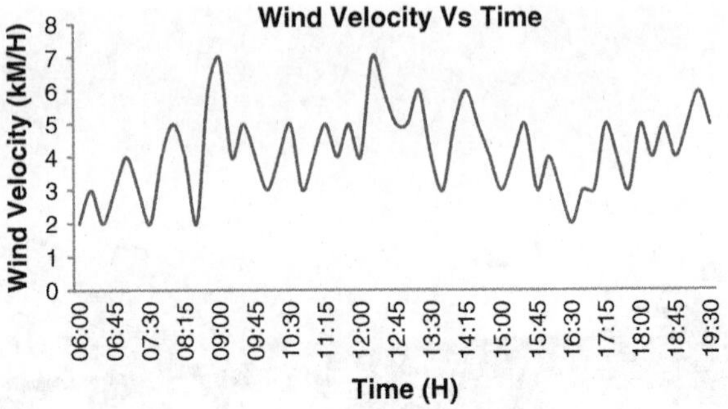

Fig. 5.6 Variations in wind velocity versus hours

5.3 Development of Neuron Model

5.3.1 Artificial Neural Network Model

ANNs imitate the learning process of the human brain and can process problems involving nonlinear and complex data even if the data are imprecise and noisy. Thus, they are ideally suited for the modeling of solar power forecasting which are known to be complex and often nonlinear. It is a biologically inspired technique. Artificial neural network function likes human brain network. The artificial neural network method is slightly different from artificial intelligence method. The brain neural network is shown in Fig. 5.7 [7].

The following steps are necessary for model development:

- Selection of input parameter
- Selection of neural network
- Selection of perfect training algorithm
- Selection of training parameter

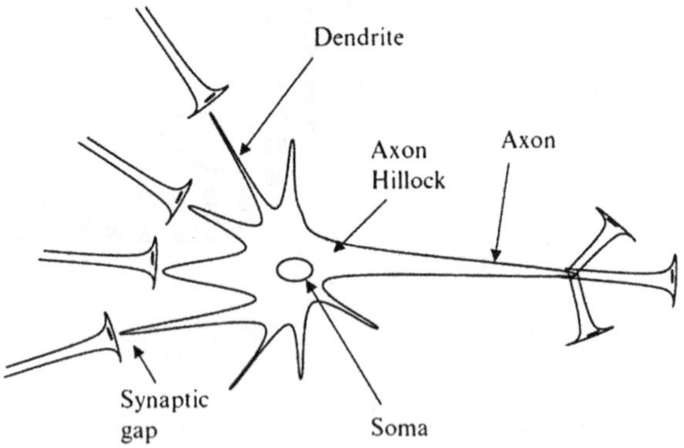

Fig. 5.7 Structure of brain neuron network

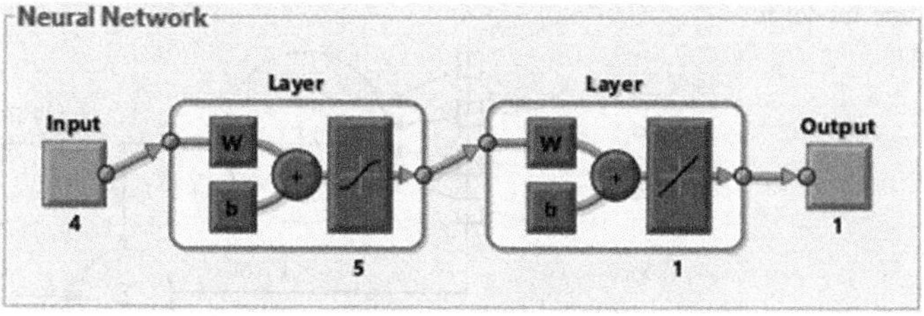

Fig. 5.8 Neural network for modeling of solar power forecasting

The data have been acquired at Indian Institute of Technology Jodhpur, India, in energy Lab. Solar power depends on the system parameters and meteorological parameters such as module temperature, solar radiation, ambient temperature, and wind velocity, which is shown in Fig. 5.8. These are four input parameters for the ANN model development. The selected input parameters have used for ANN training [8].

In this paper, multilayered feed-forward neural network is used for solar power forecasting modeling. And following specific selection of neurons and layers provides better outcomes (Table 5.1).

In this paper, backpropagation training algorithm is used with learning and momentum factors. During the training, sum-squared error is fed back to change the weight as shown in Fig. 5.9 [9] (Table 5.2).

In Fig. 5.10, actual data are compared with testing performance of ANN model. And ANN error deviation with solar radiation is indicated in Fig. 5.11. Error reduces in nine epochs, and ANN learning tool is shown in Fig. 5.12. It depicts the comparison with actual solar power curve and ANN model-based solar power curve (Table 5.3).

Solar power forecasting is basically nonlinear and complex problem so adaptive neurofuzzy interface system is able to rectify the following problem of ANN [10–12].

- ANN takes a lot of time to compute the algorithm.
- The training time depends on the nature of the data or sequence of presentation of data.
- The time of preparation depends on the mathematical function of the input output form.

Table 5.1 Structure of neural network

S. No.	Network parameters	Value
1	Number of input variables	4
2	Number of outputs	1
3	Number of input layers	1
4	Number of hidden layer neurons	5
5	Number of hidden layers	1

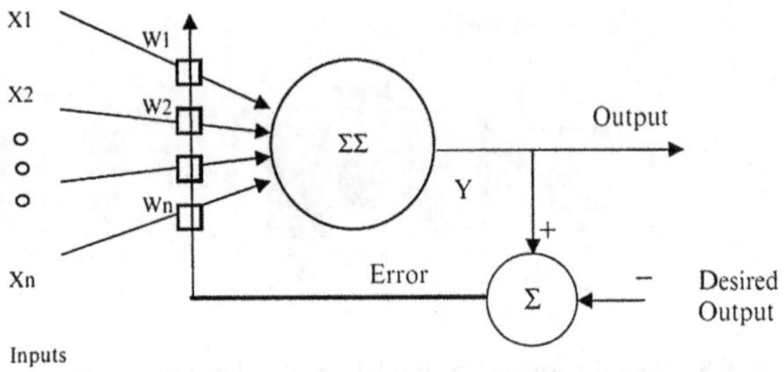

Fig. 5.9 Aldine model for model learning

Table 5.2 Values of training parameters

S. No.	Parameters	Value
1	Number of epochs	250
2	Error tolerance	0.001
3	Learning rate	0.003
4	Momentum factor	0.1

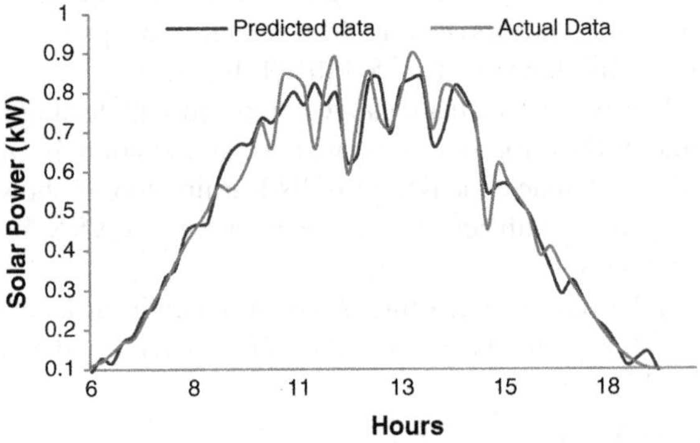

Fig. 5.10 Solar power output using ANN

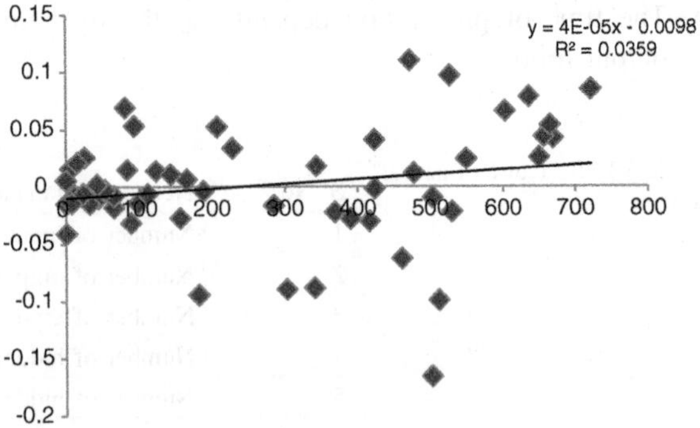

Fig. 5.11 Error with respect to the actual solar radiation

Fig. 5.12 Mean square error in ANN model

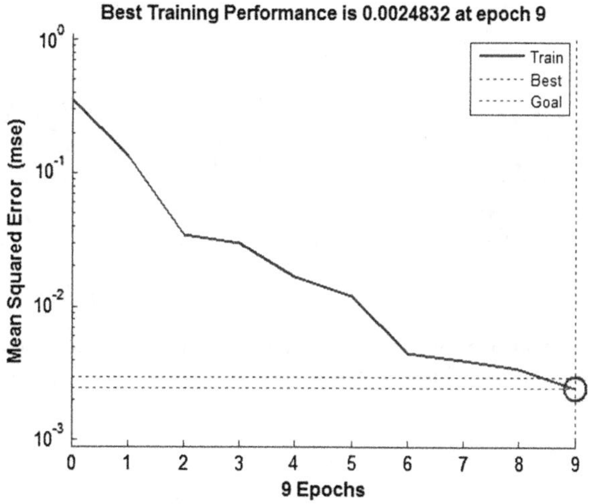

Table 5.3 RMS error analysis

S. No.	Model name	Max error	Min error	RMSE error
1	ANN	0.0868	−0.2121	0.1019

5.3.2 Generalized Neural Network Model

To rectify the problems of conventional neural network, generalized neural network is used as modified neural network which introduced by the author [8]. All training and testing parameters are same in ANN and GNN models. The combination of summation (Σ) and product (π) as aggregation function is used in generalized neural network. Use of the sigmoid threshold function and ordinary summation or product as aggregation functions in the existing models fails to cope with the nonlinearity involved in real-world problems. To deal with these, the proposed model has both sigmoid and Gaussian functions with weight sharing. The generalized neuron model has flexibility at both the aggregation and threshold function levels to cope with the nonlinearity involved in the case of applications dealt with. The neuron has both Σ and π aggregation functions. The Σ aggregation function has been used with the sigmoid characteristic function, while the π aggregation function has been used with the Gaussian function as a characteristic function. And GNN is more flexible than ANN and ANFIS. It is more appropriate for nonlinear and complex problems such as forecasting. This combination of aggregation function increases the accuracy of the algorithm. Figures 5.13 and 5.14 show the aggregation functions and block diagram for GNN model development, respectively [8].

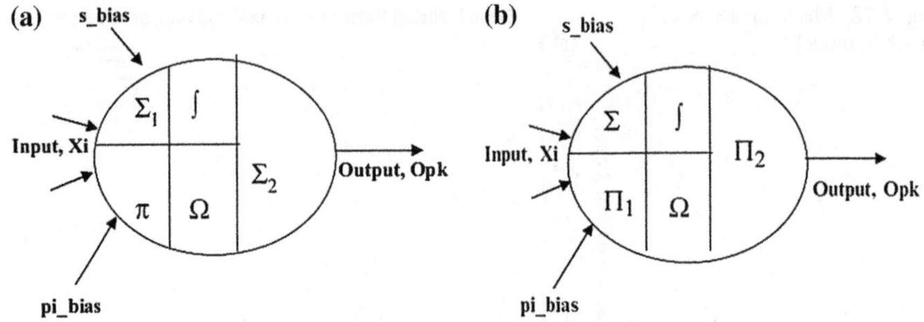

Fig. 5.13 **a** Symbolic representation of summation-type generalized neuron model. **b** Product-type generalized neuron model

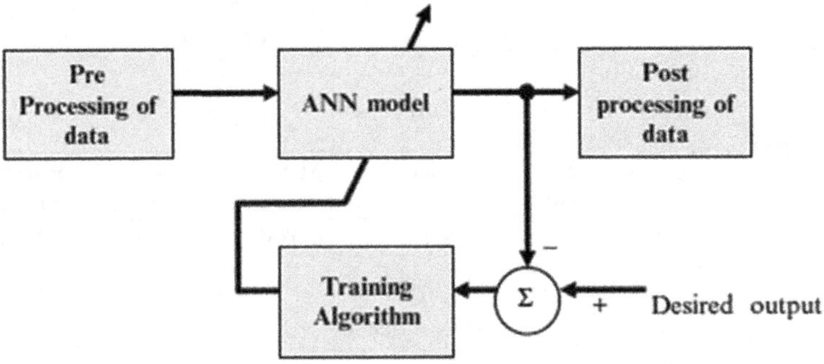

Fig. 5.14 Block diagram for GNN model development

The end product of these aggregation functions of generalized neuron written as follows:

(a) Summation case:

$$\text{GN Output} = O_\Sigma \times W + O_\Pi \times (1 - W)$$

where

O_Σ output of the summation part of the neuron Σ_1
W weight associated with O_Σ
O_Π output of the product part of the neuron (Π)

(b) Product case:

$$\text{GN Output} = O\Sigma^W \times O_\Pi^{(1-W)}$$

Advantages of Generalized Neural Network:

- Less number of unknown weights
- Less training time
- One neuron is able to solve the problem
- GN model is less complex

5.4 Results and Discussion

Due to generalization of neural network, it found in result that the generalized neural network has less RSME compared to ANN modeling. Figure 5.15 shows the test results comparing actual data with ANN and GNN model outputs. Figure 5.16 shows the error variation during day time. Solar power forecasting is very complex, and nonlinear and generalized neural network performed better and gives accurate result in less time. The RMS error during the testing phase of ANN and GNN

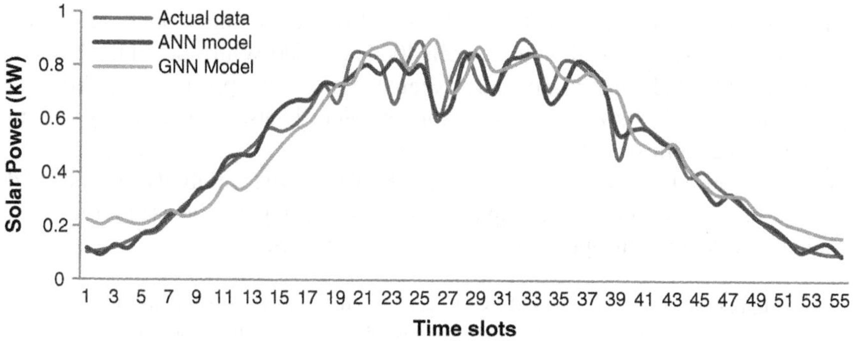

Fig. 5.15 Comparison of actual data, ANN, and GNN models

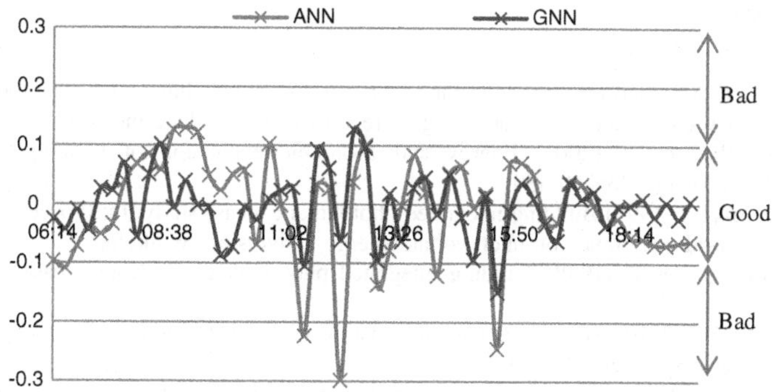

Fig. 5.16 Error classification

Table 5.4 Testing performance of ANN and GNN models for solar power forecasting

S. No.	Model name	RMSE error	Computational time (s)
1	ANN	0.1019	35
2	GNN	0.0903	15

models for solar power forecasting is shown in Table 5.4. According to the results obtained from comparative analysis, it is found that in case of GNN model error occured are less as compared to ANN model.

5.5 Conclusion

In this paper, neural network methodology is implemented in the form of ANN and GNN. The forecasting results of ANN and GNN have been compared for solar power variability forecasting. As is shown in Table 5.4 compared to the ANN with the GNN due to RMSE, computational time criteria give better results.

Figure 5.11 shows the error deviation during the morning time, especially in the lower range of solar radiation (0–400 W/m^2). Error deviation is increasing with specific range of solar radiation in between 400 and 1,000 W/m^2. So on the basis of this study, solar radiation is not only parameter for solar power variability and higher ambient temperature, and module temperature is also playing a key role in solar power variability.

In the proposed future work, season-based solar power variability analysis and its forecasting study with nature-inspired hybrid computing will be used for training of neurons to yield better performance of neural network.

Acknowledgments The authors would like to acknowledge the IIT Jodhpur for providing PV generation data and Central Power Research Institute, Bangalore, for financial support.

References

1. Osborne M (2009) Gartner posts long range forecast for photovoltaic's industry. http://www.pvtech.org/news/a/gartner, posts long range forecast for photovoltaic industry
2. Oliver M, Jackson T (2001) Energy and economic evaluation of building integrated photovoltaic. Energy 26:431–439
3. Kiss G, Kinked J (1994) Building integrated photovoltaic a case study. In: Proceedings of the IEEE first world conference on photovoltaic energy conversion, Waikoloa
4. Yoo S, Lee E, Lee K (1998) Building integrated photovoltaic's: a Korean case study. Sol Energy 64:151–161
5. Chaturvedi DK (2010) Short-term load forecasting using soft computing techniques. Int J Commun Netw Syst Sci 3(1):270–279
6. Barra OA (1981) La conversion fototermica dell'energia solareorogettazionericerca di impianti e sistemi solari. ETAS libri

7. Singh VP, Vaibhav K, Chaturvedi DK (2012) Solar power forecasting using soft computing approach. In: IEEE 3rd Nuicone, pp 1–5
8. Chaturvedi DK (2008) Soft computing techniques and its applications in electrical engineering. Springer, Berlin
9. Rui Y, Jin P (1994) The modeling method for ANN-based forecaster. In: CDC'94, China
10. Fausett L (1994) Fundamentals of neural networks, architecture, algorithms, and applications. Prentice-Hall, Englewood Cliffs
11. Young M (1989) The technical writer's handbook. University Science, Mill Valley
12. Abraham A (2004) Meta-learning evolutionary artificial neural networks. Neurocomputing 56:1–38

Chapter 6
Soft Calibration Technique with SVM for Intelligent Flow Measurement

K.V. Santhosh and B.K. Roy

Abstract Design of an adaptive soft calibration circuit with support vector machine (SVM) for flow measurement using orifice flowmeter is reported in this paper. The objectives of the present work are the following: (i) to extend linearity range of measurement to 100 % of full-scale input range and (ii) to make measurement technique adaptive to variations in ratio of orifice hole to pipe diameter (β), liquid density, and liquid temperature. SVM model is trained to map output voltage of data conversion unit to produce linear characteristics with input flow rate, and independent of variations in ratio of orifice hole to pipe diameter, liquid density, and liquid temperature all within certain range. Further, the proposed technique is tested with simulated data. Results show that the proposed technique has fulfilled its set objectives.

Keywords Adaptation · Calibration · Flow measurement · Orifice · Support vector machine

6.1 Introduction

Measurement of flow of liquids is a critical need in many industrial plants. In some operations, the ability to conduct accurate flow measurements is so important that it can make the difference between making a profit or taking a loss. With most liquid flow measurement instruments derives its principle from Bernoulli laws; the flow

K.V. Santhosh (✉)
Department of Instrumentation and Control Engineering, Manipal Institute of Technology, Manipal, India
e-mail: kv.santhu@gmail.com

B.K. Roy
Department of Electrical Engineering, National Institute of Technology, Silchar, Silchar, India
e-mail: bkr_nits@yahoo.co.in

V. Vijay et al. (eds.), *Systems Thinking Approach for Social Problems*, Lecture Notes in Electrical Engineering 327,
DOI 10.1007/978-81-322-2141-8_6

rate is determined inferentially by measuring the liquid's velocity or the change in kinetic energy. Velocity depends on the pressure differential that is forcing the liquid to flow through a pipe or conduit. Since pipe's cross-sectional area is known, the average velocity is an indication of flow rate. Orifice finds very wide applications because of its high sensitivity and ruggedness. However, the problem of offset, high nonlinear response characteristics, dependence of output on diameters of pipe and orifice hole, density of liquid, and temperature of liquid under measure have restricted its use and further impose some difficulties.

Calibration is needed for linearization of sensor output. Several techniques are used for the purpose of calibration. Analog circuits are used for the purpose of calibrating output of orifice sensor [1, 2]. Electromagnetic flowmeters are calibrated using analog circuits [3]. Head-type flowmeters are used for online measurement of powder flow rate in a pneumatic conveying system [4]; calibration of flow sensor is achieved by analog circuits. These conventional designs of calibration using hardwire analog circuits have drawbacks like the following: These are tedious and sometimes need to be replaced every time there exists a requirement for modification in process parameters or sensing parameters.

Some reported works have discussed the use of soft calibration techniques. A semicylindrical capacitive sensor with an interface circuit is used for flow rate measurement [5]. Curve fitting algorithms are used for calibration of flow sensor [6]. Numerical equations are used to create inverse characteristics of flow sensor [7, 8], which on cascade nullify the nonlinearity. Fuzzy logic algorithms are used to represent measured output in terms of fuzzy variables.

Neural network algorithms are used to produce inverse characteristics of sensor and so on cascade to sensor compensates nonlinearity. Back-propagation neural network algorithm is used for linearization of slotted orifice flowmeter [9]. Flow measurement using capacitance sensor is discussed in [10, 11]; linearization of sensor output for a certain range is achieved using neural network algorithms. Though several works are reported using neural network, these are limited only to a portion of full scale.

Several papers discuss the affect of sensor and process parameters on the measurement technique like in [12]; the relation of orifice plate dimension and pressure difference in head-type flow measurement is reported. Calculation of discharge coefficient for flow measurement using orifice is discussed in [13]. Effect of liquid viscosity and temperature on flow rate measurement is reported in [14]. Paper [15] discusses the effect of liquid density on flow measurement. Effect of temperature on flow measurement is also discussed in [16].

To overcome the above-discussed drawbacks, a technique is proposed using SVM. A technique is proposed in this paper to design an adaptive calibration technique for intelligent measurement of flow using orifice to achieve the following objectives: (i) to extend the linearity range to 100 % of full scale and (ii) to make calibration technique adaptive to variation in orifice hole to pipe diameter ratio (β), liquid density (ρ), and liquid temperature, all within certain range. The proposed technique is simulated using MATLAB.

6.2 Flow Measurement Using Orifice

6.2.1 Orifice Plate Flowmeter

An orifice plate, shown in Fig. 6.1, is a device used for measuring the volumetric flow rate. It uses the Bernoulli's principle which states that there is a relationship between pressure of fluid and velocity of fluid. When velocity increases, pressure decreases and vice versa. An orifice plate is a thin plate with a hole in the middle. It is usually placed within a pipe in which fluid flows. When the fluid reaches the orifice plate with hole in the middle, the fluid is forced to pass through the small hole. The point of maximum convergence actually occurs shortly downstream of the physical orifice, at the so-called vena-contracta point. As it does so, the velocity and the pressure changes. Beyond the vena-contracta, the fluid expands and the velocity and pressure changes once again. By measuring the difference in fluid pressure between the normal pipe section and at the vena-contracta, the volumetric and mass flow rates can be obtained from Bernoulli's equation [17, 18].

$$Q = C_d \cdot A_b \cdot \sqrt{\frac{1}{(1-\beta^4)}}\sqrt{\frac{2\Delta P}{\rho}} \quad (6.1)$$

where

C_d Discharge coefficient
A_b Area of the flowmeter cross section
β Ratio of D_b to D_a
P_a Pressure at study flow
P_b Pressure at vena-contracta
ρ Density of liquid

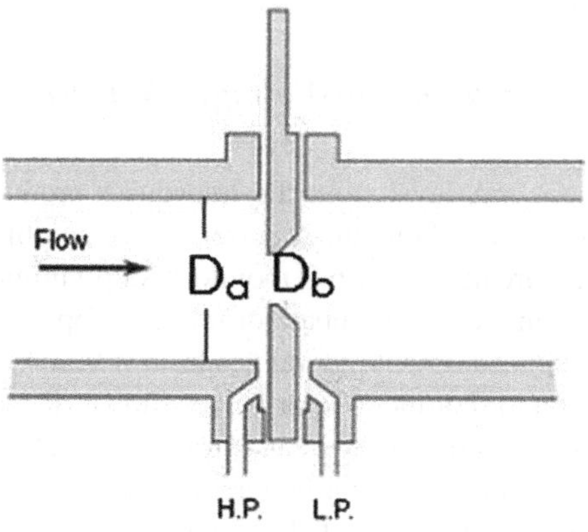

Fig. 6.1 Flow measurement using orifice

Effect of temperature on liquid density [14–16] can be given by

$$\rho_t = \frac{[\rho_{to}/(1+\alpha(t-t_o))]}{\left[1-\frac{P_t-P_{to}}{k}\right]} \tag{6.2}$$

where

- ρ_t Density of liquid at temperature 't °C'
- ρ_{to} Density of liquid at temperature 't_o °C'
- P_t Pressure at temperature 't °C'
- P_{to} Pressure at temperature 't_o °C'
- k Bulk modulus of liquid
- α Temperature coefficient of liquid

6.2.2 Pressure Transmitter

A pressure transducer is a transducer that converts pressure into an analog electrical signal. Although there are various types of pressure transducers available, one of the most common types is the strain gage-based transducer. The conversion of pressure into an electrical signal is achieved by the physical deformation of strain gages which are bonded onto the diaphragm of the pressure transducer and wired in a Wheatstone bridge configuration. Pressure applied to the pressure transducer causes the diaphragm to deflect which introduces strain to the gages. The strain produces a change in electrical resistance. The output of pressure transmitter is 4–20 mA.

Since a 4–20-mA signal is least affected by electrical noise and resistance in the signal wires, these transducers are best used when the signal need to be transmitted for long distances [19].

6.2.3 Current-to-Voltage Converter

The 4 mA level from the transducer produces 0 V output, and the 20 mA level produces 5 V output. A current sense amplifier generates this analog 0–5 V output. The circuit in Fig. 6.2 monitors loop current with a current sense amplifier (IC1) and employs a comparator/reference/op-amp device (IC2) to generate a ground-referenced output that ranges from 0 V at 4 mA to 5 V at the full-scale loop current (20 mA). For the resistor values shown (R2–R6), IC1 produces an output at pin 8 of approximately 1.25 V at 4 mA, and 6.25 V at 20 mA. In turn, the IC2 op-amp (configured as a unity-gain difference amplifier) generates an output range of 0.05–5.045 V. The IC2 comparator can be used to monitor input voltage or flag a preset loop current [20].

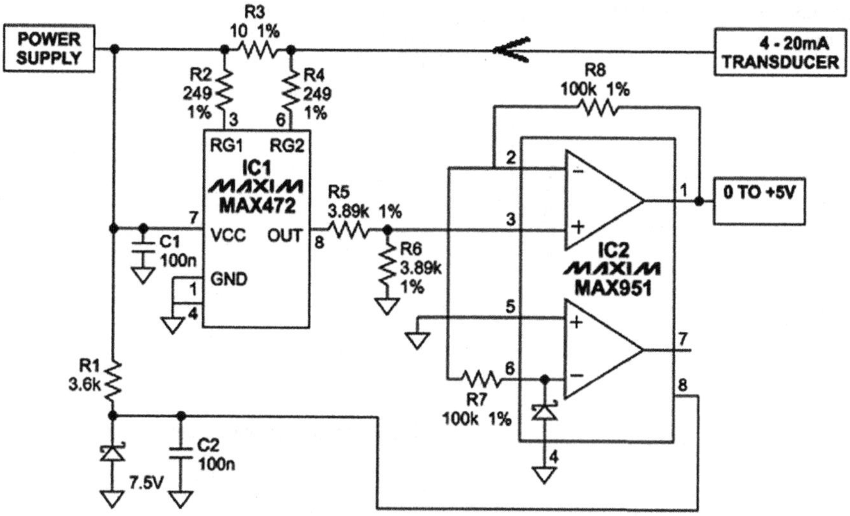

Fig. 6.2 Current-to-voltage converter

6.3 Problem Statement

In this section, characteristics of orifice are simulated to understand the difficulties associated with the available measurement technique. For this purpose, simulation is carried out with three different ratios of diameter between orifice hole and pipe considered. These are $\beta_1 = 0.3$, $\beta_2 = 0.6$, and $\beta_3 = 0.9$. Three different specific densities as $\rho_1 = 0.5$ kg/m^3, $\rho_2 = 1.0$ kg/m^3, and $\rho_3 = 1.5$ kg/m^3 are chosen. Four different temperatures, such as $t_1 = 25$ °C, $t_2 = 50$ °C, $t_3 = 75$ °C, and $t_4 = 100$ °C, are used to find the output differential pressure of orifice with respect to various values of input flow considering a particular discharge coefficient, ratio of diameter between orifice and pipe, density of liquid, and liquid temperature. These output pressure data are used as input for pressure transmitter, and output currents are generated. Finally, voltage signals are produced using current-to-voltage converter (I–V).

Figures 6.3, 6.4 and 6.5 show the variation of voltage with the change in input flow rate considering different values of diameter ratio, liquid density, and liquid temperature.

It has been observed from the graphs (Figs. 6.3, 6.4 and 6.5) that for a particular voltage value, there exist three different flows prompting available measuring technique to give erroneous readings. Datasheet of orifice suggests that input range of 10–60 % of full scale is used in practice as linear range. These are the reasons which have made the user to go for an intelligent calibration technique in comparison with conventional techniques, which have drawbacks that these are time-consuming and need to be calibrated every time whenever there is any change in β, ρ, and t. Further, use is restricted only to a portion of full range of input scale.

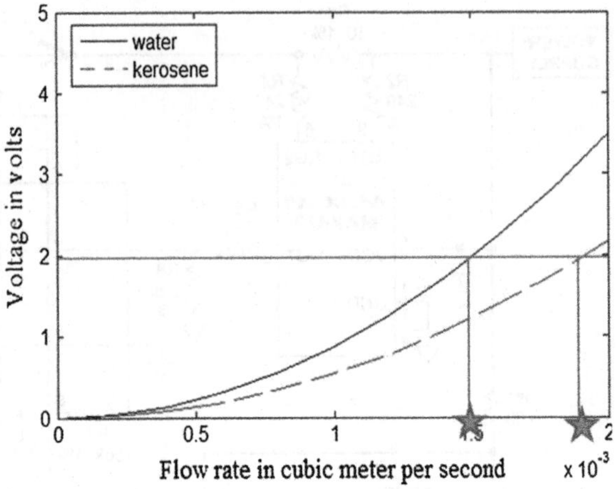

Fig. 6.3 Affect of β on flow measurement with orifice

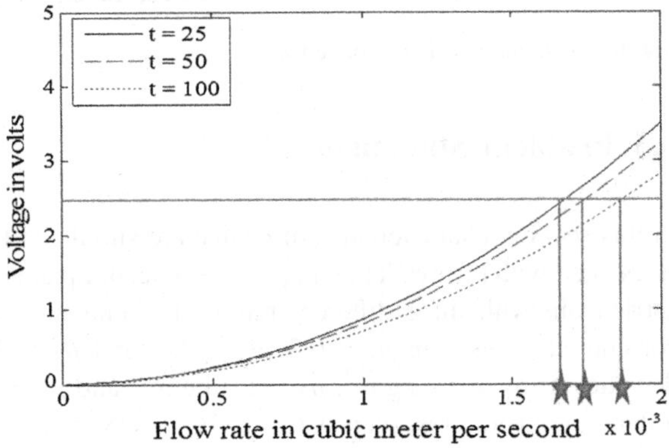

Fig. 6.4 Affect of ρ on flow measurement with orifice

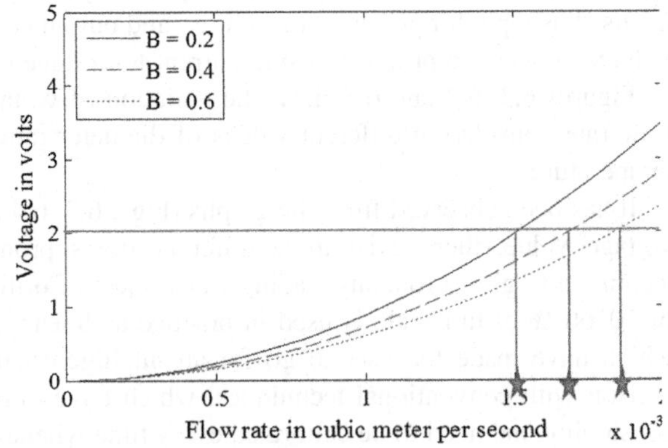

Fig. 6.5 Affect of liquid temperature on flow measurement with orifice

6.4 Problem Solution

The drawbacks discussed in the earlier section are overcome by adding a SVM model in cascade with data converter unit, in place of conventional calibration circuit. This model is designed by using MATLAB.

When we establish a calibration model of orifice sensor based on SVM, we should solve a regression problem in deed. Support vector machines provide a framework for regression problem, and it can be applied to regression analysis. There is only one kind of sample in SVM regression analysis, and the optimal hyperplane is not to separate the two kinds of samples, but to minimize the margin between all samples and optimal hyperplanes [21, 22].

The calibration principle based on SVM makes use of input parameters mapped to high-dimensional space by nonlinear transformation function; thus, regression analysis can be performed in high-dimensional space, and finally, the input/output function can be obtained [23–26].

Considering a set of training data about orifice flow sensor input and output $\{x_{ik}, y_{ik}\}$, $i = 1, \ldots, n$, where $x_{ik} \in R^N$ is input parameter SVM (i.e., orifice flow sensor data conversion output for variations in level, at different tank diameter, liquid permittivity, and liquid temperature). $y_{ik} \in R$ is output parameter of SVM (i.e., target output of measurement technique which has a linear relation with input flow and independent of diameter ratio, liquid density, and liquid temperature). The regression function based on LSSVM is denoted as in Eq. (6.3):

$$f(x) = \omega \cdot \Phi(x) + b \tag{6.3}$$

where $\omega \cdot \Phi(x)$ is the inner product of ω and $\Phi(x)$ and ω is the vector in high-dimensional space. $b \in R$ is the bias. By using the relaxation variable $\zeta, \zeta^* \geq 0$, the value of ω and b in Eq. (6.3) can solve an optimization problem:

$$\min_{\omega,\zeta} \left[\frac{1}{2} \|\omega^2\| + C \sum_{i=1}^{n} (\zeta + \zeta^*) \right] \tag{6.4}$$

First, the calibration model should be trained to get the corresponding parameters of the calibration model, such as kernel function, chastisement parameter, and error bias ε and so on. Second, the calibration model should make use of the training sampling data to obtain the values of α_i and b. If the output error is satisfied, the training ends. Otherwise, the calibration model parameters should be adjusted according to the error. Finally, the verifying sampling data should be used to verify the calibration model to determine the parameters of the calibration model (Table 6.1).

Table 6.1 Summary of SVM model proposed

Parameters of the SVM model					
Database		20 % training for CV		18	
		20 % training for test		18	
Projection algorithm				*K*-mean clustering	
o/p dimension				50 %	
Input optimization				Back-elimination	
Input		Flow	β	ρ in kg/m^3	*T* in °C
	Min	0.0×10^{-3} m^3/s	0.3	0.5	0
	Max	2.0×10^{-3} m^3/s	0.9	1.5	100

Table 6.2 Response of the proposed technique due to simulated data

AF in lpm	β	ρ in kg/m^3	*T* in °C	MF in lpm	% error
0.0	0.30	500	20	0.0013	0.000
0.0	0.35	550	50	0.0016	0.000
1.0	0.40	600	65	0.9988	0.120
1.0	0.45	650	75	0.9992	0.080
2.0	0.50	700	62	1.9986	0.070
2.0	0.55	750	90	2.0035	−0.175
3.0	0.60	800	100	2.9972	0.093
3.0	0.65	850	80	3.0041	−0.137
4.0	0.70	900	65	3.9976	0.060
4.0	0.75	950	75	4.0043	−0.107
5.0	0.80	1,000	85	4.9972	0.056
5.0	0.85	1,050	90	5.0037	−0.074
6.0	0.9	1,200	18	5.9964	0.060
6.0	0.35	1,300	22	6.0042	−0.070
7.0	0.40	1,400	38	6.9974	0.037
7.0	0.50	1,500	41	7.0042	−0.060
8.0	0.55	1,600	56	7.9973	0.034
8.0	0.60	1,700	61	8.0051	−0.064
9.0	0.65	1,800	66	8.9984	0.018
9.0	0.70	1,600	71	8.9961	0.043
12.0	0.75	1,500	76	11.9963	0.031
12.0	0.80	1,400	81	12.0041	−0.034
14.0	0.85	1,300	86	14.0032	−0.023
14.0	0.90	1,200	91	14.0051	−0.036
17.0	0.85	1,100	110	16.9972	0.016
18.0	0.80	1,050	125	17.9986	0.008
18.0	0.75	950	130	18.0044	−0.024

AF Actual flow rate in lpm
MF Measured flow rate in lpm

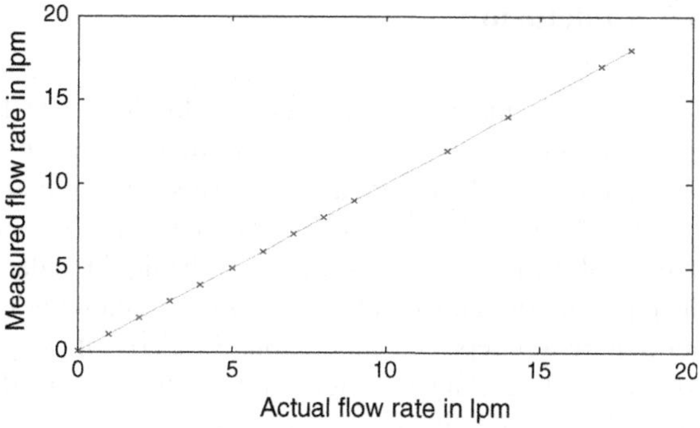

Fig. 6.6 Actual versus measure flow in simulation of proposed technique

6.5 Results and Analysis

The proposed SVM is trained, validated, and tested with the simulated data. Once the training is over, the system with orifice along with other modules, as shown in Fig. 6.2, is subjected to various test inputs corresponding to different flow rate with a particular diameter ratio, liquid density, and temperature, all within the specified ranges. For testing purposes, the range of flow rate is considered from 0 to 18 lpm, range of diameter ratio is from 0.3 to 0.9, range of liquid density is from 0.5 to 1.5 kg/m^3, and liquid temperature ranges from 0 to 100 °C. The outputs of proposed technique are noted. These are listed in Table 6.2. The input–output results are plotted, and the output graph is shown in Fig. 6.6, and graph of variation in percentage error of proposed technique is shown in Fig. 6.7.

It is evident from Figs. 6.6, 6.7 and Table 6.2 that the proposed measurement technique has incorporated intelligence by making system adaptive to variations in diameter ratio, liquid density, and liquid temperature. Also it has increased linearity range of flow measurement.

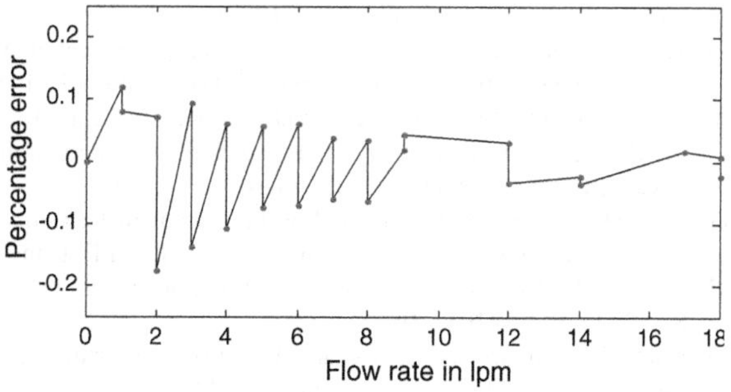

Fig. 6.7 Percentage error in simulation of proposed technique

6.6 Conclusion

The proposed measurement technique for flow rate has advantages over similar reported works. It uses the full scale of input range against limited use of input range in most of the reported works. Repeated calibration is avoided in the present approach by making it adaptive to variations in β, ρ, and t, which is not considered in most of the reported works. Thus, if the liquid under measure, and/or diameter of the pipe, and/or diameter of orifice is/are varied/replaced, the proposed system does not require any repeated calibration. Similarly, if there is a change in environment conditions, like change in temperature, the system does not require any further calibration to give the accurate reading.

References

1. Yoder J (2002) Flow calibration: how, why, and where. White paper on control for the process industries. Putman Publishers, New York
2. Technical Report (2010) Fundamentals of orifice meter measurement. Emerson Process Management, Bloomington
3. Hemp J (2001) A technique for low cost calibration of large electromagnetic flow meters. Flow Meas Instrum J 12:123–134
4. Huang Z, Wang B, Li H (2002) An intelligent measurement system for powder flow rate measurement in pneumatic conveying system. IEEE Trans Instrum Meas 51(4):700–703
5. Chiang CT, Huang YC (2006) A semi cylindrical capacitive sensor with interface circuit used for flow rate measurement. IEEE Sens J 6(6):1564–1570
6. Meriboua M, Al-Rawahia NZ, Al-Naamany AM, Al-Bimani A, Al-Busaidia K, Meribout A (2010) A multisensor intelligent device for real-time multiphase flow metering in oil fields. IEEE Trans Instrum Meas 59(6):1507–1519
7. Liu Y, Suna L, Qi L, Li S, Wei Y (2012) Development of a gas flow and velocity calibration facility. In: Proceedings of international conference on consumer, electronics, communications and networks, Three Gorges, China
8. Yanfeng G, Jinwu Z, Gang S (2006) Measurement of two-phase flow rate based on slotted orifice couple and neural network ensemble. In: Proceedings of international conference on information acquisition, Shandong, China
9. Terzic E, Nagarajah R, Alamgir M (2011) A neural network approach to fluid quantity measurement in dynamic environments. J Mechatron 21:145–155
10. Pathmanthan E, Ibrahim R (2010) Development and implementation of fuzzy logic controller for flow control application. In: Proceedings of international conference on intelligent and advanced systems, Manila, Philiphines
11. Bera SC, Mandal H (2012) A flow measurement technique using a noncontact capacitance type orifice transducer for a conducting liquid. IEEE Trans Instrum Meas 61(9):2553–2559
12. Owens CL (1965) Ionization gauge calibration system using a porous plug and orifice. J Vac Sci Technol 2(3):104–108
13. Harris MJ, Sattary JA, Spearman EP (1995) The orifice plate discharge coefficient equation-further work. J Flow Meas Instrum 6(2):110–114
14. Nakatani Noboru (2000) Measurements of the velocity and temperature in a turbulent flow by the laser photo thermal effect with the new compulsorily phase locked interferometer. J Rev Sci Instrum 71(5):1971–1974

15. Kolahi K, Shroder T, Rock H (2006) Model-based density measurement with coriolis flow meter. IEEE Trans Instrum Meas 55(4):1258–1262
16. Matharu RS, Perchoux J, Rakic AD (2011) Influence of ambient temperature on the performance of vesel based self-mixing sensors: flow measurements. In: Proceedings of IEEE sensors, Limerick, Ireland
17. Doebelin EO (2003) Measurement systems—application and design, 5th edn. Tata McGraw Hill Publishing Company, Noida
18. Liptak Bela G (2003) Instrument engineers' handbook: process measurement and analysis, 4th edn. CRC Press, London
19. Walt Boyes (2010) Instrumentation reference book. Butterworth-Heinemann Publishers Co., UK
20. Coughlin Robert F, Driscoll Frederick F (1998) Operational amplifiers and linear integrated circuits. Prentice Hall, New Delhi
21. Chang CC, Lin CJ (2001) LIBSVM-A library for support vector machines. Available at http://www.csie.ntu.edu.tw/~cjlin/libsvm
22. Xuegong Z (2000) Statical learning theory and support vector machines. Acta Automatica Sinica 26(1):32–42
23. Suykens JAK, Vandewalle J (1999) Least squares support vector machine classifier. Kluwer Academic Publisher, Amsterdam
24. Schäokopf B, Smola A (2002) Learning with kernels: support vector machines, regularization, optimization and beyond. MIT Press, Cambridge
25. Suykens JAK, Gestel TV, Brabanter JD, Moor BD, Vandewalle J (2002) Least squares support vector machines. World Scientific Publishers Co., Singapore
26. Gestel TV, Suykens JAK, Baesens B, Viaene S, Vanthienen J, Dedene G, De Moor B, Vandewalle J (2004) Benchmarking least squares support vector machine classifiers. J Mach Learn 54(1):5–32

Chapter 7
Load Frequency Control Considering Very Short-term Load Prediction and Economic Load Dispatch Using Neural Network and Its Application

Kalyan Chatterjee, Ravi Shankar and T.K. Chatterjee

Abstract The paper presents a new technique for the load frequency control (LFC) of interconnected power system. The LFC system monitors, at a minimum, power system frequency, generator output, net interchange schedule and tie-line power flows. It compares the actual frequency and tie-line values to the desired or scheduled values and generates an error value called the area control error (ACE). ACE is the instantaneous estimate of load demand in the area. However, due to relatively fast area load demand fluctuations and the relatively slow area generation response rate, only the instantaneous estimate of load demand cannot provide us with a good dynamic response. For this purpose, a look-ahead load forecasting feature is needed for effective LFC. To improve performance for LFC strategies neural network use for very short-term load prediction. From this forecasted load every 5-min ahead find out load change or error of the system can be recorded. This total load change divided all the units by concept of the unit's participation factor. Because of load is always changing so a fixed integral controller is not suitable for LFC, a fuzzy logic controller is used for this purpose.

Keywords Load frequency control · Neural network · Economic load dispatch · Load predication · Fuzzy logic controller

7.1 Introduction

Load frequency is very important in power system operation and control for supplying sufficient and reliable electric power with good quality. The generic functions of LFC include the following aspects: (1) Keeping frequency and tie-line power constant and

K. Chatterjee (✉) · R. Shankar · T.K. Chatterjee
Department of Electrical Engineering, Indian School of Mines (ISM), Dhanbad 826004, India
e-mail: kalyanbit@yahoo.co.in

V. Vijay et al. (eds.), *Systems Thinking Approach for Social Problems*, Lecture Notes in Electrical Engineering 327,
DOI 10.1007/978-81-322-2141-8_7

(2) considering economic load dispatch. Area load changes and abnormal conditions such as outages of generation lead to mismatches in frequency and scheduled power interchanges between areas. LFC is an essential mechanism in electric power system, which balances generated power and demand in each control area in order to maintain the system frequency at nominal value and the power exchange between areas at its scheduled value. Due to the relatively fast load demand fluctuations and the relatively slow area generation response rate, only the instantaneous estimate of load demand cannot provide us with the satisfactory performance. A look-ahead feature is needed if a more effective LFC is expected [1, 15, 17, 20, 27, 28].

In this view, the key features of this paper are the development of a look-ahead estimation for LFC and economic load dispatch. Automatic generation control will act to adjust the output generation every instant of time, while economic load dispatch will adjust the participation factors every few minutes to minimize the overall generating cost with respect to needed load demand while keeping the frequency within the normal range. The idea is to estimate or predict the load 10–15 min ahead of real time on intervals of 1 or 2 min. This predicted load will be distributed the entire generator by minimizing the cost function of the total load demand by adjusting the unit's participation factors of the generating units [7, 8, 18, 29].

In this paper, we have consider two control area in which first control area contains the combination of hydro-, thermal- and gas-generating unit, and in second control area, it contains the combination of the thermal and hydro-generating units. The short-term load forecasting using neural networks was proposed [4, 12, 20]. The predicted hourly load is being used to provide the future load estimates to the LFC system. As load is constantly varying, the values of gain of PI controller and the singular bounded hyper plane are required to change to get the better response. But the amount of control required is easy for the operator to describe. This required control can easily be described in qualitative terms and symbolic form. This automatically leads the necessity of soft computing-based intelligent controller such as fuzzy logic controller [5, 9, 11, 26, 30]. The fuzzy controller gives the opportunity to describe the control action in qualitative terms. For this purpose, in this paper, fuzzy logic controller is used for LFC.

7.2 Modeling of the Interconnected Power System

For the system, studies consider the combination of three and two generating units in first and second control area, respectively, of thermal, hydro, and gas and the combination of thermal- and hydro-generating unit with their system regulation. Each control area is connected through tie-line for their net balance interchanged tie-line power. The transfer function model is shown in Fig. 7.1 [14, 22].

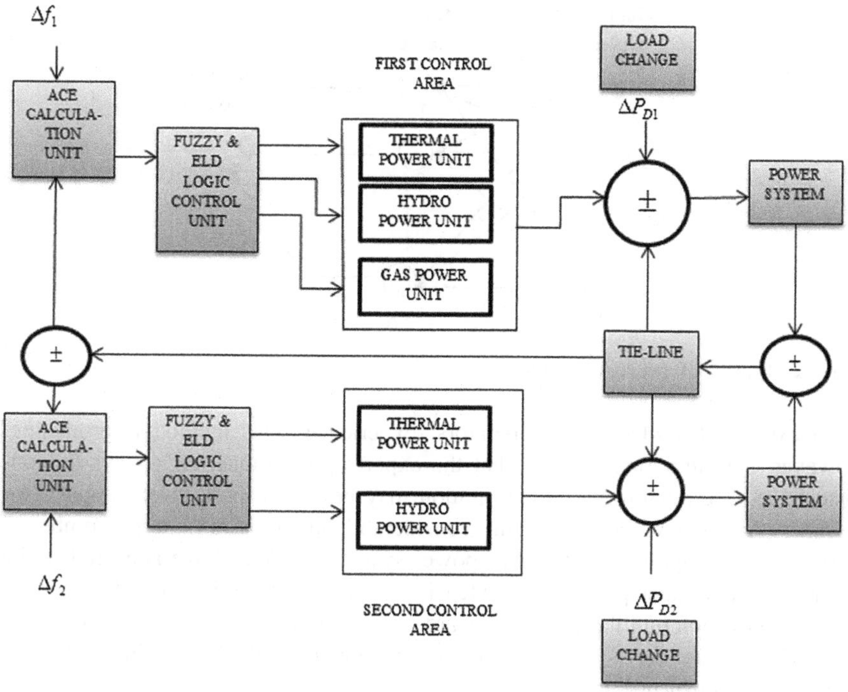

Fig. 7.1 Block diagram of the interconnected power system

7.3 Short-term Load Forecast Using Neural Network

The total demand of a system is equal to sum of all the consumers demand at the particular time. The objective of short-term load forecasting (STLF) is to forecast the future system load. A good understanding of the power system characteristics helps to design reasonable forecasting models and select appropriate models operating in different situations. Because of its importance, load forecasting has been extensively researched and a large number of models were proposed during the past several decades, such as Box-Jenkins models, ARIMA models, Kalman filtering models, and the spectral expansion techniques-based models [2–4, 6, 10, 12, 16, 19–21, 23–25]. Generally, the models are based on statistical methods and work well under normal conditions; however, they show some deficiency in the presence of an abrupt change in environmental or sociological variables which are believed to affect load patterns. Also, the employed techniques for those models use a large number of complex relationships, require a long computational time, and may result in numerical instabilities. Therefore, some new forecasting models were introduced recently. The development of artificial intelligence (AI), expert system (ES), and artificial neural networks (ANN) has been applied to solve the STLF problems [2–4, 12, 19–21]. An ES forecasts the load according to rules extracted

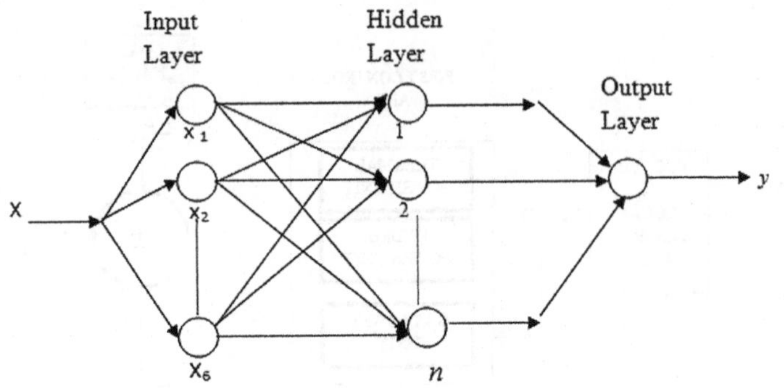

Fig. 7.2 Typical BP network structure

from experts' knowledge and operators' experience. This method is promising; however, it is important to note that the expert opinion may not always be consistent, and the reliability of such opinion may be in question. Over the past two decades, ANNs have been receiving considerable attention and a large number of papers on their application to solve power system problems have appeared in the literature. This paper presents an extensive survey of ANN-based STLF models. Back-propagation multilayer feed-forward designing the NN model. The typical BP network structure for STLF is a three-layer network with the nonlinear sigmoid function as shown in Fig. 7.2.

The BP network is a kind of array which can realize nonlinear mapping from the inputs to the output variables. Therefore, the selection of input variables of a load forecasting network is of great importance. The input variables can be classified into six classes: (i) temperature, (ii) hour of day, (iii) day of the week, (iv) holiday (weekend indicator (0, 1)), (v) previous 24-h average load, and (vi) 168-h (previous week) lagged load. The number of output variable is 1, i.e., load at particular hour.

7.4 Five-Minute Ahead Error Predictor

For improving the LFC performance, a look-ahead control algorithm is to be designed. In this purpose, hourly estimates load from NN model can interpolate it to a 5-min sample rate. So from this interpolation, 24-h forecast load data can be generated at 5-min interval. From these data, look-ahead error can be predicated. Suppose at time t, the estimated load is $y(t)$ and after time $(t + 5)$, the estimated load is $y(t + 5)$. The error or load change within this 5 min is

$$\Delta P_d(t) = y(t+5) - y(t). \tag{7.1}$$

If this load changes consider at time t, then 5-min look-ahead error incorporates in the LFC system and correspondingly designs the control algorithm. This total load change divided all the units by concept of the unit's participation factor.

7.5 Generation Allocation of the Generating Unit

Every control area of the interconnected power system may contain single or many generating units of different driving sources (hydro, thermal, nuclear, gas, etc.), which insure to provide the stable frequency and net balance interchanged tie-line power flow between the different control area via coordination with the calculation of automatic generation control or control mechanism and economic load dispatch. This coordination insuring to provide the information and logic that how much each generating unit will participate or take the load sharing out of the calculation of total load demand in context with their economics. This logic leads to introduce the concept of the unit's participation factor. Participation factor defined as the rate of change of each unit's output with respect to a change in total generation of that control area. From the definition and properties of the participation factor, its summation is equal to unity for each respective control area. When the economic load dispatch calculation is performed, then the sum of the present unit generation equal to the total generation and then it will be assigned as the base-point generation. This base-point generation will be the most economic output of the every generating unit. Thus, we will assign the generation allocation of the multiple generating units with respect to total change in generation output with the help of the conjunction of the load frequency control (LFC) mechanism and the concepts of economic load dispatch such control system is shown in (Fig. 7.3).

$$P_{i\text{des}} = P_{\text{ibase}} + pf_i \times \Delta P_{\text{total}} \tag{7.2}$$

$P_{i\text{des}}$ Desired output from unit i,
ΔP_{total} Change in total generation,
$P_{i\text{base}}$ Base-point generation for unit i,
pf_i Participation factors for unit i,

$$\sum pf_i = 1 \tag{7.3}$$

To find out the participation factors of each unit, we use the concepts of economic load dispatch calculation and the following equations accomplish the execution task. The quadratic cost function of the power system is given as

$$F_i(P_{Gi}) = \alpha_i + \beta_i P_{Gi} + \chi_i P_{Gi}^2 \tag{7.4}$$

α_i, β_i and γ_i are the coefficient of the cost function and P_{Gi} is power generation of the ith unit.

$$\min F(T) = \sum F_i(P_{Gi}) \tag{7.5}$$

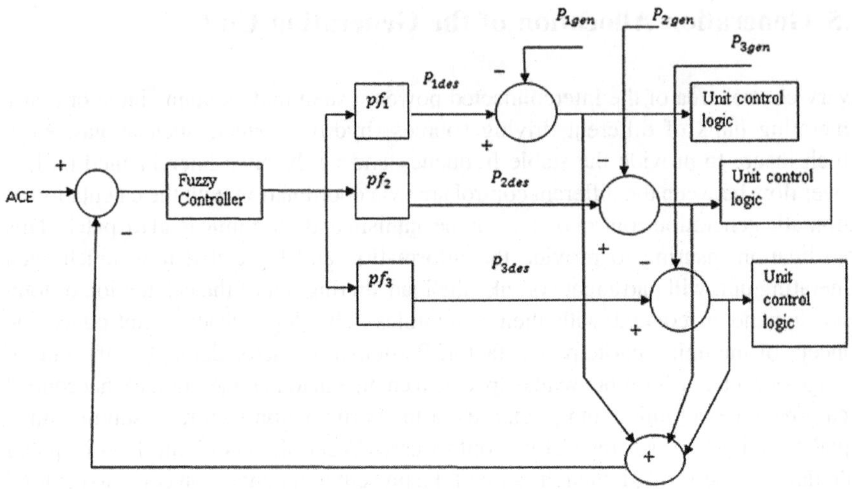

Fig. 7.3 Generation allocation block diagram of the generating unit

Power balance constraint is given as

$$\sum P_{Gi} = P_D + P_L \tag{7.6}$$

where, P_D = load demand and P_L = transmission losses and inequality is given as

$$P_{Gi}^{\min} \leq P_{Gi} \leq P_{Gi}^{\max} \tag{7.7}$$

$P_{Gi}^{\min}, P_{Gi}^{\max}$, are their minimum and maximum power generation of the ith unit.

7.6 Application of PI-like Fuzzy Controller for Load Frequency Control

The selection of the control variables (controller input and controller outputs) depends on the nature of the controlled system and the desired output. In this work, we have designed PI-like fuzzy knowledge based controller. The basic structure of the conventional PI controller is

$$u = \kappa_p e + \kappa_I \int e \, \mathrm{d}t \tag{7.8}$$

where κ_p and κ_I are the proportional and integral gains, respectively, and e is the error signal (i.e., e = process set point − process output variable). Taking the

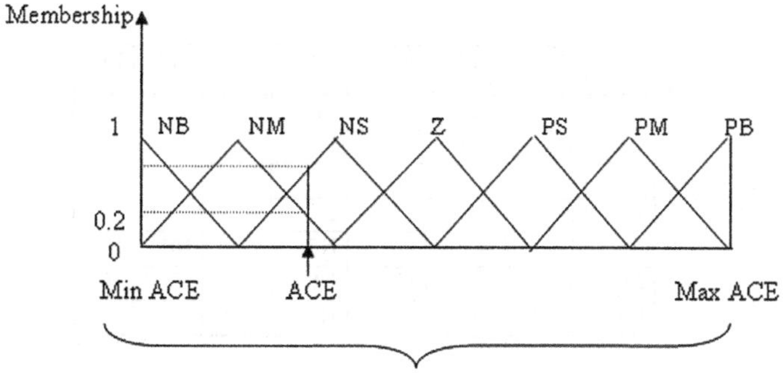

Fig. 7.4 Membership function for the fuzzy variables of ACE

derivative with respect to time, the above expression (7.8) is transformed into equivalent expression

$$\dot{u} = \kappa_p \dot{e} + \kappa_i e \tag{7.9}$$

For the LFC problem, the inputs to the fuzzy controller for *i*th area at a particular instant '*t*' are $\mathrm{ACE}_i(t)$ and $\Delta\mathrm{ACE}_i(t)$, where $\mathrm{ACE}_i(t) = \Delta P_{tiei} + B_i \Delta f_i$ and $\Delta\mathrm{ACE}_i(t) = \mathrm{ACE}_i(t) - \mathrm{ACE}_i(t-1)$ and output of the fuzzy controller is Δu. This is in accordance with the Eq. (7.9) for PI-like controller. The inputs and output are transformed to seven linguistic variables NB, NM, NS, Z, PS, PM, and PB, which stand for negative big, negative medium, negative small, zero, positive small, positive medium and positive big, respectively. Symmetrical triangular (expect of the two outermost ones which have a trapezoidal shape) membership function is considered here for all the three variables of ACE, ΔACE, and Δu. The membership function of ACE over the operating range of minimum and maximum value of ACE is shown in Fig. 7.4. The membership function would perform a mapping from the crisp value to a fuzzified value. As shown in Fig. 7.4, one particular crisp input ACE is converted to fuzzified value, i.e. $\frac{0.8}{\mathrm{NS}} + \frac{0.2}{\mathrm{NM}}$, where 0.8 and 0.2 are membership grades corresponding to the linguistic variable NS and NM in FNN system. The membership grades are zero for all other linguistic values except NS and NM. The crisp value input to the system in this way is converted to a fuzzified value consisting of several membership grades corresponding to each linguistic variable. In the same manner, the other input, ΔACE, and the output, Δu, are fuzzified.

7.6.1 Procedure for Framing the Rules

As each of the three fuzzy variables are quantized to seven fuzzy sets, so total or 49 rules are required to generate an fuzzy output relating two input fuzzy sets as shown in Table 7.1. Fuzzy rules play a major role in FLCs and have been investigated extensively.

ACE \ ΔACE	NB	NM	NS	Z	PS	PM	PB
NB	PB	PB	PM	PM	PS	PS	Z
NM	PB	PM	PM	PS	PS	Z	Z
NS	PM	PS	PS	PS	Z	Z	Z
Z	Z	Z	Z	Z	Z	Z	Z
PS	Z	Z	Z	NS	Z	NS	NS
PM	Z	Z	NS	NS	NM	NM	NB
PB	Z	NS	NS	NM	NM	NB	NB

Table 7.1 Fuzzy rules for two-area system

However, fuzzy rules usually can be generated using knowledge and operation experience with the system or by understanding of the system dynamics [26]. Table 7.1 shows the 49 rules that are generated through the behavior of the system response after a load change. The entire rule base may be divided into regions A, B, C, and D depending on the dynamics of the system as shown in Fig. 7.4. From Fig. 7.5, if e is positive and Δe is negative, the system will reduce the error itself, which means that the magnitude of the applied control action should be zero. This corresponds to the region labeled A in the Fig. 7.5. In case of negative e and negative Δe, the system tends to go to instability which requires an opposite (positive control) control action. This is interpreted as region B in the figures. In region C, the error is still negative while the rate of change of error is positive. This implies that the error is decreasing and that the control action should be kept to a minimum (zero).

When both the error and rate of change of error go positive, the system again tends to go to instability region again, and hence, it is required to apply opposite (negative) control action to compensate for this tendency toward instability. The output control action increases from zero (Z) control in rule region A to a minimum

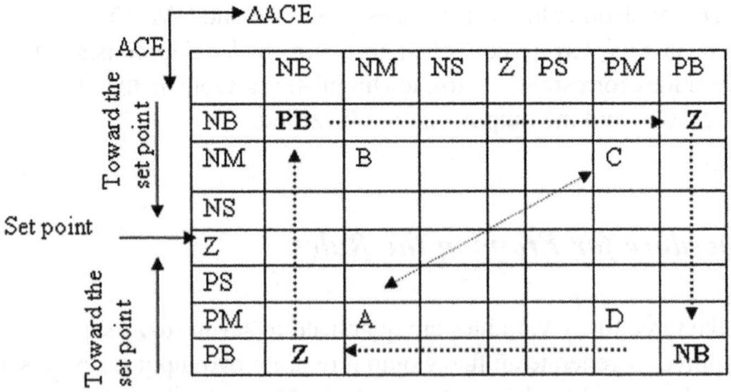

Fig. 7.5 Rules generation by understanding the system dynamics

positive control (PB) in rule region B, as indicated in Fig. 7.5. Then, the control action decreases from PB in rule region B to minimum control (Z) in rule region C. Also, when both e and Δe approach their maximum positive value, the control action goes to maximum negative (NB). The firing strength of a fuzzy control rule will be ultimately determined using Mamdani correlation minimum encoding inference. For example, suppose that at a certain instant t, the fuzzy logic employs a rule: If ACE(t) is NS and ΔACE(t) is NM, then control action Δu is PS. The firing strength of such fuzzy control rule will be,

$$(\mathrm{ACE}(t),\ \Delta\mathrm{ACE}(t)) = \min(\mu\mathrm{ACE}(\mathrm{ACE}(t)), \Delta\mathrm{ACE}(\Delta\mathrm{ACE}(t)))$$

where A: 'NS', B: 'NM' and ΔACE represent the membership functions of the linguistic values of error and rate of change of error. Suppose, e belongs to NS with a membership of 0.3 and Δe belongs to NM with a membership of 0.7, then the rule consequence (w_i) will be 0.3, the minimum between 0.3 and 0.7.

7.6.2 *Defuzzification*

The output of the inference mechanism is a fuzzy value, so it is necessary to convert this fuzzy value into a real value, since the physical process cannot deal with fuzzy value. This operation that is the inverse of fuzzification is known as defuzzification. The well-known center of gravity defuzzification method has been used because of its simplicity. The control output Δu is determined using the center of gravity by the following expression,

$$\Delta u = \frac{\sum (\text{membership of input} \times \text{output corresponding to be membership of input})}{\sum (\text{membership of input})}$$

$$\Delta u = \frac{\sum_{j=1}^{49} \mu_j u_j}{\sum_{j=1}^{49} \mu_j} \tag{7.10}$$

where μ_j is the membership value of the linguistic variable recommending the fuzzy controller action, and u_j is the precise numerical value corresponding to that fuzzy controller action. This Δu is added with the existing previous signal which will be the actual output signal u which goes to the governor. Figure 7.2 is the block diagram of two-area system where area control error (ACE) of each area is fed to the corresponding fuzzy logic-based digital controller. The accurate control signal is generated for every incoming ACE.

7.6.3 Normalization

Generally, the universes of discourse of input and output variables of FLC are the real line. In practice, each universe is restricted to an interval that is related to the maximal and minimal possible values of the respective variable, that is, to the operating range of the variable. The rules will not be properly framed if the operating range is not suitably selected. For simplification and unification of the design of the FLC and its computer implementation, however, it is more convenient to operate with normalized universe of discourse of the input/output variables of the FLC [6]. Figure 7.6 shows block diagram representation of fuzzy logic controller with normalization and denormalization factors. The normalized universes are well-defined domains; the fuzzy values of the input/output variables are fuzzy subsets of these domains. In general, the normalized universe can be identical to the real operating ranges of the variable, but in most application, they coincide with the closed interval of (−1, 1). In this work, the inputs of ACE and ΔACE(*t*) are normalized between −1 and 1, the real operating universe to normalized universe. Scaling factor is dependent on the operating range considered for the corresponding input. As for example, $K_{\mathrm{ACE}} = \frac{1}{0.02} = 50$, as the operating range of ACE is considered as −0.02 to 0.02. Essentially, the scaling factors K_{ACE} and $K_{\Delta\mathrm{ACE}}$ map the real measured values ACE (*t*), ΔACE(*t*) to the values of $\mathrm{ACE}^*(t)$, $\Delta\mathrm{ACE}^*(t)$ from these normalized universes by linear mapping such as $\mathrm{ACE}^* = K_{\mathrm{ACE}}\mathrm{ACE}(t)$; $\Delta\mathrm{ACE}^*(t) = K_{\Delta\mathrm{ACE}}\Delta\mathrm{ACE}(t)$. Defuzzified value $\Delta u^*(t)$ obtained by the application of the FLC algorithm belongs to the normalized universe (−1, 1), and it is required to get the real change of control variable $\Delta u(t)$ from these operating range (−1, 1) through the scaling factor $K_{\Delta u}$. Therefore, the output $\Delta u(k) = K_{\Delta u}\Delta u^*(t)$. Thus, the scaling of the normalization value of the output variable is effectively denormalization, that is, bringing the output of the FLC from the normalized universe to its actual operating range (−0.01, 0.01) through the scaling factor $K_{\Delta u} = \frac{0.01}{1} = 0.01$.

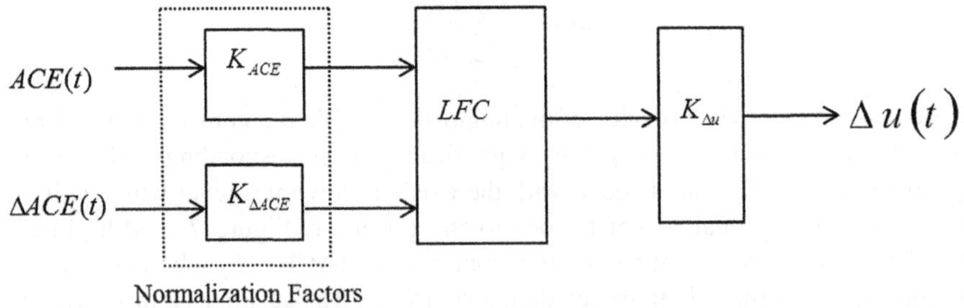

Fig. 7.6 Fuzzy logic controller with normalization and denormalization factors

7.7 Results and Discussion

The proposed design in conjunction with economic load dispatch (i.e., unit's participation factor) has been applied for typically multigenerating units of the interconnected power system fuzzy logic controller is shown in Fig. 7.7. The proposed controller is simulated in MATLAB SIMULATION TOOLBOX. A neural network-based very short-term load forecasting system is developed in this paper. The models are trained on hourly data and 35,064 data sets [14] are used in each epoch. The training parameters were set with initial learning rate, momentum constant, and error convergence 0.1, 0.01, and 10^{-7}, respectively. Total number of epochs for training is 500. A number of 1,754 sample data sets use for testing the network. The models are shown to produce less average error during the forecasting the load. The forecasted load pattern is dashed line and actual load pattern in solid line as shown in Fig. 7.8. From this forecasted load curve, a 5-min look-ahead load change is generated. The error or load change curve is shown in the Fig. 7.9. This load change now fed to the area 1. Each generating unit now "participates" in the load change by using the participation factor of each unit in context with economic load dispatch. The dynamic response or frequency deviations of area 1 and area 2 are shown in the Figs. 7.10 and 7.11, respectively. The proposed fuzzy logic controller gives better dynamic response with respect to consider a fixed gain integral controller.

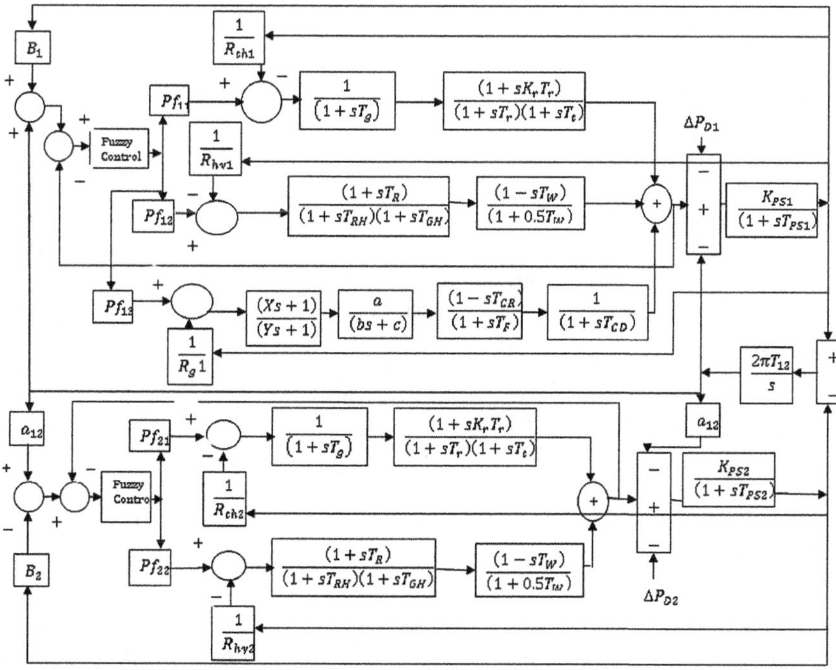

Fig. 7.7 Actual and forecasted load pattern (load vs. hour)

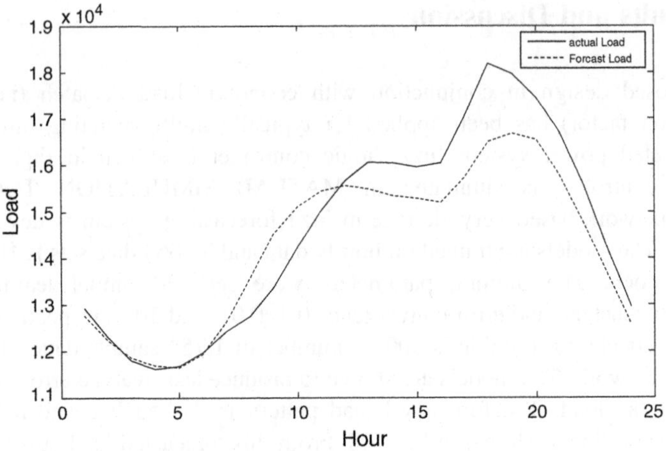

Fig. 7.8 Load change (error) signal fed to area 1

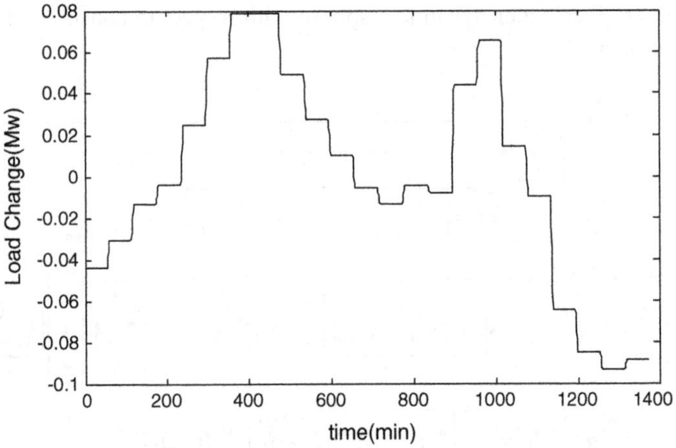

Fig. 7.9 Frequency deviations in area 1

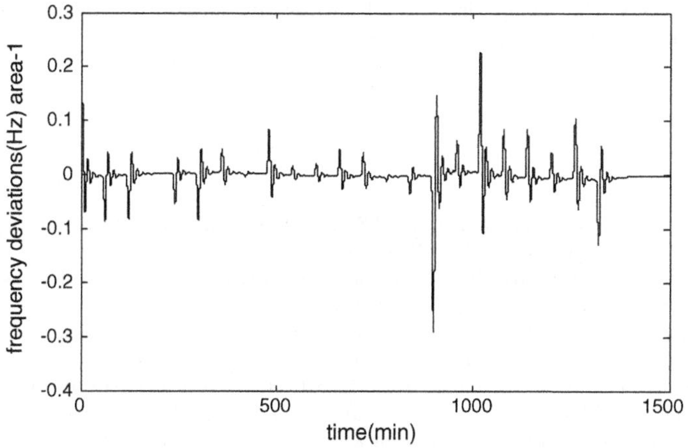

Fig. 7.10 Frequency deviations in area 2

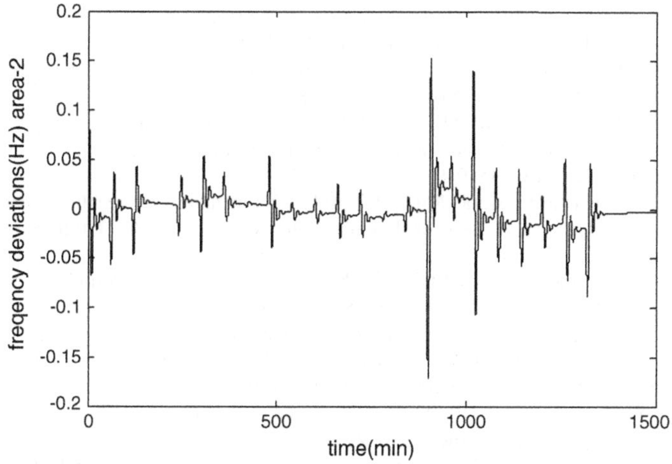

Fig. 7.11 Block diagram of the multigenerating interconnected power system

7.8 Conclusion

This paper describes a methodology for estimating the look-ahead error in 5 min steps for a two-area interconnected power system. Each unit shares the load according to their participation factors with the total load change of the control area. This paper builds a fuzzy rule base with the use of the ACE and rate of change of the error. The simulation results for continuous load change have been justified that the proposed fuzzy logic controller yields more improved control performance than a fixed gain integral controller.

Acknowledgments The authors sincerely acknowledge the financial support provided by the ISM, Dhanbad, for carrying out the present work.

References

1. Ahamad TP, NagendraRao PS, Sastry PS (2002) A reinforcement learning approach to automatic generation control. Electr Power Syst Res 63:9–26
2. Akirtzis AG, Theocharis JB, Kiartzis SJ, Satsios KJ (1995) Short term load forecasting using fuzzy neural network. IEEE Trans Power Syst 10(3):1518–1524
3. Asar A, Mcdonald JR (1994) A specification of neural network application in the load forecasting problem. IEEE Trans Control Syst Technol 2(2):135–141
4. Bakirtzis RC, Petridis V, Klartiz SJ, Alexladls MC (1996) A neural network short term load forecasting model for greek power system. IEEE Trans Power Syst 11:858–863
5. Chown GA, Hartman RC (1998) Design and experience with a fuzzy logic controller for automatic generation control. IEEE Trans Power Syst 13(3):965–970
6. Danishi H, Daneshi A (2008) Real time load forecasting in power system. In: DPRT 2008, Nanjing, 6–9 Apr 2008
7. Elgerd OE (1982) Electric energy systems theory, 2nd edn. McGraw Hill, New York, pp 315–389
8. Elgerd OL, Fosla CE (1970) Optimum megawatt frequency control of multi-area electric energy systems. IEEE Trans Power Apparatus Syst PAS-89(4):556–563
9. El-Hawary ME (1998) Electric power applications of fuzzy systems, Chap. 5. Series on power engineering. IEEE Press, New York, pp 112–148
10. El-Naggar MK, Al-Rumaih AK (2007) Electric load forecasting using genetic based algorithm, optimal filter estimator and least error squares technique: comparative study. World Acad Sci Eng Technol 6:945–949
11. Farook S, Sangamasmeswara Raju P (2011) Optimization of feedback controller in restructured power system using evolutionary genetic algorithm. Int J Eng Sci Technol 3 (5):4074–4083
12. Hippert HS, Pedriera CE, Souza RC (2001) Neural networks for short-term load forecasting: a review and evaluation. IEEE Trans Power Syst 16(1):44–55
13. http://iso-ne.com/markets/hstdata/znl_info/hourly/index.html
14. IEEE Committee Report (1973) Dynamic models for steam and hydro turbines in power system studies. Trans Power Apparatus Syst 92(6):1904–915
15. Karnavas YL, Papadopoulos DP (2002) AGC for autonomous power system using combined intelligent techniques. Electr Power Syst Res 62:225–239
16. Khan MR, Abraham A (2003) Short term load forecasting models in Czech Republic using soft computing paradigm. Int J Knowl Based Intell Eng Syst 7(4):172–179
17. Kothari ML, Nanda J, Kothari DP, Das D (1988) Discrete-mode automatic generation control of a two-area reheat thermal system with new area control error. IEEE Trans Power Syst 4 (2):1988–1994
18. Kundur P (1994) Power system stability and control. McGraw Hill, New York
19. Lu CN, Vemuri S (1993) Neural network for short-term load forecasting. IEEE Trans Power Syst 8(8):336–342
20. Lui K, Subbarayan S, Shoults R, Manery MT, Kwan C, Lewis FL, Naccarion J (1996) Comparison of very short-term load forecasting techniques. IEEE Trans Power Syst 11:877–882
21. Moghram I, Rahman S (1989) Analysis and evaluation of five short term load forecasting techniques. IEEE Trans Power Syst 4(4):1484–1491
22. Ramakrishna KS, Sharma P, Bhatti PS (2010) Automatic generation control of interconnected power system with diverse sources of power generation. Int J Eng Sci Technol 2(5):51–65

23. Rothe JP, Wadhwani AK, Wadhwani S (2009) Short term load forecasting using multi parameter regression. Int J Comput Sci Inf Secur 6(2):303–306
24. Senjyu T, Uazato T, Higa P (1998) Future load curve shaping based on similarities using fuzzy logic approach. IEE Proc Gener Transm Distrib 145(4):375–380
25. Shulte RP (1996) An automatic generation control modification for present demands on interconnected power systems. IEEE Trans Power Syst 11:1286–1294
26. Talaq J, Al-Basri F (1999) Adaptive fuzzy gain scheduling for load frequency control. IEEE Trans Power Syst 14:145–150
27. Tripathy SC, Hope GS, Malik OP (1982) Optimization of load-frequency control parameters for power systems with reheat steam turbines and governor dead band nonlinearity. IEE Proc Gener Transm Distrib 129(1):10–16
28. Trudnowski J, McReynold L, Jonson J (2001) Real-time very short-term load prediction for power-system automatic generation control. IEEE Trans Control Syst Technol 9(2):254–260
29. Wood AJ, Woolenberg BF (1998) Power generation operation and control. Wiley, New York
30. Yager RR, Filev DP (1994) Essentials of fuzzy modelling and control. Wiley, New York

Chapter 8
A Systems View of Pathological Tremors

Viswanath Talasila, Ramkrishna Pasumarthy, Sindhu S. Babu and Sudharshan Adiga

Abstract In this paper, we consider a specific case of movement disorders, i.e., resting tremors and attempt to formulate a simple mathematical description of these tremors. A novel aspect of the paper is that it is a first attempt at using standard tools from systems theory, such as state space and Lyapunov stability analysis, to model resting tremors. We formulate tremor control as a disturbance rejection problem, and derive conditions under which disturbance rejection is achievable.

Keywords Movement disorders · Tremors · State space model · Disturbance rejection

8.1 Introduction

Movement disorders often occur due to a defective central nervous system. One particular symptom of movement disorders is tremors, and there are 120 kinds of tremors categorized [1]. Tremors can have a debilitating affect on the normal life of subjects. A common reason for tremors is due to lack of sufficient dopamine produced in the substantia nigra, which affects the functioning of the motor cortex [2]. In this paper, we are interested in analyzing the behavior of one particular tremor, called the resting tremor. This is a tremor where the muscles are at rest and

V. Talasila (✉) · S.S. Babu · S. Adiga
MSRIT, Bangalore, India
e-mail: viswanath.talasila@msrit.edu

S.S. Babu
e-mail: sindhu732@gmail.com

S. Adiga
e-mail: sudarshanadiga1993@gmail.com

R. Pasumarthy
IIT-Madras, Chennai, India
e-mail: ramkrishna@ee.iitm.ac.in

V. Vijay et al. (eds.), *Systems Thinking Approach for Social Problems*,
Lecture Notes in Electrical Engineering 327,
DOI 10.1007/978-81-322-2141-8_8

supported against gravity. Other types of tremor include kinetic tremor—which is caused due to goal-directed movements and has the same frequency as resting tremor.

The tremor waveform is roughly sinusoidal with characteristic frequency for each of the 120 kinds of tremors. Hence, frequency is important in tremor classification, and it is often used to characterize between different types of tremors, whereas the amplitude is not consistent between different tremors and is not used for characterization, e.g., the amplitude widely vary under controlled conditions since there are many factors (psychological, pathological, environmental, etc.) which can influence it [3].

Tremors are classified as physiological tremor, generally occurring in the 8–12 Hz frequency range and pathological tremor occurring in the 4–8 Hz frequency range; there is often a small overlap between the two frequency ranges. Physiological tremors are normal tremors which all humans have, and they do not interfere with motion. There are various conditions under which pathological tremors are generated. Often, underlying neurological (e.g., Parkinsonian) conditions can make a patient susceptible to tremors.

Currently, there exist few system theoretic formulations of tremors, especially from a control theoretic viewpoint. Notions of stability, disturbance rejection, reachability, and so on could find important use in tremor analysis and control. In this paper, we attempt to formulate resting tremors as a control problem, specifically in a disturbance rejection framework. Thus, we shall argue that subjects experiencing resting tremors have a poor disturbance rejection control mechanism.

8.2 Formal Definitions of Tremors

In this section, we will provide formal notions of resting tremors, where a tremor is essentially a function mapping an input signal space into an output signal space. Tremors are usually generated under the influence of certain events, and these event (trigger)-driven dynamics can be modeled from a system theoretic viewpoint. For this paper, let us consider a single body part—the arm for our study; without loss of generality. Consider two systems $\Sigma_{\text{phystremor}}$ and $\Sigma_{\text{pathtremor}}$. $\Sigma_{\text{phystremor}}$ is a system describing the dynamics[1] of an arm experiencing only physiological tremors but not resting tremors; $\Sigma_{\text{pathtremor}}$ is a system describing the dynamics of an arm experiencing pathological tremors (in our case: resting tremors).

The tremor signals themselves can be modeled as external disturbances acting on the arm, and we are essentially interested in studying the output behavior of the systems $\Sigma_{\text{phystremor}}$ and $\Sigma_{\text{pathtremor}}$ subject to the disturbances. Figure 8.1 shows a plant G—modeling the human arm—experiencing a disturbance d at the output, with a reference signal r. Such a system provides a simple starting model of resting

[1] We shall study these dynamics later.

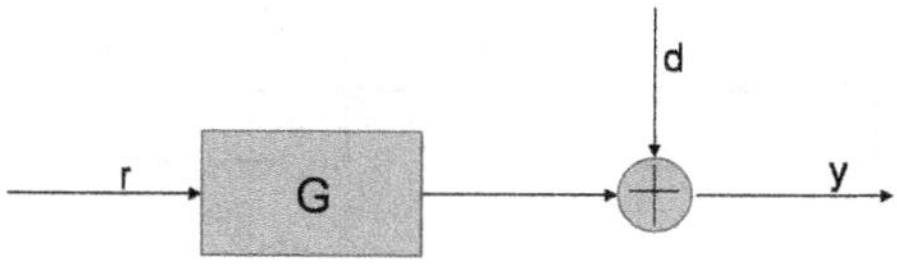

Fig. 8.1 Plant with external disturbance

tremors, where r can be considered to be zero since the arm (or system G) is at rest and supported against gravity. The disturbance d, in our case, is a sinusoidal signal and is defined as follows:

$$d := \{d_{\mathrm{lf}}, d_{\mathrm{hf}}\}, \text{ or any linear combination of } \{d_{\mathrm{lf}}, d_{\mathrm{hf}}\}$$

where d_{hf} is a high-frequency sinusoidal input with frequency in the range 8–12 Hz, and d_{lf} is a low-frequency sinusoidal input with frequency in the range 4–8 Hz.

The block diagram in Fig. 8.1 indicates that feedback is not present—though is not true for an actual human arm. Instead, we make a simplifying assumption that the plant G is modeled as an open-loop system with passive damping present, subject to external disturbances. A model of such a system is presented in the following section. Let us first formally define the physiological tremor system and the pathological tremor system.

Definition 1 (*Physiological Tremor System*) A physiological tremor system, denoted $\Sigma_{\mathrm{phystremor}}$, is formally defined by the maps

$$\Sigma_{\mathrm{phystremor}} : d_{\mathrm{hf}} \longrightarrow y_{\mathrm{hf}}, \quad \Sigma_{\mathrm{phystremor}} : d_{\mathrm{lf}} \longrightarrow 0$$

where y_{hf} is the output sinusoidal response corresponding to the sinusoidal disturbance signal d_{hf}; we define the output of $\Sigma_{\mathrm{phystremor}}$ to be zero when the input is d_{lf}. This definition simply says that the system $\Sigma_{\mathrm{phystremor}}$ is capable of rejecting the low-frequency disturbance signals and passes through the high-frequency disturbance signals.

Definition 2 (*Pathological Tremor System*) A pathological tremor system, denoted $\Sigma_{\mathrm{pathtremor}}$, is formally defined by the maps

$$\Sigma_{\mathrm{pathtremor}} : d_{\mathrm{hf}} \longrightarrow y_{\mathrm{hf}}, \quad \Sigma_{\mathrm{pathtremor}} : d_{\mathrm{lf}} \longrightarrow y_{\mathrm{lf}}$$

where y_{hf} is the output sinusoidal response corresponding to the sinusoidal disturbance signal d_{hf}; we define the output of $\Sigma_{\mathrm{phystremor}}$ to be another sinusoid y_{lf} when the input is d_{lf}. This definition says that the system $\Sigma_{\mathrm{pathtremor}}$ is unable to reject both the low- and high-frequency disturbance signals. The above two definitions provide a simple mathematical description of the two types of tremors. The following block diagram illustrates a simple systems approach to elucidate the above definitions. Figure 8.2 attempts to explain the onset of tremors under the influence of certain

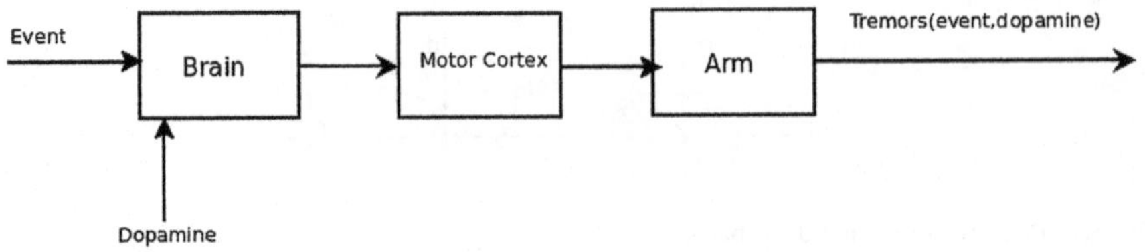

Fig. 8.2 An open-loop system representation of tremors on the human arm

events and the availability of dopamine in the brain. The figure is a massive simplification of a highly complex process but is sufficient for our purpose here. When certain events[2] occur (e.g., stress or sudden loss of dopamine producing neurons for various reasons) and if sufficient dopamine is not available in the brain[3] then abnormal synchronization activities, [4], occur in the brain and affect the functioning of the motor cortex and can induce pathological tremors, see [5, 6]. If sufficient dopamine is present, the motor cortex functions normal and pathological tremors are not observed. It is well known, [7], that Parkinson patients often rely on other signals, apart from proprioceptive feedback,[4] to control their motions during tremor. If we consider tremors in the arm, one possibility of control would be to essentially increase the damping of the arm (by stiffening the muscles) so as to reduce the tremors. In the absence of such sensory feedback, patients cannot control the tremors. Also, often the available sensory measurement is weak or faulty (e.g., in proprioception) and hence sufficient active damping is not produced to minimize the tremors.

8.3 Simple Dynamical Model of the Human Arm

In the previous section, we have provided formal definitions for systems modeling tremors—physiological and pathological. We defined tremor systems by functions mapping input spaces (signal spaces) into output spaces (signal spaces). In this section, we attempt to define simple models for these systems focusing on a specific part of the human anatomy (a single segment of the arm), and we use Lyapunov theory to show that pathological tremors essentially represent poor disturbance rejection. There exist many models to capture the behavior of the human arm [8, 9]. We adopt the simple mass spring damper (MSD) model [10, 11] for our study.

[2] Note that events may be instantaneous or may build up overtime.

[3] Observe that we are not localizing the specific part of the brain here, e.g., the substantia nigra is affected in Parkinson's case. For the purposes of this paper, we will use the generic term *brain* instead of focussing on the specific part responsible.

[4] Proprioceptive feedback often degenerates during dopamine loss.

We also do not explicitly model the feedback for motion control of the arm; instead, we implicitly assume that the damping term in the MSD model is controlled appropriately[5] when sufficient dopamine is present and is poorly controlled in the absence of sufficient dopamine. Thus, we will argue that the available damping, for patients with loss of dopamine, is not sufficient to control the tremors.

The MSD system under the influence of an external force, u, is governed by the differential equation $m\ddot{q} + c\dot{q} + kq = u$. In the state space form, this can be rewritten as

$$\begin{bmatrix} \dot{x}_1 \\ \dot{x}_2 \end{bmatrix} = \begin{bmatrix} 0 & 1 \\ -\frac{k}{m} & -\frac{c}{m} \end{bmatrix} \begin{bmatrix} x_1 \\ x_2 \end{bmatrix} + \begin{bmatrix} 0 \\ \frac{1}{m} \end{bmatrix} u \tag{8.1}$$

where $x = [x_1\ x_2]^T; x_1 = q,\ x_2 = \dot{q}$; q is the measured displacement of the mass. Consider the unforced dynamics of this system, i.e., when $u = 0$. If we assume that $x = [0\ 0]^T$ is a stable equilibrium of the system (8.1), then there correspondingly exists a Lyapunov function, $V(x)$ which is positive definite and satisfying $\frac{dV(x)}{d_t} \leq 0$. One of the standard choices of a Lyapunov function for the unforced ($u = 0$) MSD system is $V(x) = \frac{1}{2}kx_1^2 + \frac{1}{2}mx_2^2$. For the unforced MSD system, we then obtain $\dot{V}(x) = -cx_2^2$ which obviously satisfies $\dot{V}(x) \leq 0, \forall t$.

Remark 1 One could think of the unforced system as corresponding to the human arm, where there are no tremors generated (as sufficient dopamine is present in the brain) and thus we have the arm (modeled as a simple MSD system) at rest. In the absence of the tremors, we would expect the arm position to be stable; indeed, this is what we observe in the simple model used above.

8.3.1 Tremor Control as a Disturbance Rejection Problem

Pathological tremors are often classified as either resting or action [12]. Tremor is an involuntary and (often) rhythmic motion of a body part in a fixed plane and resting tremor occurs in a body part, [12], which is supported against gravity and is at rest. Thus, there is no *intentional* movement of the limb, and any observed dynamics is purely resting tremors. Resting tremors are most commonly caused by Parkinsonian disease and sometimes occur in severe essential tremors as well. We are specifically concerned with the tremors occurring in the 4–6 Hz range.

For the purpose of disturbance rejection, we can treat resting tremors as caused due to an external disturbance (a sinusoid) acting on the MSD system (8.1), also refer to Figs. 8.1 and 8.2. Thus, we have the following dynamics:

[5] Through some complex feedback mechanism, not discussed here.

$$\Sigma := \begin{bmatrix} \dot{x}_1 \\ \dot{x}_2 \end{bmatrix} = \begin{bmatrix} 0 & 1 \\ -\frac{k}{m} & -\frac{c}{m} \end{bmatrix} \begin{bmatrix} x_1 \\ x_2 \end{bmatrix} + \begin{bmatrix} 0 \\ \frac{A\sin(\omega t)}{m} \end{bmatrix} \tag{8.2}$$

Note here that the MSD system is completely supported against gravity, thus the only force acting on it is the external sinusoid (modeling the tremor signal). Our objective is to compute the lower bound on the active damping in the system (Σ) to be able to reject the (low frequency) disturbance.

Lemma 1 *The system, Σ, in* (8.2) *achieves disturbance rejection if* $A\sin(\omega t) \ll cx_2$.

Proof It is easily seen that if $A\sin(\omega t) \ll cx_2$ (strong damping force) then for any initial condition sufficiently close to the origin $x = [0\ 0]^T$, the resulting trajectory will also be sufficiently close to the origin (note that the chosen Lyapunov function has a global minima at the origin). However, since the output response of a linear system to an input sinusoid is again a sinusoid of varying amplitude and phase—the trajectory will be a sinusoid but will be in a neighborhood of the origin and has very small amplitude. □

Remark 2 Lemma 1 describes a system experiencing resting tremors if $A\sin(\omega t) > cx_2$. If $A\sin(\omega t) \ll cx_2$ then Lemma 1 describes a system where the resting tremors are negligible.

8.4 Conclusions

This paper attempts a system theoretic formulation of resting tremors, a type of movement disorder. We model the pathological tremors occurring in the 4–8 Hz range acts as a disturbance on the system, and this disturbance can be rejected provided sufficient active damping is present in the system. We derive bounds on the damping factor so as to reject this disturbance.

Acknowledgments The first author wishes to thank Dr. Nanda Kumar (MS Ramaiah Medical College) for detailed discussions on gait analysis and providing insight into an engineering approach to understanding movement disorders.

References

1. Deuschl G, Krack P, Lauk M, Timmer J (1996) Clinical neurophysiology of tremor. J Clin Neurophysiol 13(2):110–121
2. Helmich RC, Hallett M, Deuschl G, Toni I, Bloem BR (2012) Cerebral causes and consequences of parkinsonian resting tremor: a tale of two circuits? Brain 135(11):3206–3226
3. Findley LJ (1996) Classification of tremors. J Clin Neurophysiol 13(2):122–132

4. Franci A, Chaillet A, Panteley E, Lamnabi-Lagarrigue F (2012) Desynchronization and inhibition of Kuramoto oscillators by scalar mean-field feedback. Math Control Signals Systems 24(1–2):169–217
5. Fahn Sanley (2003) Description of Parkinson's disease as a clinical syndrome. Ann N Y Acad Sci 991:1–14
6. Damier P, Hirsch EC, Agid Y, Graybiel AM (1999) The substantia nigra of the human brain. II. Patterns of loss of dopamine-containing neurons in Parkinson's disease. Brain 122 (8):1437–1448
7. Jobst EE, Melnick ME, Byl NN, Dowling GA, Aminoff MJ (1997) Sensory perception in parkinson disease. Arch Neurol 54(4):450–454
8. Kutz M (2003) Standard handbook of biomedical engineering and design. McGraw Hill, New York
9. Shadmehr R, Wise SP (2005) Computational neurobiology of reaching and pointing: a foundation for motor learning. MIT Press, Cambridge
10. Fu M, Cavusoglu MC (2012) Human arm-and-hand dynamics model with variability analyses for a stylus-based haptic interface. IEEE Trans Syst Man Cybern Part B 42(6):1633–1644
11. Nowak DA, Hermsdörfer J, Marquardt C, Topka H (2003) Moving objects with clumsy fingers: how predictive is grip force control in patients with impaired manual sensibility? Clin Neurophysiol 114(3):472–487
12. Crawford P, Zimmerman E (2011) Differentiation and diagnosis of tremor. Am Fam Physician 83(6):697–702

Chapter 9
Recognition of Medicinal Plants Based on Its Leaf Features

E. Sandeep Kumar and Viswanath Talasila

Abstract Ayurveda is one of the oldest forms of medicine, and the use of medicinal plants is a crucial aspect in Ayurvedic treatment. In this paper, we develop an automated system to identify the vast number of medicinal plants relevant for Ayurveda. We focus on the use of image processing and pattern recognition algorithms for plant identification. A unique feature identifier is computed, and this feature algorithm is tested on ten different medicinal plants for accurate identification. The main result in this paper is a demonstration that the features attributed to the leaf of each plant are Gaussian distributed.

Keywords Ayurvedic medicinal system · Image processing · Gaussian distribution

9.1 Introduction

Ayurveda is a medicinal system which originated in India. Medicinal plants form the fundamental requirement of this system of medicine. Although allopathic medicine has a significant impact on disease treatment, recent trends indicate a rising importance of Ayurvedic medicine and thus the increasing importance of medicinal plants. The identification of many medicinal plants often requires an expert, and such experts are of shortage in India. Furthermore, the conversion of

E. Sandeep Kumar (✉) · V. Talasila
Department of TCE, M.S. Ramaiah Institute of Technology, Bangalore, Karnataka, India
e-mail: sandeepe31@gmail.com

V. Talasila
e-mail: viswanath.talasila@msrit.edu

V. Vijay et al. (eds.), *Systems Thinking Approach for Social Problems*, Lecture Notes in Electrical Engineering 327,
DOI 10.1007/978-81-322-2141-8_9

rural lands and forests into commercial developments has an unknown impact on the number of medicinal plants which may be threatened. To safeguard medicinal plants requires a concrete knowledge of the availability and geographical spread of such plants across India, and how commercial development is affecting the survival and availability of medicinal plants.

Thus, there is a need to develop mechanisms for the analysis of medicinal plant availability in India, as well as a means of characterizing the health of the plants. One specific aspect of this would involve the design and development of an automated medicinal plant identification system, which is the focus of this paper.

This work is an improvement of Sandeep [1] where the author builds a system which identifies the medicinal plant species based on area, color, and edge features. This paper extends the results and computes additional features such as the roundness, rectangularity, and elongation of the leaf. Both this present work and [1] use the calculation of the area specified by Sanjay et al. [2]. In [3], survey was made on many edge detection algorithms and Canny edge detection algorithm was used for leaf edge detection. The use of neural networks for classification of medicinal plants was attempted in Anami et al. [4], in which they take the image of the entire plant and they consider only the color, texture, and edges. Arun et al. [6] classify medicinal plants based on texture features.

Senthilkumaran et al. [7] proposed a method of edge detection based on soft-computing techniques and compared their results with the existing edge detection algorithms like Canny, Roberts and Prewitt. Kaushal et al. [8] proposed an adaptive thresholding technique for edge detection in images and compare the obtained results with the existing techniques. Mani et al. [9] compared various edge detection techniques of image processing and concludes the work with a comparison study. Arai et al. [11] propose a method of identifying ornamental plants based on discrete wavelet transformation approach. Zhang et al. [12] proposed a method of extracting features from facial images of humans based on the discrete wavelet transforms. They discussed the use of artificial neural networks as a classifier for facial images. Sandeep et al. [14] proposed an image-processing algorithm for identifying leaf spot diseases in groundnut plants. Wu et al. [15] proposed a method of classification of plants using leaf images and probabilistic neural networks. They discussed the accuracy of the applied method with the different classes. Beghin et al. [16] proposed a method of identifying plants based on shape and texture features from the leaf images. Papers [17], [18] discuss the classification of plants based on leaf images using artificial neural networks. Herdiyeni [19] propose a method of recognizing medicinal plants using a mobile application. The smart app uses combination of fuzzy techniques for the plant identification.

In this proposed work, experimental analysis was carried out with 10 plant species. This includes *Vinca rosea* pink and white species, *Hibiscus rosa senesis*, *Malvaviscus arboreus*, *Leucas aspera*, *Mentha arvensis*, *Phyllantus acidus*, *Santalum album*, *Murayya keonigii*, and *Azadirachta indica*. Fifty leaves were taken from each plant species, and clustering was done. Also specially from *V. rosea* pink plant species, 50 leaves each from 6 plants from different locations and 50 leaves each from 5 plants from different locations were taken for studying their environmental behavior and changes. This proposed work embeds itself within some complicated methodology with more accuracy. The graphical plots shown at the end of this paper proves this method is successful.

This proposed work provides some interesting results, for example, proving that the distribution of the leaf features in a plant is Gaussian, and if we consider several such plants belonging to the same family, even then the distribution is Gaussian. *V. rosea* plants were used for this study, which is solely a new result regarding medicinal plants and possibly supports the notion that a majority of the plants have Gaussian-type features.

Hence, this sole project uses some new concepts toward handling botanical data (leaf) for the identification of medicinal plants. This work is one of the few attempts done on medicinal plants.

9.2 Proposed Methodology

This section gives a detailed description of the work done.

9.2.1 Image Acquisition

The leaf was plucked from the plant and was placed on a white background, and the image was captured at a distance of 10 cm. A Sony cyber-shot camera with a resolution of 16.2 M pixels was used. The images were captured in natural daylight. Camera calibration was not performed before image capture since it was assumed that any sensor errors induced would be negligible in this specific experimental setup.

9.2.2 Image Samples

Ten plant species were taken with 50 leaf images from each plant. In addition, to analyze the statistical features of the plant leaves, the *V. rosea* plant species with 50 leaves each from 6 pink-flowered plants and 50 leaves each from 5 white-flowered plants were considered. The ten plant species considered are *V. rosea* pink and white species, *H. rosa senesis, M. arboreus, L. aspera, M. arvensis, P. acidus, S. album, M. keonigii,* and *A. indica.* It is interesting to note that *V. rosea* pink- and white-flowered plants fall in the same plant family as the *H. rosa* and *M. arboreus.* The samples are shown in Fig. 9.1.

Fig. 9.1 **a** *Phyllantus acidus,* **b** *Hibiscus rosa senesis,* **c** *Leucas aspera,* **d** *Vinca rosea* (white), **e** *Mentha arvensis,* **f** *Murayya keonigii,* **g** *Azadirachta indica,* **h** *Santalum album,* **i** *Malvaviscus arboreus,* and **j** *Vinca rosea* (pink)

9.2.3 Method

Figure 9.2 illustrates the system block diagram of the proposed method. The images are captured in the image acquisition stage, and the captured image is then pre-processed. Morphological features are extracted in the feature extraction stage leading to the identification of the medicinal plants.

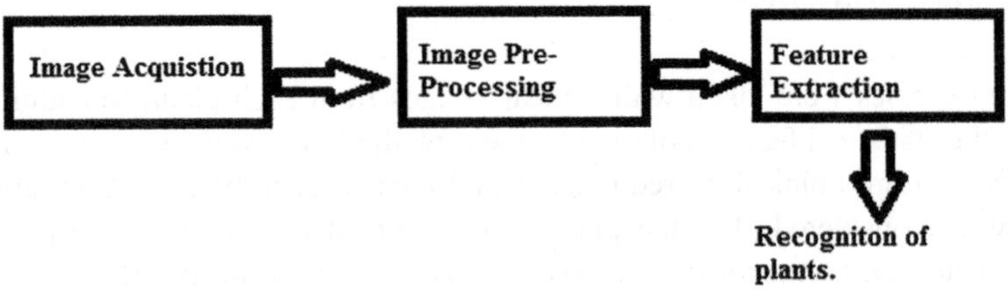

Fig. 9.2 System block diagram

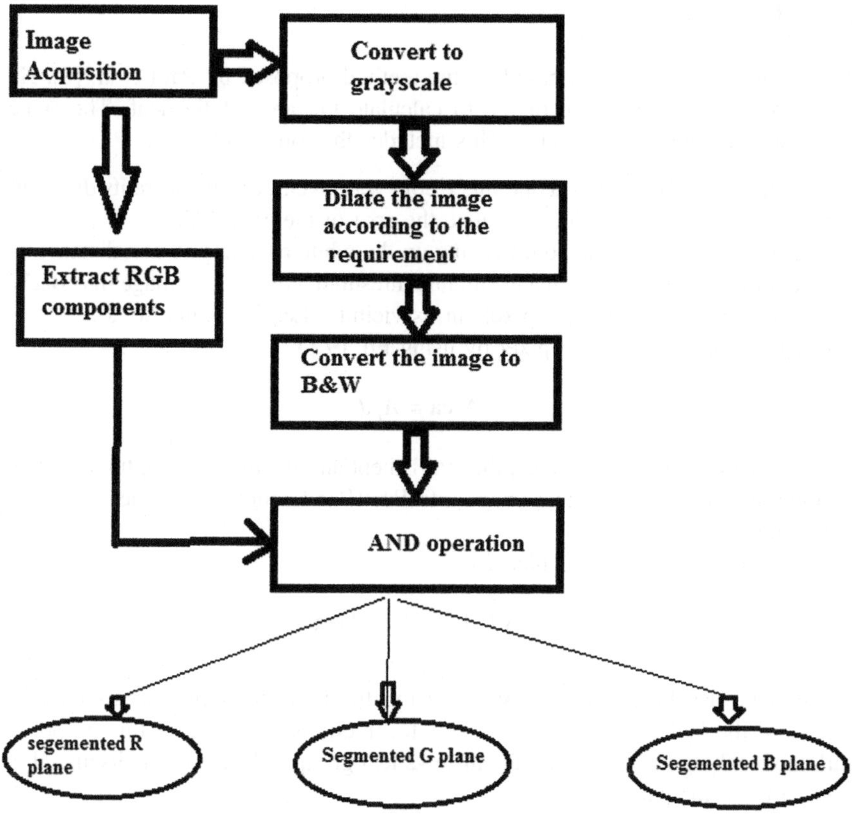

Fig. 9.3 Segmentation flowchart

9.2.4 Image Segmentation

This is an important step in any image preprocessing. Image is captured and converted to B&W. The color image is split into its plane, and later, masking of the background is done using a simple 'and' operation. This leads to focus only on the leaf portion in the image. The segmentation algorithm is depicted in the Fig. 9.3.

9.2.5 Feature Extraction

The following features were extracted from the leaf images for the identification.

9.2.5.1 Leaf Area

The leaf area was calculated based on the method proposed by Patil et al. [2]. He used a coin as the reference object to calculate the area of the leaf. The same technique was followed here also. This includes the following steps:

- Take a coin (in this work, a one rupee coin was taken), and measure its diameter manually. Using this value, calculate the area of the coin, 'A'.
- Capture the image of the coin keeping on the white background at a distance of 10 cm from the object. By using suitable threshold, convert the image into B&W and measure the number of pixels in its vicinity. Let this count be 'J'.
- Calculate the area of each pixel by using Eq. (9.1).

$$\text{Area} = A/J \tag{9.1}$$

- At the same reference distance, the leaf is kept and the image is captured. Using some threshold, convert this image into B&W and count the number of pixels. Let this be 'N'.
  The area of the leaf is calculated by

$$\text{A.O.L} = N \times \text{Area} \tag{9.2}$$

Area can also be calculated by just using the number of pixels count in the vicinity of the leaf and normalizing. This technique is also used in this proposed work for breaking the cluster-1 and cluster-2 leaf groups which will be discussed in latter parts of this paper.

9.2.5.2 Leaf Roundness

$$R = (4 \times \text{A.O.L})/(\pi \times \text{length} \times \text{length}) \tag{9.3}$$

The length of the leaf was calculated by finding the Euclidean distance between the end pixel coordinates. The Euclidean distance is calculated by the following formulae

$$\text{Distance} = \text{sqrt}((X_2 - X_1)^2 + (Y_2 - Y_1)^2) \tag{9.4}$$

9.2.5.3 Rectangularity

$$\text{R} = (\text{A.O.L})/(\text{length} \times \text{width}) \tag{9.5}$$

where the length is calculated as in the case of the roundness, whereas the width is calculated between vertical end points as shown in the Fig. 9.4.

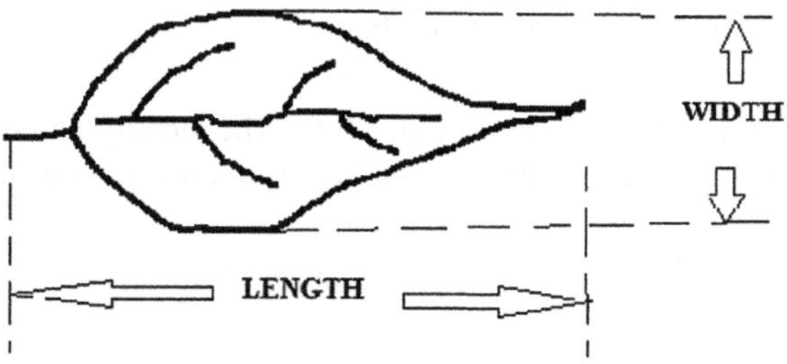

Fig. 9.4 Extraction of length and width

9.2.5.4 Edges and Textures

The method used to extract the edge and texture information is using another image as reference. The reference image is an either entire white image or an entirely black image. In our work, we have used entire black image as the reference image. Let the black reference image be denoted as 'B'.

Let 'S' denote the image containing only the leaf image after applying the segmentation algorithm. The segmentation algorithm is dealt in detail in the later section. Consider another image which is entirely dark image with all the pixel values to be zeros. Now, apply Canny edge detection algorithm to the image 'S'.

After the application of the canny method, we get the edges and the texture information of the leaf as a B&W image. Calculate the normalized image histogram of this image. Let this image be 'B'. Now, find the difference between these image plane histograms i.e., (between S and B) using Euclidean distance Eq. (9.4). This will be a single number.

9.2.5.5 Color

The color JPEG image of the leaf is split into three planes R, G, and B. Now, we apply the segmentation algorithm to each plane to extract only the leaf region leaving the background. A similar method of edge and texture feature estimation is applied to calculate the redness number, blueness number and the greenness number of the leaf. Here also, entire black image is taken as the reference. These features are helpful because the leaf colors vary from one plant to another. Even though many of them are green, they have varying intensity in greenness itself.

9.2.5.6 Elongation

$$E = \text{Width}/\text{length} \tag{9.6}$$

The elongation factor gives us the aspect ratio of the leaf. This is a number which indicates how much the leaf is elongated.

These eight features make a powerful feature group to identify the leaf.

9.2.6 System Algorithm

Let the features extracted be f1–f8. The identity number for each plant is obtained by using the weighted averaging technique. The equation is given by

$$I = \sum_{j=1}^{N} (W_j \times f_j)/N \tag{9.7}$$

where

N number of features extracted
W_j weights given to the features
f_j feature values extracted

Step 1 Capture image.
Step 2 Segment the leaf from the background.
Step 3 Extract features.
Step 4 Calculate unique ID number from Eq. (9.7).
Step 5 Check for the uniqueness of the number. If unique, go to Step 6, else adjust weight of features.
Step 6 Stop.

9.3 Distribution of Leaf Features

This work is unique and novel as per the study of medicinal plants which gives a result of great interest in designing classifiers, especially if attempted for a parametric classifier.

To find the distribution of the leaf features, we adapted a popular graphical test. This plot is formed by plotting the sorted feature values versus their expected z-values that are the values of z which divide the area under the standard normal curve into $n + 1$ equal area, where n is the number of samples. Consider an example sample set 5, 1, 3, 6, and 9. These samples are sorted in increasing order. For these sorted lists, cumulative distribution function (CDF), $C(z)$, is calculated: 1/6, 3/6, 5/6, 6/6, and 9/6. Once the CDF is calculated, now find the corresponding z values from the standard normal chart for each value of the sample. Now, plot the graph of sorted sample feature values versus the expected z values. If the plot matches with any one of the patterns given in Fig. 9.5, we can conclude that it is that specific type of distribution. The equation behind this standard normal chart is

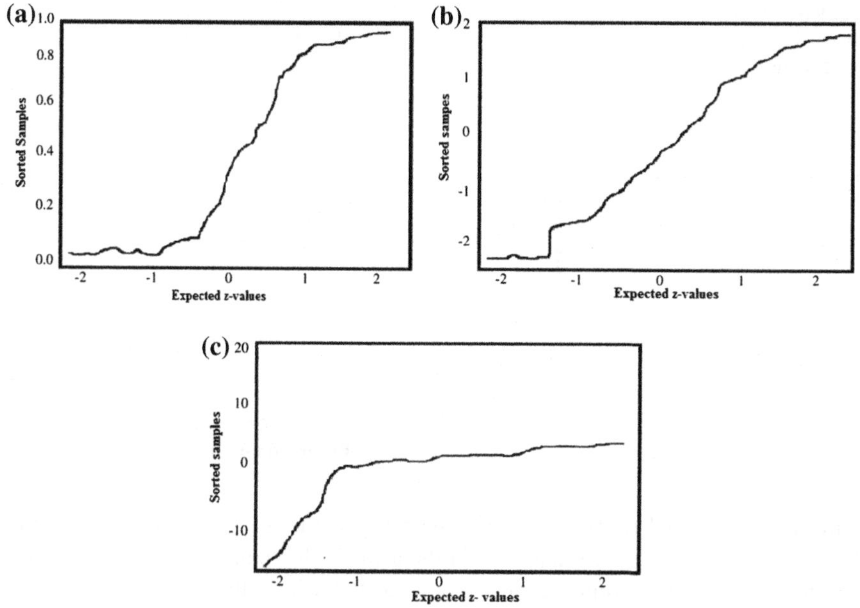

Fig. 9.5 Normal plots for **a** uniform distribution, **b** normal distribution, and **c** Cauchy distribution

$$C(z) = \int_{-\infty}^{z} \left(\frac{1}{\sqrt{2\pi}}\right) e^{-\left(\frac{z^2}{2}\right)} dz \tag{9.8}$$

In our work, we took *V. rosea* pink-flowered plants 50 leaves. We calculated the identity number as per the Eq. (9.7) by extracting various features from the leaves. We got a set of 50 identity numbers. Now, we divided each identity number by 51. This resulted in 50 $C(z)$ values. Find the corresponding expected z values, and plot these values leading to the normal plot. By inspection, one can come to a conclusion that this is a normal distribution since the graph obtained for *V. rosea* matches with the normal plot of the Fig. 9.5 and is shown in Fig. 9.6. To strengthen the result obtained, we used another test which is **Method of Moments (M.O.M)**. From the graphical method, we came to a conclusion that is surely not Cauchy distribution, but to confirm it is neither uniform, we use the M.O.M method. For the 50 feature values, calculate the mean and the variance. For the Vinca plant, we got $\mu = 42.6707$ which is the mean and $\sigma^2 = 78.3640$ which is the variance. For the uniform distribution, $(\widehat{a} + \widehat{b})/2 = 42.6707$ and $\frac{(\widehat{b-a})^2}{12} = 78.3640$.

These two simultaneous equations are solved, and the estimated values of a and b were calculated. As per the data set of the *V. rosea* pink-flowered plant, the feature values set considered, the range values $a = 29.1750$ and $b = 63.4358$. By solving the equations, the estimated values of a and b are $\widehat{a} = 27.3380$ and $\widehat{b} = 58.0034$. By

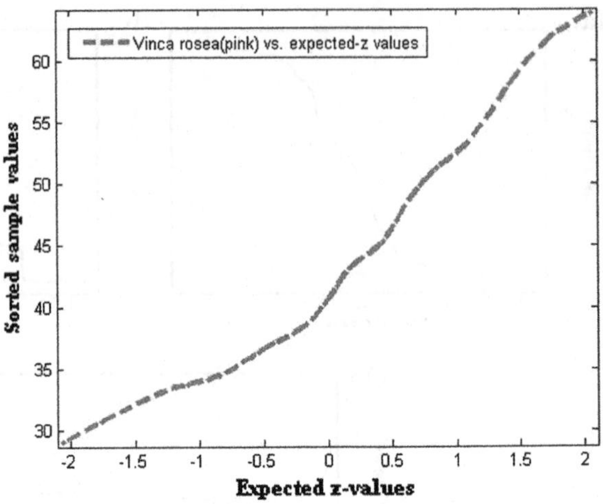

Fig. 9.6 Normal plot obtained for *Vinca rosea* plant leaf feature values

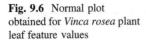

comparing the actual range values and the estimated range values, it is noticeable that the estimated value of b falls within the actual value range, but the same is not true for a. Since the estimated value does not fall within the range of the actual values, we can come to the conclusion that this is not uniform distribution.

Hence, from the graphical analysis and the Method of Moments estimate, we came to the conclusion that the medicinal plants are **Gaussian distributed**, which provides evidence for choosing a proper distribution or for assuming normal distribution for designing parametric classifiers, which is the future enhancement of this work.

9.4 Results and Discussions

The main idea behind this work was to generate a unique identity number for each plant. The analysis started with the leaf features. We used some image processing tools and algorithms to generate and prove the uniqueness of the identity numbers obtained for the plants. We considered 50 leaf sample images from each species. We considered 10 different plant species for our analysis. The algorithm was executed and tested in MATLAB 2008a.

The identity number for each plant was calculated by weighted average method. The first level of clustering based on the area was obtained. Since here automatically the area was given the more weight-age because it was not normalized. The plot of the clustering is as shown in the Fig. 9.7. From this plot, we can observe that the samples have clustered into four groups with the small area leaves at the bottom and the large area leaves at the top.

The next task was to break the cluster at the bottom say cluster-1. Here, the area was the normalized pixel count in the vicinity of the leaf and the roundness

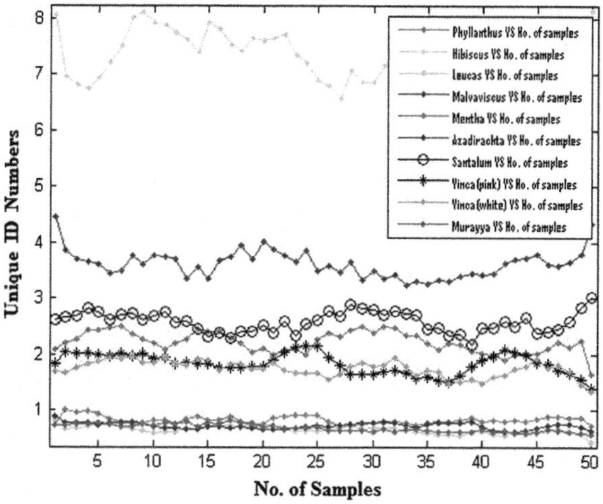

Fig. 9.7 Area-based clustering

parameter was considered. The roundness was given 1,000 weightage, and the algorithm was executed. But it was found that *Mentha* and *Leucas* got separated and *Murayya* and *Azadirachta* still packed in the cluster. Again the roundness and edges were given 1,000 weightage, and the algorithm was executed and all the four leaf groups in the cluster-1 separated. Each time, a proper decision boundary was estimated between the feature values and the algorithm was executed. The cluster separation is shown in Fig. 9.8. The next task was to separate the cluster-2. This was the most difficult task, and all the leaves in the group were almost similar. By repeated trial and error techniques, the roundness and greenness were given 1,000 weightage, and the algorithm was executed, and the cluster was separated, and this is shown in the graph of Fig. 9.9.

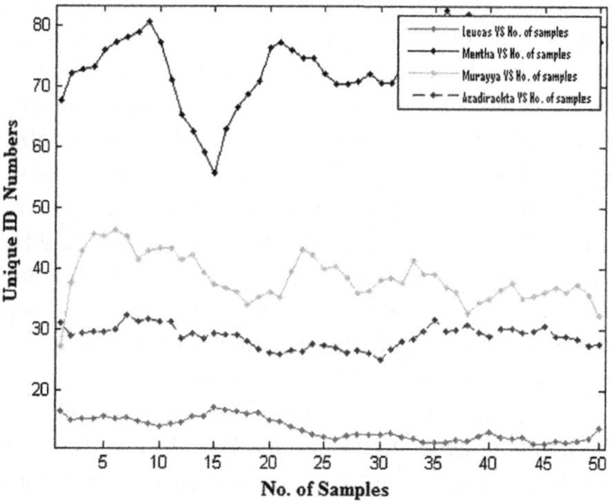

Fig. 9.8 Separation of cluster-1

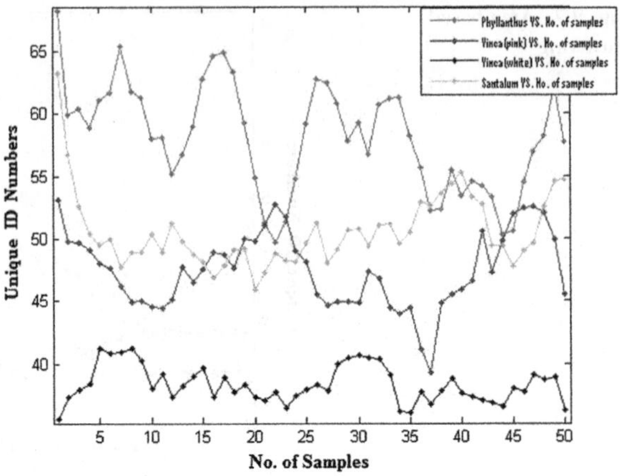

Fig. 9.9 Separation of cluster-2

Table 9.1 Gaussian parameters for the plant species

Plant species	Standard deviation	Mean	Variance
Hibiscus	**0.9769**	**7.2391**	**0.9543**
Malvaviscus	**0.5396**	**3.5914**	**0.2911**
Phyllantus	8.0896	57.9588	65.4412
Santalum	5.5645	50.4500	30.9632
Vinca (pink)	5.9960	47.5482	35.9521
Vinca(white)	4.0234	38.3952	16.1862
Leucas	**2.3500**	**13.6219**	**5.5223**
Mentha	**9.8603**	**73.1065**	**97.2245**
Murayya	**6.8338**	**38.7227**	**46.7011**
Azadirachta	**3.6006**	**28.8582**	**12.9644**

Hence, all the leaf groups were separated. In the graph of Fig. 9.9, there is slight perplexity in the classification of leaves belonging to cluster-2. But since it had been proved that these features are Gaussian in nature, we extract the parameters mean, variance, and standard deviation. It was noticed that each plant varies greatly than the other. Table 9.1 depicts the values calculated as per 50 samples in each species. Bold fonts in Table 9.1 depict different clusters. Bottom most four plants belong to cluster-1, which is a group of small sized leaves, middle with non-bold names indicate cluster-2 with medium sized leaves, Malvaviscus and Hibiscus belong to the group of large sized leaves, but since Malvaviscus leaves are smaller sized compared to Hibiscus leaves these two plants can be put to two separate clusters, where Malvaviscus in cluster-3 and Hibisucs in cluster-4. Hence we can see the formation of four clusters depending on the area feature.

Hence, from Table 9.1, it was concluded that all the plant species are unique and the numbers generated as per the Eq. (9.7) was also unique in nature; that is, if the numbers generated are unique for their cluster groups, then it is concluded to be

Table 9.2 Environmental variations of plant species

Plant species	Standard deviation	Mean	Variance
Vinca-pink 1	6.5957	43.5039	26.2387
Vinca-pink 2	5.7644	43.6554	33.2286
Vinca-pink 3	6.5957	39.8497	43.5039
Vinca-pink 4	4.4970	43.7813	20.2226
Vinca-pink 5	8.8523	42.6707	78.3640
Vinca-pink 6	5.9960	47.5482	35.9521
Vinca-white 1	4.0508	43.9247	16.4088
Vinca-white 2	4.6214	36.5139	21.3572
Vinca-white 3	4.0234	38.3952	16.1862
Vinca-white 4	6.0754	40.7858	36.9104
Vinca-white 5	6.6822	47.6514	44.6514

unique. In Table 9.1, even though 38 are appearing in the mean column twice, they both belong to the different clusters which are well separated by choosing a proper decision boundary. Mean is the actual unique ID generated for a plant species. The next task was to find the changes in the plant belonging to same species but taken at different locality. For this work, six Vinca (pink) plants and five Vinca (white) plants were taken and the Gaussian parameters were found. It was found that these two plant species, since they belong to the same family, jumble with each other as the locality changes.

The plants were taken from different localities belonging to Shimoga district, Karnataka, India. The result was tabulated and is shown in Table 9.2. It was observed that the plants start mixing with each other due to varying environmental factors. Hence, this quandary can be solved by taking additional features from the plants.

Both these plants are well known for their vincamine which is common in both the species. For a botanist, these plants are same, but as per the traditional Ayurvedic practitioners, this in-depth differentiation of plants is necessary. This problem may be common to the plant species belonging to the same plant family. Also separation of cluster-2 is difficult solely by the leaf species if number of plant species increases. Hence, to increase the authenticity, we can opt for flowers and fruits as features.

The normal plot analysis was conducted for the mean values of Vinca pink plants. It was observed that even the variations of the features in different plants belonging to the different locality are also Gaussian, and it is shown in Fig. 9.10.

The results obtained in this paper were derived from a data set comprising of 10 species. An open issue is whether the features from these 10 species (which are unique within these 10) remain unique when a larger number of medicinal plant species are considered. Furthermore, it is quite possible that when we consider all non-medicinal plants as well, these features will not remain unique. In such cases, an awareness of the situational context in which the images are captured becomes critical.

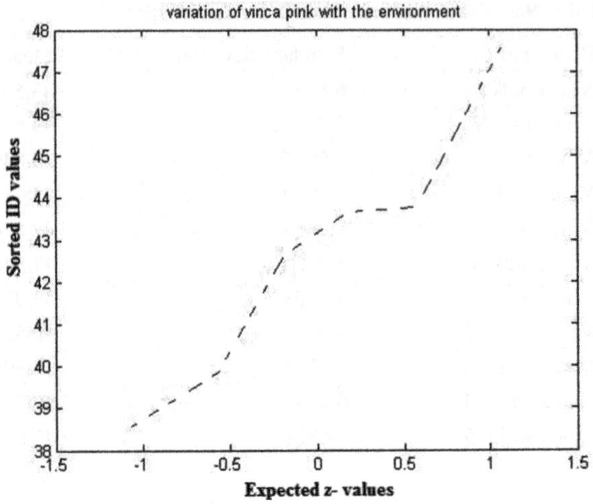

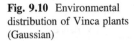

Fig. 9.10 Environmental distribution of Vinca plants (Gaussian)

9.5 Conclusions

The work presented in this paper is an attempt to perform automatic recognition of medicinal plants, as well as to analyze the statistical nature of the image features used for recognition. Ten plant species including plants belonging to the same families were considered. It was proved that even some of the species though they belong to the same family are different due to slight variations in the features. A novel result in this paper is the experimental demonstration that the various image-related features have a clear Gaussian distribution. This work is a contribution toward serving horticulture in India and an attempt toward a faster and better means of identifying medical plant species.

Acknowledgments Authors would like to thank Dr. Vijaya Kumar B.P., Head and Prof. of Information Science and Engineering, M.S. Ramaiah Institute of Technology, for his constant support and encouragement throughout this work.

References

1. Sandeep Kumar E (2012) Leaf color, area and edge features based approach for identification of Indian medicinal plants. Indian J Comput Sci Eng (IJCSE) 3(3):436–442
2. Patil SB, Bodhe SK (2011) Betel leaf measurement using image processing. Int J Comput Sci Eng (IJSCE) 3(7):2656–2660
3. Canny Edge Detection, 09gr820, 23 Mar 2009. http://www.cse.iitd.ernet.in/~pkalra/csl783/canny.pdf
4. Anami BS, Suvarna SN, Govardhan A (2010) A combined color, texture and edge feature based approach for identification and classification of Indian medicinal plants. Int J Comput Appl 6(12):45–51 (0975-8887)

5. Solomon C, Breckon T (2011) Fundamentals of digital image processing. Wiley Black-well, Oxford
6. Arun CH, Sam Emmanuel WR, Christopher Dhurairaj D (2013) Texture feature extraction for identification of medicinal plants and comparison of different classifiers. Int J Comput Appl 62 (12):1–9 (0975-8887)
7. Senthilkumaran N, Rajesh R (2009) Edge detection techniques for image segmentation-a survey of soft computing approaches. Int J Recent Trends Eng 1(2):250–254
8. Kaushal M, Singh A, Singh B (2010) Adaptive thresholding for edge detection in gray scale images. Int J Eng Sci Technol 2(6):2077–2082
9. Mani R, Aggarwal H (2010) Study and comparison of various image edge detection techniques. Int J Image Process (IJIP) 3(1):1–11
10. Gonzalez RC, Woods RE (2002) Digital image processing. Prentice Hall, Upper Saddle River
11. Arai K, Abdullah IN, Okumura H (2013) Identification of ornamental plant functioned as medicinal plant based on redundant discrete wavelet transformation. Int J Adv Res Artif Intell 2(3):34–38
12. Zhang P, Guo X (2012) A cascade face recognition system using hybrid feature extraction. Digit Sig Process 22(6):987–993
13. Gose E, Johnsonbaugh R, Jost S (2009) Pattern recognition and image analysis. PHI publications, New Delhi
14. Sandeep Kumar E, Benak Patel MP (2012) Analysis of early leaf spot disease and damage by *Helicoverpa armigera* pest attack on groundnut leaves using image processing. IJSRCSAMS 1(2):1–5
15. Wu S, Bao FS, Xu EY, Wang Y-X, Chang Y-F, Xiang Q-L (2007) A leaf recognition algorithm for plant classification using probabilistic neural network. In: IEEE international symposium on signal processing and information technology, pp 11–16
16. Beghin T, Cope JS, Remagnino P, Barman S (2010) Shape and texture based plant leaf classification, advanced concepts for intelligent vision systems. Springer, Berlin, pp 345–353
17. Chaki J, Parekh R (2011) Plant leaf recognition using shape based features and neural network classifiers. Int J Adv Comput Sci Appl 2(10):41–47
18. Zhang L, Kong J, Zeng X, Ren J (2008) Plant species identification based on neural network. In: IEEE natural computation-2008, Oct 2008, ICNC-08
19. Herdiyeni Y, Wahyuni NKS (2012) Mobile application for Indonesian medicinal plants using fuzzy local binary pattern and fuzzy color histogram. IEEE, ICACSIS

Chapter 10
CLOSENET—Mesh Wi-Fi in Areas of Remote Connectivity

Sriragh Karat, Sayantani Goswami, Aparna Sridhar, Aakash Pathak, D.S. Sachin, K.S. Sahana, Viswanath Talasila and H.S. Jamadagni

Abstract The availability of high-speed network connectivity in Indian villages can have significant socioeconomic impact. This can have a positive influence on many aspects of development such as telemedicine, smart agriculture, and education among others. In this paper, we describe our attempts to develop and deploy CLOSENET, a wireless mesh network (WMN) in rural areas facing insufficient network coverage. The network topology is chosen based on a case study on a specific village (C.K.Pura in Tumkur, Karnataka), taking several constraints into account such as terrain interference, population of users, and available power sources. We analyze factors such as normalized routing load, packet delivery ratio, data throughput, and data latency in order to demonstrate the efficiency, reliability, and durability of the network through a simulation model developed using the NS2 simulation tool. Advanced reactive routing protocols such as the ad hoc on-demand distance vector (AODV) have been chosen in order to guarantee good performance in terms of scalability, reliability, throughput, load balancing, congestion control, and efficiency. Furthermore, the terrain is mapped onto an appropriate attenuation model implemented in NS2; this helps decide the throughput and latencies at each node, as well as the number of nodes required to obtain acceptable data transfer rates. The network is self-adaptive in nature. The work presented in the paper is a part of a planned long-term initiative to develop connectivity in remote areas, which will have significant benefits to society.

Keywords Wireless mesh networks · Self-adaptive and self-organizing systems · Ad hoc on-demand distance vector · Teledentistry

---

S. Karat · S. Goswami · A. Sridhar · A. Pathak · D.S. Sachin (✉) · K.S. Sahana · V. Talasila · H.S. Jamadagni
M. S. Ramaiah Institute of Technology, Bangalore, India
e-mail: sachin.ds92@gmail.com

S. Karat
e-mail: sriragh.karat@gmail.com

S. Goswami
e-mail: sayantani.goswami20@gmail.com

V. Vijay et al. (eds.), *Systems Thinking Approach for Social Problems*,
Lecture Notes in Electrical Engineering 327,
DOI 10.1007/978-81-322-2141-8_10

10.1 Introduction

In the current era of globalization, where the Internet is considered to be the prime medium of communication by a majority of world's population, that a very minor portion of the world has access to it is very appalling. In India, there is a lack of connectivity in the rural areas. The penetration of telecommunication services such as telephony and Internet access is low and in some regions non-existent. This is probably due to the fact that service providers assume that the investment is uneconomical owing to the lack of industrial development in these areas. However, it is of utmost importance that the potential of rural areas be realized and equal facilities be given to them by providing them with better connectivity with the rest of the world. This will be of great use since numerous applications suitable to these regions can be implemented for their benefit.

The problem of providing cost-effective solutions can be resolved by wireless mesh networks (WMNs) [2, 3]. WMNs adopt a self-adaptive, self-organizing (SASO) mechanism resulting in reliability, wide area coverage, spectral efficiency, congestion control, and scalability with low incremental costs. One of the main advantages that CLOSENET provides is privacy. One can be assured that the content he transmits over the network is secure and private to oneself. This is possible since every node behaves as an access point, and nobody needs to go to a public portal to do his work, which is of essence for the applications.

The area of implementation of our model is Chennakeshavapura (C.K.Pura), which is a small village in Pavagada taluk of Tumkur district, and is a drought-affected semiarid region. The village comprises 1,350 households apart from a few schools, a bank, a post office, and a Gram Panchayat office. The power supply at C.K.Pura is very irregular; the residents get about 10 h of electricity every day. A few mobile networks are available for regular usage, but the residents cannot afford the high mobile data rates. There are five wired Internet connections which give minimal bandwidth. Good medical facilities are limited to a few doctors in the nearest hospital which is 105 km away. There is a veterinarian for the cattle in the village, as agriculture is the main occupation of the lower strata of the society. The water available at C.K.Pura has high fluoride content, which has adverse effects on the general health of the people. Dental fluorosis is commonly seen as an effect of this and needs to be attended to with great care.

Thus, in this paper, we propose to provide cost-effective and efficient high-speed Internet connectivity by establishing a dynamic, self-adapting, and self-organizing wireless mesh network in a rural area which has an Ethernet backbone, since the lack of connectivity has resulted in the isolation of rural areas from the rest of the technologically advanced world, deterring the holistic development of various fields such as agriculture, health, and social welfare.

10.2 Wireless Mesh Network

10.2.1 Why WMN?

WMNs have the capabilities of self-organizing, self-healing, and self-forming within the network, which is very essential when there are multiple static and mobile nodes in the architecture [9]. The wireless mesh network is formed automatically once the mesh nodes have been configured and activated, which reduces the setup time and maintenance cost, owing to its self-forming nature. Due to the self-healing nature of WMNs, a disjoint node rejoins the network seamlessly and does not require reconfiguration once restored. WMNs are also self-organizing, i.e., if there is an obstacle present in the path between the source and the destination nodes, the next reliable shortest path between the source and the destination is calculated using routing protocols and transfer of data packets takes place through this alternate path. This solves the line of sight (LoS) problem and minimizes the loss of data packets. The network automatically incorporates a new node into the existing structure without any need for adjustments by a network administrator and is thus self-configuring.

Mesh routers have minimal mobility and perform dedicated routing and configuration, which significantly decreases the load of mesh clients and other end nodes, thus preventing congestion [17]. Alternate paths can be chosen which allow the traffic load to be balanced in the network. Load balancing and minimizing the bottleneck via alternate routing can significantly increase network reliability in WMNs. The technology is independent of the underlying radio technology. Each customer can enjoy guaranteed bandwidth when the routers are placed tactically along with sufficient gateways that are available at C.K.Pura, which also enhances the heterogeneity of the network that we have created. The nodes of the WMN can be built with extremely low power requirements using autonomous energy sources, which are very much feasible for our area of deployment, since there is a lot of solar energy available which can be capitalized upon. All these factors have led us to adopt WMN as the best available option for the provision of connectivity in rural areas (Fig. 10.1).

10.2.2 Network Architecture for C.K.Pura

Some of the important design aspects to be taken into consideration while mapping network architecture to a specific region are discussed here. Various radio techniques such as directional and smart antennas, multiple input multiple output (MIMO), and multiradio/multichannel systems exist, among which one can be chosen based on the load of the network. Scalability is an important factor in WMN to gauge its impact at the region of implementation in the long run. Reactive routing protocols are used to ensure efficient mesh connectivity to avoid congestion and

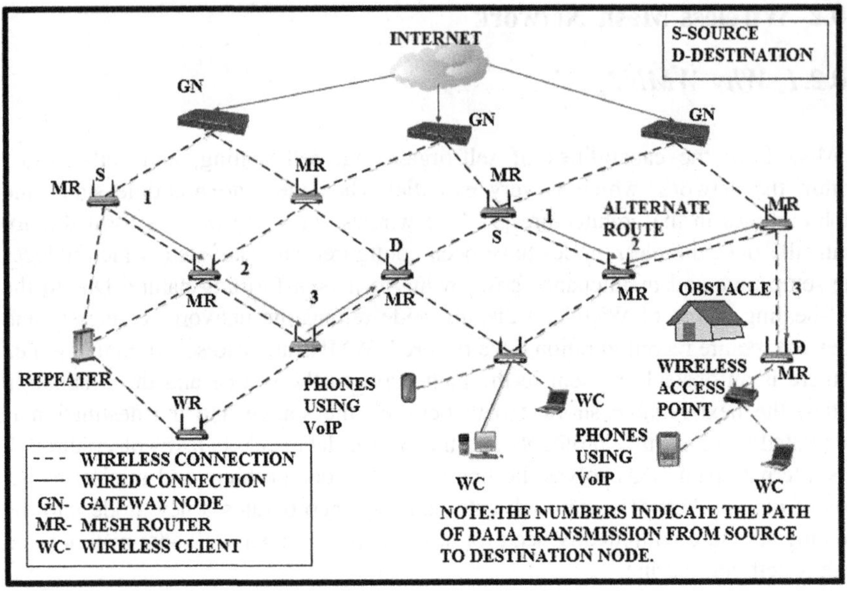

Fig. 10.1 Wireless mesh network architecture

fading. In addition to end-to-end transmission delay and fairness, more performance metrics, such as delay jitter, aggregate and per-node throughput, and packet loss ratios, are considered by communication protocols. Security schemes must be selected based on various parameters such as load and type of data that is to be transmitted over the network apart from the type of threats that could be possible (Fig. 10.2).

After thorough land survey and referring to topography sheets of C.K.Pura, it was seen that the land is mostly plain with minimal physical obstructions for signal transmission. This precludes high signal reflection and hence good signal quality. All the routers to be set up in the WMN will be new dual-band routers complying with 802.11n. Thus, there is no issue of backward compatibility of devices, ensuring high speeds and full utilization of network resources.

10.3 AODV Algorithm and Protocols

Ad hoc on-demand distance vector is a reactive protocol which creates routes when demanded by the source host, and the routes are maintained and used when needed. Hello messages are used to detect and monitor links with neighbors. Each active node periodically broadcasts a Hello message that all its neighbors receive. In case a node fails to receive several Hello messages from a neighbor which are broadcasted, a link break is found [4].

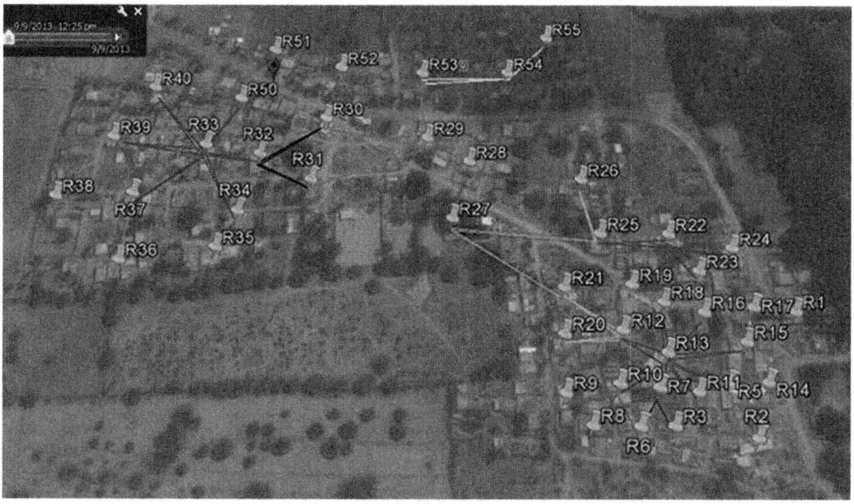

Fig. 10.2 Placement of routers at C.K.Pura after consideration of all parameters along with AODV algorithm

In an AODV network, when a data packet is to be transmitted between two nodes and the source node must send data to an unknown location, it initiates the route discovery process in order to obtain a route toward the destination node by broadcasting a route request (RREQ) message for that destination. When an RREQ is received, at every intermediate node, a route to the source is created. If the receiving node has not received this RREQ before and is not the destination or does not have a current route to the destination, then it rebroadcasts the RREQ. If the receiving node is the destination or has a current route to the destination, it generates a route reply (RREP). As the RREP propagates, each intermediate node creates a route to the destination. When the source receives the RREP, it records the route to the destination and can begin sending data. If multiple RREPs are received by the source, the route with the shortest hop count is chosen.

Each node along the route updates the timers associated with the routes to the source and the destination as the data flow, maintaining the routes in the routing table. If a route is not used for some period of time, the node removes the route from its routing table as it is not sure if it is valid. If data are flowing and a link break is detected, a route error (RERR) is sent to the source of the data in a hop-by-hop fashion. As the RERR propagates toward the source, each intermediate node invalidates routes to any unreachable destinations. When the source of the data receives the RERR, it invalidates the route and reinitiates route discovery if necessary [5].

AODV mainly has three distinct features which are very important to understand its functionality. Firstly, every request is assigned a sequence number. These are assigned so that the nodes do not repeat route requests that have already been passed on. Secondly, there is a feature called time to live which is present for every

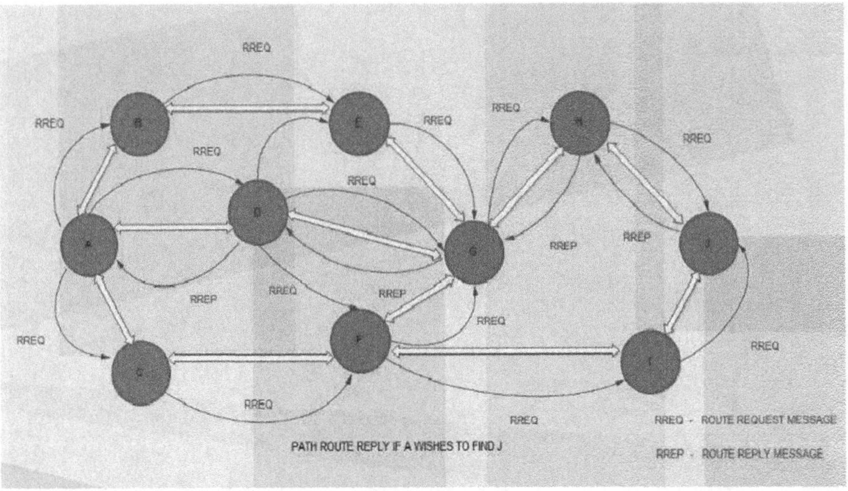

Fig. 10.3 Transmission of data between source and destination as per AODV routing algorithm

route request, which limits the number of times they can be retransmitted. Also, in a situation where a route request fails, we cannot assign another route request unless and until twice as much time has passed since the time-out of the previous request.

AODV has a few advantages over other routing protocols such as optimized link state routing (OLSR), which makes it one of the most preferred protocols. It does not require a central administrator to manage. The control traffic messages are reduced, but it is at the cost of increased latency in finding new routes. Minimal routing is practiced as the route information exists in the routing table, which shows the active routes in the network. It reacts quickly to topological changes and updates any host affected by the change in router error message (RRER) [10] (Fig. 10.3).

AODV offers a large number of advantages in our model. AODV performs very well in a low-traffic and low-mobility environment like C.K.Pura, where the reactive protocol scales perfectly to a larger network with low bandwidth and storage overhead. Since the AODV protocol is resource critical in nature, the usage of resources and their shortage acts as a major factor in environments similar to C.K.Pura's and is thus suitable for CLOSENET. AODV requires less protection as compared to others as only RREP and RRER messages require protection, thus reducing the routing overhead. Security is also implemented by encrypting all the messages in the network using public key cryptography [15].

This algorithm is flashed onto the routers using open firmware programs such as OpenWRT [14], along with the other necessary protocols. The other protocols being used are IEEE 802.11n, IEEE 802.11s, and WPA2/AES. IEEE 802.11n is a wireless networking standard that provides standardized support for MIMO, frame aggregation, and security improvements among other features. It can be used in the 2.4 or 5 GHz frequency bands. IEEE 802.11s is an IEEE 802.11 amendment for mesh networking, which defines how wireless devices can interconnect to create a

WLAN mesh network. It basically helps routers form mesh links with each other, over which mesh paths can be established using a routing protocol and in this case the AODV protocol.

10.4 Privacy and Security

Wireless mesh network (WMN) is a new wireless networking paradigm which is characterized by fast, easy, and inexpensive deployment solutions. In a WMN, the client should have end-point-to-end-point security assurance [16]. Wireless networks provide great flexibility and mobility but are not as secure as wired networks. However, to provide applications to residents via mobile applications and for peer-to-peer communication, wireless mesh network is important. A WMN is vulnerable to various types of attacks, which is why appropriate security measures must be provided with utmost priority, especially when there is a need for maintaining privacy for each user. In CLOSENET, it is extremely essential that confidentiality be maintained and there is no leakage of data in order to support applications such as teledentistry and Internet banking.

Some of the common security threats and attacks that can be faced in a WMN are briefly discussed in this section [11, 19]. Physical threat is the threat posed by the physical location of the routers in cases where they can be replaced by malicious routers so that incorrect routing information is communicated with the network. Confidentiality and integrity of the information sent out by the mesh routers is very crucial to protect from being tampered with or intercepted. This could be realized by employing encryptions in various layers. However, finding a viable encryption policy for protecting confidentiality and integrity while minimizing the algorithm complexity and cost becomes the foremost problem. The existing wired equivalent privacy (WEP) is not suitable due to its inherent flaws. A strong authentication mechanism must be in place so that all nodes joining the network are able to verify the identity of the other nodes. The current mechanism of public key authentication is not the most efficient in our model since it causes a time delay apart from accelerating the depletion of battery life. Apart from these threats, the routing security might be a risk since, by attacking the routing policy of the WMN, the attacker could affect the performance of the network by altering the topological information in the route packet.

WMNs are vulnerable to various kinds of attacks. Redirection by modified route sequence numbers is possible attack against the AODV protocol. Redirection with modified hop count attack is targeted against the AODV protocol in which a malicious node can increase the chances that they are included on a newly created route by resetting the hop count field of a RREQ packet to zero. Apart from these, repudiation, masquerading, and denial of service are also threats to WMNs. There are two main attacks that AODV routing protocol is prone to. They are as follows:

1. *Wormhole attack* Two distant points in the network are connected by a malicious connection using a direct low-latency link called the wormhole link [18]. Once the wormhole link is established, the attacker captures wireless transmissions on one end, sends them through the wormhole link, and replays them at the other end. The malicious nodes could also drop packets and cause network disruption. The attacker can also spy on the packets going through and use large amount of information gained to launch other types of attacks and compromise the security of the network.
2. *Blackhole attack* While receiving the routing request, the attacker claims to have a link to the destination node even if there is not and then forces the source to send the packet through it without forwarding the data packet to the next hop [1].

AODV is a vital part of the WMN for it directly determines the implementation of network function and its efficiency, but is not sufficient to tackle the security issues. A more secure extension of AODV known as secure ad hoc on-demand vector (SAODV) that can be used to protect the route discovery mechanism and provide security features such as integrity, authentication, and non-repudiation can be employed. SAODV assumes that each node has a signature key pair from a suitable asymmetric cryptosystem and that each node is capable of securely verifying the association between the address of a given node and the public key of that node. Two mechanisms are used to secure the AODV messages—digital signatures to authenticate the non-mutable fields of the messages and hash chains to secure the hop count information (the only mutable information in the messages) [20]. This is because for the non-mutable information, authentication can be performed in a point-to-point manner, but the same kind of techniques cannot be applied to the mutable information. Route error messages are protected in a different manner as they have a big amount of mutable information. In addition, it is not relevant which node started the route error and which nodes are just forwarding it. The only relevant information is that a neighbor node is informing to another node that it is not going to be able to route messages to a certain destination anymore. Therefore, every node (generating or forwarding a route error message) uses digital signatures to sign the whole message and any neighbor that receives it verifies the signature. Thus, we have chosen SAODV as a suitable security mechanism in order to support AODV in our WMN model.

10.5 Simulation Results

A few standard parameters such as throughput, end-to-end delay, constant bit rate (CBR), normalized routing load, and congestion window have been simulated on NS2 network simulator [13] along with NAM for better visualization in order to verify the efficiency of our design. Fifty static nodes placed at strategic locations in C.K.Pura and 50 mobile nodes have been taken into consideration in order to assume heavy load on the network. The reason behind this is that it is necessary to

Table 10.1 Scenario for NS2 topology

Performance measurement parameter	Value
Number of simulated nodes	50 static and 50 mobile
Area size of topography (m^2)	3,13,500
Packet size (bytes)	512
Pause time (s)	20
Traffic type	CBR
Simulation time (s)	140
Simulated routing protocol	AODV

analyze the performance of the network under situations of high pressure. The nodes for the simulation were placed as per the C.K.Pura layout in order to gauge the results when deployed. The parameters that have been considered for the NS2 topology simulation are as shown in Table 10.1.

Packet delivery ratio is the ratio of the number of data packets received to the number of data packets sent and is useful in determining the CBR as well as data packet loss ratio. In our model, the packet delivery ratio is calculated to be 0.9920 as shown in Fig. 10.4a, which is an excellent received-to-sent ratio, and indicates that the data packet loss ratio is very minimal, thus ensuring security of the data transmitted [8, 12]. Normalized routing load is defined as the total number of routing packets such as RREQ, RREP, RRER, and HELLO, transmitted per data packet, counted by bit/s, and is found to be 0.262, which can be improved and is shown in Fig. 10.4b. The average end-to-end data delay includes all possible delays caused by buffering during routing discovery latency, queuing at the interface queue, and retransmission delays at the media access control (MAC), propagation, and transfer times. The delay is found to be 187.837 ms as shown in Fig. 10.4c. Average data throughput is the average of the data packets generated by every

Fig. 10.4 Simulation results of **a** CBR, **b** normalized routing load, **c** delay, and **d** throughput

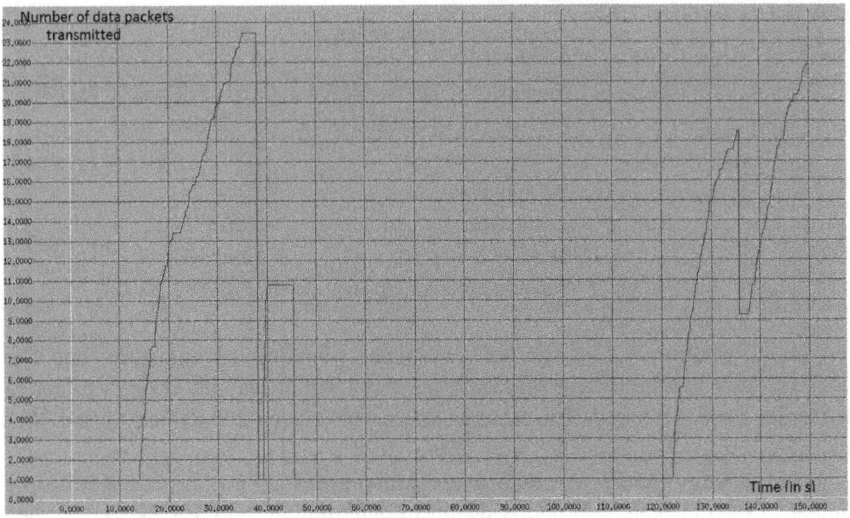

Fig. 10.5 Graph of number of data packets versus time

source, counted by bit/s. In our network simulation, we are able to achieve about 448 bits/s, which is almost equivalent to the standard AODV network simulation results, which is about 510 bits/s. This is shown in Fig. 10.4d.

The congestion window is one of the factors that determine the number of bytes that can be outstanding at any time. This is a means of stopping the link between two places from getting overloaded with too much traffic. The size of this window is calculated by estimating how much congestion is there between the two places. The congestion window for our network is as shown in Fig. 10.5. Thus, it can be determined that CLOSENET can be deployed and will provide an efficient infrastructure to support the applications beneficial to the residents.

10.6 Applications

Various applications can be provided through CLOSENET at C.K.Pura, which will be extremely helpful to the residents there. A few of them include teledentistry, smart street lighting, resolution of human–animal conflict, Internet telephony, e-education, precision agriculture, and IP radio.

10.6.1 Teledentistry

Teledentistry is a branch of telemedicine which utilizes both information and communication technologies and includes the electronic exchange of diagnostic image files, including radiographs, photographs, video, optical impressions, and photomicrographs of patients for consultation with authorized healthcare professionals [6]. As mentioned before, dental fluorosis, which is a developmental disturbance of dental enamel caused by excessive exposure to high concentrations of fluoride (more than 2 ppm) during tooth development, is a common health problem that persists among the residents of C.K.Pura. Although dental care is extremely important, it is neglected and is not considered a matter of priority, as a result of which people are not ready to travel too far to get appropriate treatment.

With the establishment of CLOSENET in C.K.Pura, this issue can be handled efficiently and in a hassle-free manner, which is convenient for the residents since they can enjoy good dental care while sitting at home. The medical history and complaints of the patient, i.e., data [still photographs, digital radiographs, and paper or electronic data], are sent across to the doctor. This involves the usage of cameras, microphones, headset, or external speakers which are installed in the centers that serve as the medium of communication. The data can be sent in two ways, which are real-time video conferencing (asynchronous) and store-and-forward communication (synchronous) (data are stored and sent later when required). Dental management software (dental records, dental billings, appointment scheduling, reporting requirements) can be used to manage the data of the patients and send diagnosis reports and reminders about appointments to the patient's mobile phone as a notification through CLOSENET [7]. The doctor or specialist at the other end diagnoses the ailment of the patient and indicates the further steps to be taken for the treatment of the patient.

One of the major concerns involved in this kind of medication is the confidentiality of patient information. It arises from the transfer of medical histories and records as well as from general security issues of electronic information stored in computers. CLOSENET ensures the security of the data being transferred. To make this more confidential, the patient can be addressed by using initials, a pseudonym, or a reference number. Some practitioners use secure socket layer (SSL) to encrypt the information that flows between the Web browser at the user end (center of the patient) with the Web server receiving the referral (center of the doctor). An encrypted Web-based referral is more secure than a conventional, paper-based referral. Thus, teledentistry with the guarantee of safety and security in C.K.Pura is of prime importance, which can be provided by CLOSENET.

10.6.2 Smart Street Lighting

The present-day street lamps run on a periodic on–off cycle whose period will be predetermined. Since the cycle is fixed, either the lights are on when it is not required or the lights are off when required. This can be overcome by using Wi-fi. Connecting all the street lamps in a meshlike fashion will allow officials to maintain precise operation of each lamp from one location to determine the appropriate output for the area and time of day. They will also know immediately when a light requires maintenance instead of having to rely on visual confirmation from the field, thus reducing the maintenance and energy costs over a period of time. Using the mesh network that we propose, there can be one sensor which will sense the light intensity and this information will be sent to all nodes (lamps which are interconnected using Wi-fi). In turn, all street lamps will be operated without the requirement of manual control. Since only one sensor needs to be deployed, too much additional costs are not incurred.

10.6.3 Resolution of Human–Animal Conflict

Human–animal conflict can be broadly classified into four categories—space conflict, crop raiding, predation on livestock, and death of humans. Video surveillance can be employed in rural areas which transmit data wirelessly. This will help in better monitoring of fields, barns, and livestock movement, thus reducing the risk of conflicts. Introducing low-cost and low-power video surveillance systems allows people to check what is happening in real time. This also helps in keeping a check on intruders who usually get into fenceless fields. Low-cost cameras can be installed as sensors as there is no need of high-definition cameras since its purpose is just monitoring. This proposed solution aims at making the farmers' lives easier at C.K.Pura, as majority of the population there is comprised of agriculturists.

10.7 Conclusion

The CLOSENET model provides a solution to realize the hidden potential of rural areas by providing them with better connectivity and Internet access. This can maximize the untapped economic potential of several regions in a country like India, where about 70 % of the country dwells in rural areas. The implementation of our model at C.K.Pura has the capacity to bring about many improvements in the village in terms of technology, health care, agriculture, education, and overall standard of living without compromising individual privacy. Some of the most prominent applications that would help the residents C.K.Pura are teledentistry, smart street lighting, and human–animal conflict. The deployment of this SASO network is feasible since it is a cost-effective solution and has the capability of providing uninterrupted, high-speed Internet to the entire village, as has been shown in the simulation.

The solution for security issues has been proposed as SAODV, which can be further looked into and simulated. Hybrid networks can be formed using this network as the base. The network can be scaled to include the neighboring villages and introduce intervillage communication with the help of antennas and integration with other technologies. There are also multiple upcoming projects in C.K.Pura like COMMONSENSE, which is a farmers' helpline for smart agriculture, for which our network can provide the infrastructure. CLOSENET can also be implemented at other areas of remote connectivity posing various other constraints, and its efficiency can be improved based on the results that are obtained. There is a great need for providing good connectivity solutions like this, considering the overall progress of the country in the current scenario of globalization.

Acknowledgments The authors thank the people of C.K.Pura, specifically Mr. P.R. Sheshagiri Rao, for giving a clear picture of the climate, terrain, living conditions, and problems plaguing the village. This helped us to design a specific network model with relevant applications.

References

1. Aad I, Hubaux J-P, Knightly EW (2008) Impact of denial of service attacks on ad hoc networks. IEEE/ACM Trans Networking 16(4):791–802
2. Akyildiz IF, Wang X, Wang W (2005) Wireless mesh networks: a survey. Comput Netw 47 (4):445–487
3. Akyildiz IF et al (2006) Next generation dynamic spectrum access cognitive radio wireless networks: a survey. Comput Netw 50(13):2127–2159
4. Cerri D, Ghioni A (2008) Securing AODV: the A-SAODV secure routing prototype. IEEE Commun Mag 46(2):120–125
5. Chakeres ID, Belding-Royer EM (2004) AODV routing protocol implementation design. In: Proceedings of the 24th international conference on distributed computing systems workshops, pp 698–703. IEEE, 2004
6. Chen, J-W et al (2003) Teledentistry and its use in dental education. J Am Dent Assoc 134 (3):342–346
7. Clark GT (2000) Teledentistry: what is it now, and what will it be tomorrow? J Calif Dent Assoc 28(2):121
8. Hassan YK, El-Aziz MHA, El-Radi ASA (2010) Performance evaluation of mobility speed over MANET routing protocols. Int J Netw Secur 11(3):128–138
9. Hossain E, Leung KK (eds) (2008) Wireless mesh networks: architectures and protocols. Springer, Berlin
10. Huhtonen A (2004) Comparing AODV and OLSR routing protocols. In: Seminar on internetworking, Sjkulla, pp 26–27
11. McDaniel P, McLaughlin S (2009) Security and privacy challenges in the smart grid. IEEE Secur Priv 7(3):75–77
12. Natesapillai K, Palanisamy V, Duraiswamy K (2009) A performance evaluation of proactive and reactive protocols using NS2 simulation. Int J Eng Res Ind Appl 2(11):309–326
13. Network simulator website. http://www.isi.edu/nsnam/ns/
14. OpenWRT website. https://openwrt.org/
15. Royer EM, Toh C-K (1999) A review of current routing protocols for ad hoc mobile wireless networks. IEEE Pers Commun 6(2):46–55
16. Sen J (2013) Security and privacy issues in wireless mesh networks: a survey. In: Wireless networks and security, pp 189–272. Springer, Berlin
17. Seyedzadegan M et al (2011) Wireless mesh networks: WMN overview, WMN architecture. In: International conference on communication engineering and networks IPCSIT, vol 19
18. Yi P et al (2006) Flooding attack and defence in ad hoc networks. J Syst Eng Electron 17 (2):410–416
19. Yi P et al (2010) A survey on security in wireless mesh networks. IETE Tech Rev 27(1):6
20. Zapata MG, Asokan N (2002) Securing ad hoc routing protocols. In: Proceedings of the 1st ACM workshop on wireless security. ACM, Atlanta

Chapter 11
Hybrid Backstepping Control for DC–DC Buck Converters

Tousif Khan Nizami and Chitralekha Mahanta

Abstract This paper presents a backstepping control technique in combination with the sliding-mode mechanism for simultaneous control of the capacitor voltage and inductor current in a DC–DC buck converter. The proposed hybrid controller is capable of tackling both the matched and mismatched types of uncertainties like input voltage change and load current variation. The backstepping control can reject both matched and mismatched types of uncertainties, whereas the sliding-mode control is robust against matched uncertainties only. The systematic controller design procedure of backstepping and invariance property of SMC for matched uncertainty have been utilized for robust tracking of both the capacitor voltage and inductor current simultaneously. It is found that by switching between these two different control structures, one exclusively for the matched and the other for the mismatched uncertainties, excellent transient and steady-state performances can be ensured. In the case of backstepping control, performance of the buck converter is largely dependent on design parameters. Hence, these design parameters are judiciously selected to assure optimum performance. Simulation studies have been carried out to verify the effectiveness of proposed hybrid control structure. Transient performances like peak overshoot, peak undershoot, settling time, and also steady-state error have been measured under widely varying changes in input voltage and load current. Simulation results demonstrate that as compared to existing controllers, the proposed hybrid control strategy offers superior transient and steady-state performances.

Keywords Hybrid backstepping control · DC–DC buck converters · Sliding-mode mechanism

T.K. Nizami (✉) · C. Mahanta
Department of Electronics and Electrical Engineering, Indian Institute of Technology Guwahati, Guwahati 781039, India
e-mail: tousif@iitg.ac.in

C. Mahanta
e-mail: chitra@iitg.ac.in

V. Vijay et al. (eds.), *Systems Thinking Approach for Social Problems*,
Lecture Notes in Electrical Engineering 327,
DOI 10.1007/978-81-322-2141-8_11

11.1 Introduction

The buck converter is a step-down DC–DC converter used in a variety of industrial and residential applications [1, 2]. It is inherently nonlinear in nature. Accurate and fast control of both the capacitor voltage and the inductor current in buck converters, when subjected to wide range of uncertainties, is still a challenging task. A lot of work has been carried out in the past for capacitor voltage regulation in buck converters [3–7]. However, majority of these methods ignore the inductor current control part. Both the voltage and current control are equally important for good transient and steady-state performances of the buck converter. Simultaneous control of current and voltage is also vital to reduce the size and cost of the converter and simplify its implementation.

Recently, Chiu and Shen [8] proposed finite time control of buck converters via internal terminal sliding modes. The proposed method works well for capacitor voltage regulation, but inductor current control in the presence of disturbance has not been investigated. Komurcugil [9] presented an efficient method of controlling both the voltage and current in the buck converter with adaptive terminal sliding-mode control technique. The proposed controller tracks the reference input satisfactorily but at the cost of slow start-up response.

Backstepping is an effective control tool with a systematic and recursive procedure for designing a controller for uncertain systems and has been applied successfully to DC–DC buck converters [10–12]. However, the performance of backstepping controllers is largely dependent on the design parameters which have high influence on the operation of the converter. Sureshkumar and Ganeshkumar [13] proposed backstepping with time-invariant design constants and showed the performance for the capacitor voltage. Studies pertaining to inductor current response and subjugation of the system to wide range of matched and mismatched uncertainties have not been reported.

Sliding-mode control (SMC) [14, 15] is another well-established control mechanism for tackling matched uncertainties. SMCs are widely popular due to their simplicity and robustness and have found useful application in buck converters [16–18]. Unfortunately, SMC method does not guarantee immunity of the converter to mismatched uncertainties [9].

In order to improve the robustness of buck converters applied for voltage and current tracking, a combined architecture involving backstepping and SMC is proposed. The proposed control algorithm uses switching action between the time-varying design parameter-based backstepping control and hysteresis modulation (HM)-based sliding-mode control techniques. The proposed controller operates in the backstepping mode and switches to the SMC mode only when the controller detects matched uncertainty present in the system. This results in superior transient and steady-state responses for a wide range of matched and mismatched uncertainties.

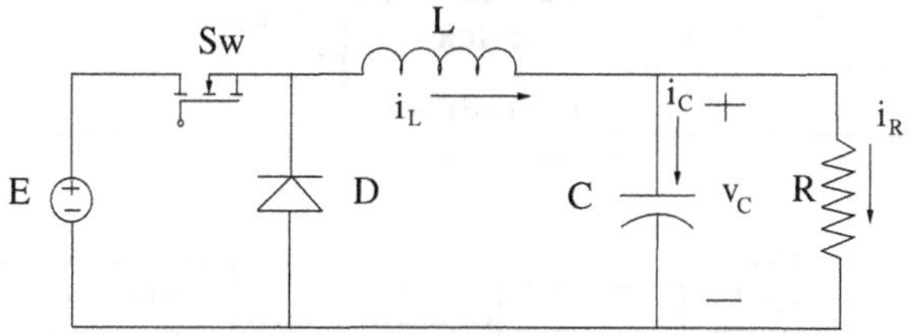

Fig. 11.1 Basic topology of buck converter

This paper is organized as follows. The buck converter is briefly described in Sect. 11.2. The proposed hybrid controller design methodology is described in Sect. 11.3. Simulation results are presented in Sect. 11.4. Conclusions are drawn in Sect. 11.5.

11.2 Buck Converter

The basic topology of buck converter required for modeling, analysis, and its subsequent control design is shown in Fig. 11.1.

The dynamics of the buck converter are described as

$$\dot{x}_1 = -\frac{x_1}{RC} + \frac{x_2}{C} \tag{11.1}$$

$$\dot{x}_2 = -\frac{x_1}{L} + \frac{uE}{L} \tag{11.2}$$

where

$$x_1 = v_C \tag{11.3}$$

$$x_2 = i_L \tag{11.4}$$

and $u = 1$ denotes the on time period and $u = 0$ denotes the off time period of the switch S_W.

11.3 Controller Design

The block diagram of the proposed hybrid backstepping control methodology for buck converters is shown in Fig. 11.2. The proposed controller uses a backstepping control method in conjunction with a sliding-mode controller (SMC) and switches between the two according to the nature of uncertainty affecting the buck converter.

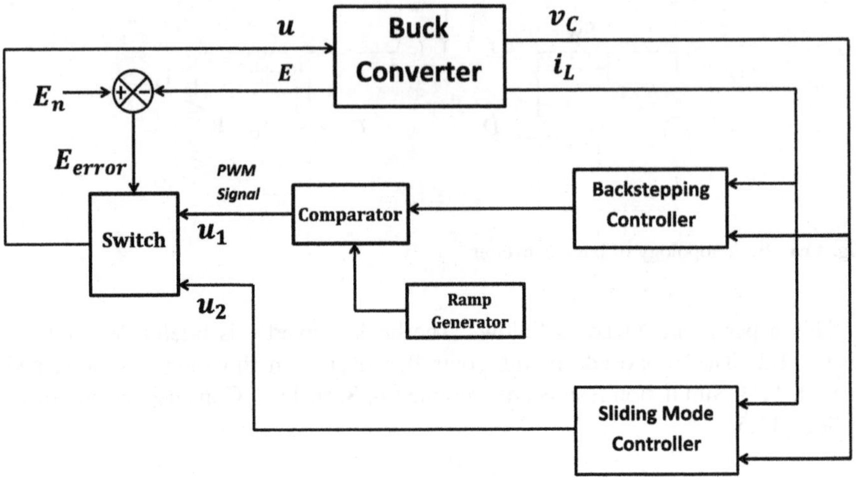

Fig. 11.2 Block diagram of the proposed hybrid backstepping controller for capacitor voltage v_C and inductor current i_L control of buck converter

The backstepping control is applied to tackle mismatched uncertainties which may arise because of change in load and other system parameters like inductance and capacitance. The SMC acts only when any matched uncertainty like input voltage change occurs in the system. Backstepping controllers are capable of rejecting both matched and mismatched types of uncertainties, but sliding-mode control is inherently immune to matched uncertainties. Moreover, transient performance of SMCs against matched uncertainties has been established as superior. Hence, the proposed controller is designed to utilize these two control strategies appropriately. The proposed control methodology is described below.

11.3.1 Design of Backstepping Control

Let u_b be the control signal produced by the backstepping controller. Let the voltage error be defined as

$$z_1 = x_1 - y_r \tag{11.5}$$

where y_r is the reference capacitor voltage. Differentiating (11.5) and using (11.1) yields

$$\dot{z}_1 = \frac{x_2}{C} - \frac{x_1}{RC} - \dot{y}_r \tag{11.6}$$

Let the desired value of x_2 be x_{2d}. Now

$$x_{2d} = C(-c_1 z_1 + \frac{x_1}{RC} + \dot{y}_r) \tag{11.7}$$

where c_1 is a design parameter. Substituting (11.7) for x_2 in (11.6) yields

$$\dot{z}_1 = -c_1 z_1 \tag{11.8}$$

and

$$\lim_{t \to \infty} \dot{z}_1 = 0 \tag{11.9}$$

Let us now define a new variable z_2 as

$$z_2 = x_2 - x_{2d} \tag{11.10}$$

Rearranging (11.10) and using (11.7) gives

$$x_2 = z_2 + C(-c_1 z_1 + \frac{x_1}{RC} + \dot{y}_r) \tag{11.11}$$

Using (11.11) in (11.6) gives

$$\dot{z}_1 = \frac{z_2}{C} - c_1 z_1 \tag{11.12}$$

Differentiating (11.10) yields

$$\dot{z}_2 = \dot{x}_2 - \dot{x}_{2d} \tag{11.13}$$

Further simplification of (11.13) using (11.11), (11.12), (11.1), and (11.2) yields

$$\dot{z}_2 = -\frac{x_1}{L} + \frac{u_b E}{L} + c_1 z_2 - C c_1^2 z_1 - \frac{x_2}{RC} + \frac{x_1}{R^2 C} - C \ddot{y}_r \tag{11.14}$$

Let us consider a Lyapunov function as

$$V = 0.5 z_2^2 \tag{11.15}$$

Differentiating (11.15) gives

$$\dot{V} = z_2 \dot{z}_2 \tag{11.16}$$

Substituting (11.14) in (11.16) yields

$$\dot{V} = -\frac{x_1 z_2}{L} + \frac{u_b E z_2}{L} + c_1 z_2^2 - C c_1^2 z_1 z_2 - \frac{x_2 z_2}{RC} + \frac{x_1 z_2}{R^2 C} - z_2 C \ddot{y}_r \tag{11.17}$$

Let the control input u_1 be

$$u_b = \frac{L}{E z_2}\left(\frac{x_1 z_2}{L} - c_1 z_2^2 + C c_1^2 z_1 z_2 + \frac{x_2 z_2}{RC} - \frac{x_1 z_2}{R^2 C} + z_2 C \ddot{y}_r - c_2 z_2^2\right) \tag{11.18}$$

Substituting (11.18) in (11.17) leads to

$$\dot{V} = -c_2 z_2^2 \tag{11.19}$$

where c_1 and c_2 are positive design parameters guaranteeing $\dot{V} < 0$ and thereby ensuring stability for the backstepping controller given by (11.18). Further the derived backstepping control signal u_b is made discontinuous signal by passing it through pulse width modulation (PWM) technique and thus obtaining u_1 as the final backstepping control signal.

11.3.2 Design of Hysteresis Modulation-Based Sliding-Mode Control

Let u_2 be the control signal produced by the SMC. The sliding surface s is designed as

$$s(x_1, x_2) = (y_r - x_1)k_1 + (p_r - x_2)k_2 \tag{11.20}$$

where p_r is the reference inductor current and k_1 and k_2 are design parameters. The control output from the SMC is produced by using the HM technique which has a hysteresis band with predefined switching boundaries to limit the switching frequency of the converter. The sliding-mode control signal is produced as per the following switching logic,

$$u_2 = \begin{cases} 1 & \text{when } s < -h \\ 0 & \text{when } s > h \end{cases} \tag{11.21}$$

where h is the switching limit of hysteresis band. The proposed controller switches between the backstepping and the SMC according to the following condition:

$$u = \begin{cases} u_1 & \text{when } |E_{\text{error}}| = 0 \\ u_2 & \text{when } |E_{\text{error}}| \neq 0 \end{cases} \tag{11.22}$$

and

$$E_{\text{error}} = E_n - E \tag{11.23}$$

where E_n and E are the nominal and measured input voltages of the buck converter.

11.4 Results and Discussion

The proposed controller is applied to a buck converter as described in Fig. 11.2 having the following parameters:

$$E = 15\,\text{V}, \quad L = 20\,\text{mH}, \quad C = 6\,\mu\text{F}, \quad R = 30\,\Omega \tag{11.24}$$

Simulation is carried out using Simulink in MATLAB with 1 μs as the sensing interval. The proposed controller is implemented by selecting time-varying design parameters c_1 and c_2, in order to give the best converter performance. The design parameters chosen are as follows:

$$c_1 = \begin{cases} 5500 + e^{100t} & \text{if } t \leq 0.002\,\text{s} \\ 1e^7 & \text{otherwise} \end{cases} \tag{11.25}$$

$$c_2 = 8e^6 \tag{11.26}$$

A fixed ramp signal with a switching frequency of 20 kHz is compared with the continuous control signal produced by the backstepping controller to generate the PWM signal, needed to operate the power electronic switch S_W. For the SMC, a hysteresis band of h = ±0.0001 and gains k_1 = 20 and k_2 = 10 are selected.

Transient and steady-state performances of the buck converter are investigated under varying source and load conditions. Dynamic behaviors of both the capacitor voltage and the inductor current are studied against changes in input voltage and load resistance. During the start-up, responses of the capacitor voltage and the inductor current are noted as quick start-up response, since it is very important in buck converters. Simulation results obtained by using the proposed controller are compared with the results obtained by applying proportional–integral (PI) and backstepping controllers reported in [13]. In [13], design constants used are k_p = 0.0015, k_i = 10, and c_1 = 100, c_2 = 390 for the PI and the backstepping controllers, respectively. The simulation results are plotted in Figs. 11.3 and 11.4.

Figures 11.3 and 11.4 show the capacitor voltage and the inductor current during start-up period of the buck converter. It is observed in Fig. 11.3 that the proposed method successfully tracks the reference signal of 5 V with the settling time of 0.96 ms, with zero peak overshoot and negligible steady-state error, and shows superior performance than PI and backstepping control. As it can be seen in Fig. 11.4, PI controller [13] exhibits large ripple in the current and long settling

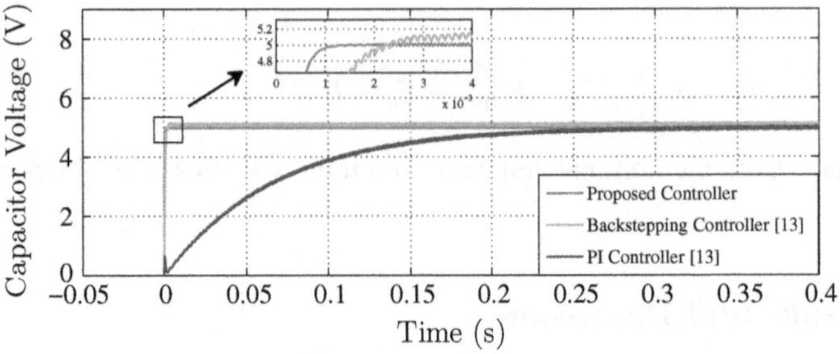

Fig. 11.3 Profile of capacitor voltage v_C during start-up

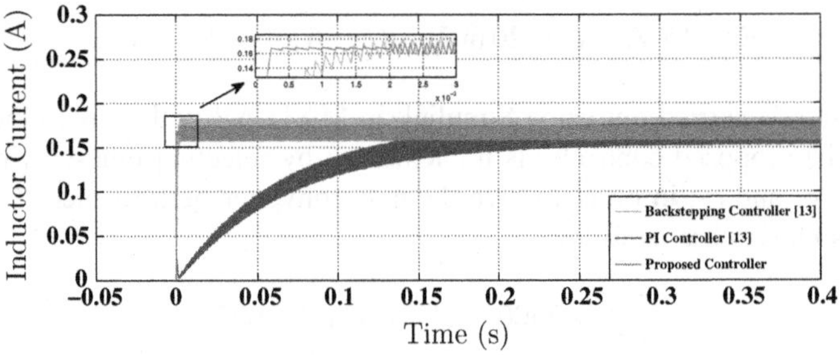

Fig. 11.4 Profile of inductor current i_L during start-up

time, whereas the proposed controller in comparison with both PI [13] and backstepping [13] controllers offers less settling time and less inductor current ripple.

Table 11.1 compares the transient parameters like rise time (t_r), settling time (t_s), peak overshoot (P_o), and steady-state parameter like steady-state error (E_{ss}) of PI, backstepping, and the proposed controller for capacitor voltage and inductor current

Table 11.1 Comparison of control schemes for capacitor voltage and inductor current during start-up

Response	Controller	Performance			
		Transient		Steady state	
		t_r (ms)	t_s (ms)	P_o (%)	E_{ss} (%)
Capacitor voltage	PI [13]	146.80	262	0	0.40
	Backstepping [13]	1.20	2.10	1.7	2.10
	Proposed	0.44	0.96	0	0.06
Inductor current	PI [13]	149.80	255	0	0.60
	Backstepping [13]	1.00	1.20	0	2.34
	Proposed	0.10	0.23	0	0.20

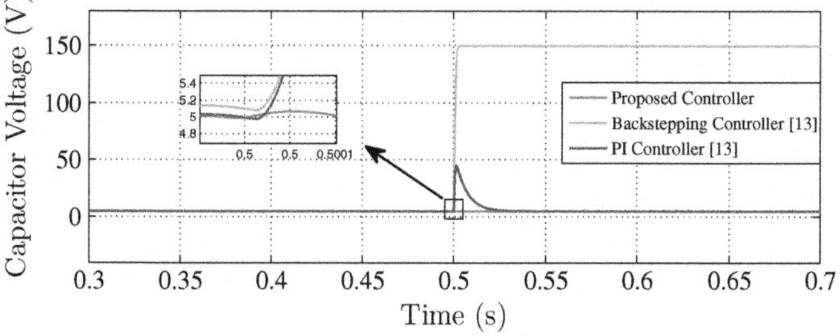

Fig. 11.5 Capacitor voltage response in the presence of matched uncertainty, i.e., change of *E* from 15 to 150 V at $t = 0.5$ s

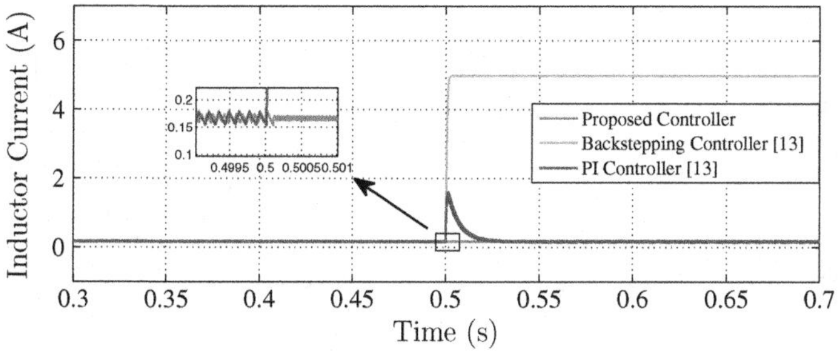

Fig. 11.6 Inductor current response in the presence of matched uncertainty, i.e., change of *E* from 15 to 150 V at $t = 0.5$ s

responses during the start-up period of the buck converter. It is evident from Table 11.1 that the proposed hybrid backstepping controller performs better.

The buck converter is now subjected to a change in input voltage from 15 to 150 V, and the corresponding system response is plotted in Figs. 11.5 and 11.6. The capacitor voltage is shown in Fig. 11.5, and the inductor current is shown in Fig. 11.6, for the case when the system is subjected to matched uncertainty. It is observed from Figs. 11.5 and 11.6 that the PI controller [13] is able to track the reference signal but at the cost of high peak overshoot and long settling time. It is worth noting that the backstepping controller proposed by [13] exhibits poor steady-state performance by producing intolerable steady-state error. In contrast, the proposed hybrid backstepping controller performs best with the least settling time and steady-state error and produces no peak overshoots or undershoots. The accurate measurements are tabulated in Table 11.2.

Further, the buck converter is subjected to change of load resistance *R* from 30 to 15 Ω, denoting a mismatched type of uncertainty. The corresponding capacitor voltage and inductor current response are plotted in Figs. 11.7 and 11.8, respectively.

As evident from the Fig. 11.7 and Table 11.3, both PI controller and backstepping controller proposed in [13] gives large undershoot and long settling time.

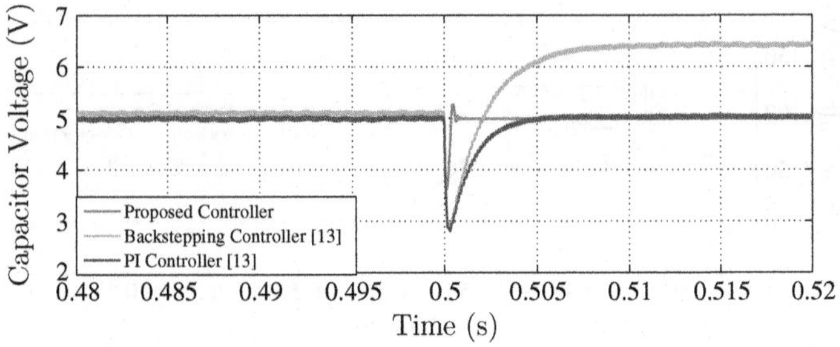

Fig. 11.7 Capacitor voltage response in the presence of mismatched uncertainty, i.e., change of R from 30 to 15 Ω at $t = 0.5$ s

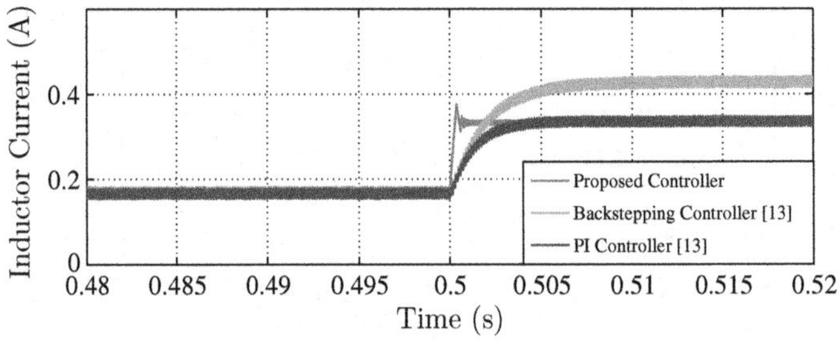

Fig. 11.8 Inductor current response in the presence of mismatched uncertainty, i.e., change of R from 30 to 15 Ω at $t = 0.5$ s

Table 11.2 Comparison of control schemes for capacitor voltage and inductor current during input voltage change

Response	Controller	Performance		
		Transient		Steady state
		t_s (ms)	P_o (%)	E_{ss} (%)
Capacitor voltage	PI [13]	37	793	0.70
	Backstepping [13]	3.5	2,880	2,890
	Proposed	0.10	1.35	0.06
Inductor current	PI [13]	38.60	860	0.36
	Backstepping [13]	2.00	0	2,892
	Proposed	1.00	7.60	0.24

Again it must be noted here that the backstepping controller exhibits significant steady-state error and settling time. In view of Table 11.3, it is clear that the proposed controller offers superior transient and steady-state performance in comparison with [13].

Table 11.3 Comparison of control schemes for capacitor voltage during load change

Response	Controller	Performance		
		Transient		Steady state
		t_s (ms)	P_u (%)	E_{ss} (%)
Capacitor voltage	PI [13]	4.50	44.20	0.50
	Backstepping [13]	10	42.5	28.4
	Proposed	6.00	27.60	0.09

Table 11.4 Comparison of control schemes for inductor current during load change

Response	Controller	Performance		
		Transient		Steady state
		t_s (ms)	P_o (%)	E_{ss} (%)
Inductor current	PI [13]	4.20	0	0.60
	Backstepping [13]	100	0	28.56
	Proposed	10	13.30	0.015

Although the proposed method gives satisfactory response for capacitor voltage under change in load resistance, yet the inductor current response in Fig. 11.8 suffers from an overshoot of 13.3 % and settling time of 10 ms as shown in Table 11.4.

It is evident from these tables that the transient and steady-state performances of the proposed controller are the best among the three controllers. However, the peak overshoot (P_o) in the inductor current is high, which is a drawback of the proposed controller. Overall, it is clear that the proposed method of combining the backstepping and sliding-mode control works satisfactorily in presence of wide range of matched and mismatched uncertainties and provides effective simultaneous control of both the capacitor voltage and the inductor current.

11.5 Conclusion

In this paper, a hybrid controller is proposed for the buck converter by combining the backstepping control method with the sliding-mode control technique. The proposed controller aims to control both the capacitor voltage and the inductor current in the buck converter in presence of matched and mismatched uncertainties like input voltage change and load change, respectively. The proposed controller switches between the two controllers according to the type of uncertainty. When the buck converter is affected by matched uncertainty due to input voltage change, the SMC acts. When mismatched uncertainty like load resistance change occurs in the buck converter, the backstepping control becomes operational. This strategy is conceived because the SMC is robust to matched uncertainty, whereas it cannot tackle mismatched uncertainties. On the other hand, backstepping controller can deal with both matched and mismatched uncertainties, but its performance against matched uncertainties is not at par with SMCs. The systematic controller design procedure of backstepping and invariance property of SMC for matched uncertainty have been utilized for robust tracking of both the capacitor voltage and inductor current simultaneously. Simulation results demonstrate that as compared to existing controllers, the proposed hybrid control strategy offers superior transient and steady-state performances.

References

1. Rashid MH (2003) Power electronics: circuits, devices and applications, 3rd edn. Pearson Publications, New Delhi
2. Wai R-J, Shih L-C (2011) Design of voltage tracking control for DC–DC boost converter via total sliding-mode technique. IEEE Trans Industr Electron 58(6):2502–2511
3. Raviraj VSC, Sen PC (1997) Comparative study of proportional-integral, sliding mode and fuzzy logic controllers for power converters. IEEE Trans Ind Appl 33(2):518–524
4. Leung KK, Chung HS (2004) Derivation of a second-order switching surface in the boundary control of buck converters. IEEE Power Electron Lett 2(2):63–67
5. Calderan AJ, Vinagre BM, Feliu V (2006) Fractional order control strategies for power electronic buck converters. J Sig Process 86(10):2803–2819
6. Yan W-T, Ho CN-M, Chung HS-H, Keith T (2009) Fixed-frequency boundary control of buck converter with second order switching surface. IEEE Trans Power Electron 24(9):2193–2201
7. Hassan Hosseinnia S, Tejado I, Vinagre BM, Sierociuk D (2012) Boolean-based fractional order SMC for switching systems: application to a DC–DC buck converter. J SIViP 6 (3):445–451
8. Chiu C-S, Shen C-T (2012) Finite time control of DC–DC buck converters via integral terminal sliding modes. Int J Electron 99(5):643–655
9. Komurcugil H (2012) Adaptive terminal sliding-mode control strategy for DC–DC buck converters. ISA Trans 51(6):673–681
10. El Fadil H, Giri F, Haloua M, Ouadi H (2003) Nonlinear and adaptive control of buck power converters. In: Proceedings of 42nd IEEE conference on decision and control, vol 5, Maui, Hawaii, USA, pp 4475–4480
11. Salimi M, Soltani J, Markadeh GA (2011) A novel method on adaptive backstepping control of buck choppers. In: Proceedings of 2nd power electronics, drive systems and technologies conference, vol 2, pp 562–567
12. Wei Z, Bao-bin L (2012) Analysis and design of DC–DC buck converter with nonlinear adaptive control. In: Proceedings of 7th international conference on computer science and education, Melbourne, Australia, July 2012, pp 1036–1038
13. Sureshkumar R, Ganeshkumar S (2011) Comparative study of proportional integral and backstepping controller for buck converter. In: Proceedings of IEEE international conference on emerging trends in electrical and computer technology, pp 375–379
14. Utkin VI (1977) Variable structure systems with sliding modes. IEEE Trans Autom Control 22 (2):212–222
15. Edwards C, Spurgeon SK (1998) Sliding mode control: theory and applications. Taylor & Francis, London

16. Tan S-C, Lai YM, Tse CK (2006) A unified approach to the design of PWM-based sliding-mode voltage controllers for basic dc-dc converters in continuous conduction mode. IEEE Trans Circuits Syst I Regul Pap 53(8):1816–1827
17. Venkataramanan R, Sabanoivc A, Cuk S (1985) Sliding mode control of dc-dc converters. In: Proceedings of IEEE conference on industrial electronics, control and instrumentation, pp 251–258
18. Tan S-C, Lai YM, Tse CK (2008) General design issues of sliding mode controllers in dc-dc converters. IEEE Trans Industr Electron 55(3):1160–1174

Chapter 12
Power Optimization of Refrigerator by Efficient Variable Compressor Speed Controlled Driver

S.D. Zilpe and Z.J. Khan

Abstract Nowadays, refrigerator is an important domestic electrical device, but at the time of load shading or when the electricity is not available at that time, because of some drawbacks, we cannot operate this device on any inverter. The main reason why refrigerator does not work on inverter is because of switching. Since the pressurized compressor of refrigerator get continuously switched between ON and OFF, in order to maintain the temperature, it draws more current because of blocked rotor induction motor, and as switching occurs frequently, compressor requires minimum 3 min to start again for proper working in refrigerator. All these things have severe effect on the system such as nonlinear behavior of the device, and also, power quality gets deteriorated which needs to be controlled at any cost and requires some practical measures on this field, and the project's theme mainly works in this area and provides a solution to this problems. The project includes an idea and provides a design of a driver circuit which operates the compressor of refrigerator with minimum frequency and voltage at minimum temperature. Using this driver, the compressor of a refrigerator can be operated using solar panel photovoltaic cell (P-V Cell) operated inverter. The various advantages of this project are that the driver circuit eliminates the difficulty in operation of compressor while continuous ON–OFF switching of power supply across it. Instead of continuous ON–OFF switching, the driver circuit controls the temperature by gradually varying the speed of single phase induction motor.

Keywords Refrigerator · Compressor · Inverter · Switching · Harmonics

S.D. Zilpe (✉)
Rajiv Gandhi College of Engineering Research and Technology (RCERT), Chandrapur, India
e-mail: zilpeshrikant@gmail.com

Z.J. Khan
Electrical Engineering (E&P), Rajiv Gandhi College of Engineering Research and Technology (RCERT), Chandrapur, India
e-mail: zjawedkhan@gmail.com

V. Vijay et al. (eds.), *Systems Thinking Approach for Social Problems*, Lecture Notes in Electrical Engineering 327,
DOI 10.1007/978-81-322-2141-8_12

12.1 Introduction

Basically, compressor contains induction motor, and because of high pressure of compressor, it acts as a blocked rotor at start. Hence, it takes six times more load current as compared to normal operation which may affects the inverter. As discussed earlier, present refrigerators face the problem of switching. Because of switching of power across compressor, the power quality of supply gets affected due to generation of harmonics. If we cut off the supply of compressor, it requires minimum 3 min to start again; otherwise, it may get blocked. This project includes an implementation of general refrigerator compressor which is operated by a driver circuit. The various advantages of this project are that the driver circuit eliminates the difficulty in operation of compressor while continuous ON–OFF switching of power supply across it. Instead of continuous ON–OFF switching, the driver circuit controls the temperature by gradually varying the speed of compressor motor. Last but not least, this designed driver circuit for compressor is energy efficient too.

12.1.1 Compressor in Regular Refrigerator

Most compressors in regular refrigerators are single speed refrigerators. They are either "ON" or "OFF" based on the temperature in the refrigerator and the setting in the thermocouple. Most compressors are designed to handle peak load conditions (for high temperatures in summers), which means that they run at peak load even in winters when the cooling requirement is less. Every time the refrigerator door is opened, heat enters the refrigerator, and the compressor has to take care of this load as well. So most regular compressors are built to take care of peak load plus the "door open shut" load which during most of the year is much more than the actual requirement.

In regular refrigerator continuously switching as per temperature varying and therefore the compressor ON at full load at maximum speed on daytime as well as nighttime, the given observation shows the regular refrigerator output, and it consumes more power without stable output.

12.1.2 Compressor with Inverter-based Drive

A compressor with inverter-based drive works very much like a car accelerator. When the speed required is more, the acceleration is more, and when it is less, the acceleration is less. This makes sure that during summer months when cooling load is more, the compressor works at peak capacity taking more electricity. But during winter months when cooling load is less, the compressor works less thereby consuming less electricity. Even in summer season, the temperature at night is less than at daytime, and thus, electricity savings happen at night when compressor can run at a lower speed [1].

12.2 Methodology

In this project, initially, the compressor runs at main supply up to set point of T_1, i.e., 15 °C, and because initially temperature is very high, the applied voltage and frequency are high. One more set point T_2, i.e., 8 °C is used for setting the temperature inside the refrigerator which can be set manually as per the requirement. When temperature inside the refrigerator is less than T_1 °C, compressor works on inverter mode. On inverter mode, the voltage and frequency required by compressor are less and they depend on the temperature change from set value T_2 and current refrigerator temperature. And the compressor takes only that much power, so as to cool the refrigerator to the set voltage T_2 °C. When the temperature drops down the set value T_2 °C, the compressor gets turn off. This process continues when the temperature rises above the set value T_2 °C.

The methodology involved in the project can be best understood with the help of flowchart.

12.2.1 Flowchart

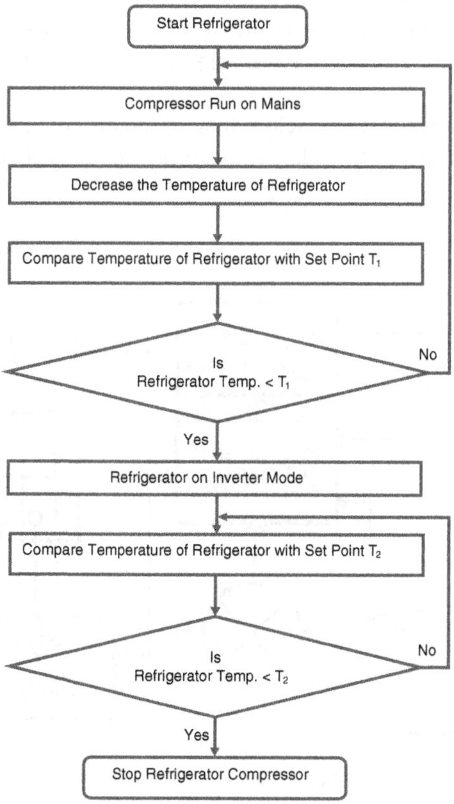

12.2.2 Project Setup

Figure 12.1 indicates the overall block diagram of the project with all the interconnections among the black boxes.

12.3 Functioning Blocks

12.3.1. Switcher
12.3.2. Power converter with voltage frequency driver
12.3.3. Temperature to voltage converter
12.3.4. Battery charging system
12.3.5. Refrigerator compressor

12.3.1 Switcher

The switcher circuit comprises of comparator IC LM358 which compares the first set point T_1 °C of temperature with the current temperature in the refrigerator sensed by thermistor. If the current temperature of refrigerator is above set point T_1 (15 °C), comparator does not generate the control pulse and the compressor runs on supply mains.

If temperature of refrigerator drops below first set point, comparator generates the control pulse which operates the relay and compressor is switched to inverter mode.

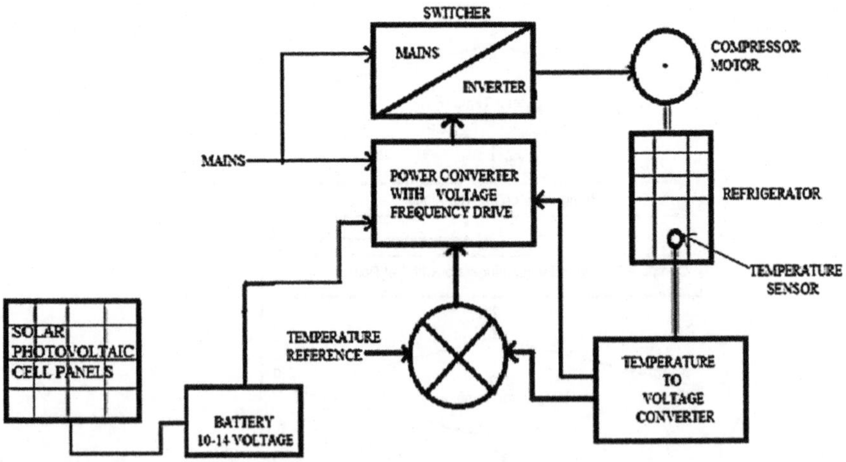

Fig. 12.1 Block diagram of inverter-based drive setup

In inverter mode, the power consumption of compressor is less because the voltage and frequency applied to the compressor are less and depend on the temperature change in current refrigerator temperature and second set point T_2 (8 °C).

12.3.2 Power Converter with Voltage Frequency Driver

Power converter circuit is a simple inverter driver which produced variable output power which is controlled by variable frequency signal generated from PWM IC SG 3524. This PWM IC generates the variable frequency signal at its output pin by sensing the variable resistance at its input which depends on the temperature in refrigerator. The temperature in a refrigerator is sensed with the help of thermistor.

12.3.2.1 Pulse Width Modulation (PWM) Techniques

There are three basic pulse width modulation techniques

Single pulse width modulation
Multiple pulse width modulation
Sinusoidal pulse width modulation

12.3.2.2 Single Pulse Width Modulation

This driver is designed with the help of single pulse width modulation. In this modulation, there is an only one output pulse per half cycle. The output is changed by varying the width of the pulses. The gating signals are generated by comparing a rectangular reference with a triangular reference. The frequency of the two signals is nearly equal. The waveform of single pulse width modulation is given in Fig. 12.2 [2].

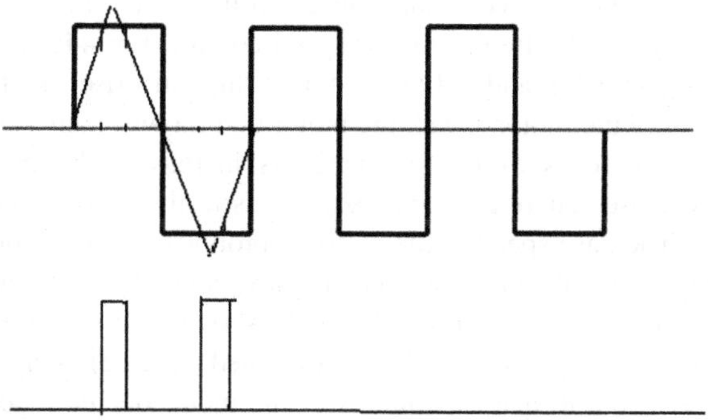

Fig. 12.2 Single pulse width modulation

The variable frequency obtained from PWM IC is fed to LM 331 which converts this variable frequency into variable voltage. This variable voltage is then compared with the reference voltage which is proportional to first set point T_1 °C and second set point temperature T_2 °C.

Two LM 358 ICs are used for different functions. One is used as a voltage follower, which strengthen the current signal with the use of increase output current of op-amp and also provide low output impedance coming from LM-331 and also for making PWM off when the temperature inside refrigerator and set point temperature are same. And second LM-358 is used to compare the refrigerator temperature with the temperature of two set points and to operate the relays.

12.3.2.3 V–F Controller for Single-phase Induction Motor

Single-phase induction motors are widely used in home appliances and industrial control. A variable frequency drive is a system for controlling the speed of a rotational or linear alternating current electric motor by controlling the frequency of the electrical power supplied to the motor. A variable frequency drive is a specific type of adjustable speed drive. Variable frequency drives are also known as adjustable frequency drives, variable speed drives, AC drives, micro-drives, or inverter drives. The multispeed operation and multipurpose operation are provided by controlling the speed of these motors [3].

In the previous days, the variable speed drives had various limitations such as larger space, poor efficiencies, and lower speed. But, the invention of power electronics devices changes the situation so now, variable speed drive is constructed in smaller size, high efficiency, and high reliability.

12.3.2.4 Volts-per-Hertz Ratio

This term describes a relationship that is fundamental to the operation of motors using adjustable frequency control. An AC induction motor produces torque by virtue of the flux in its rotating field. Keeping the flux constant will enable the motor to produce full load torque. Below base speed, this is accomplished by maintaining a constant voltage-to-frequency ratio applied to the motor when changing the frequency for speed control. For 440 and 230 V motors, the ratio is 440/50 = 8.8 and 230/50 = 4.6. If this ratio rises as the frequency is decreased to reduce the motor speed, the motor current will increase and may become excessive. If it reduces as the frequency is increased, the motor torque capabilities will decrease. There are some exceptions to this rule which are described below.

The base speed of the motor is proportional to supply frequency and is inversely proportional to the number of stator poles. So, by changing the supply frequency, the motor speed can be changed. Above base speed, this ratio will decrease when constant voltage (usually motor rated voltage) is applied to the motor. In these cases, the torque capabilities of the motor decrease above base speed [4].

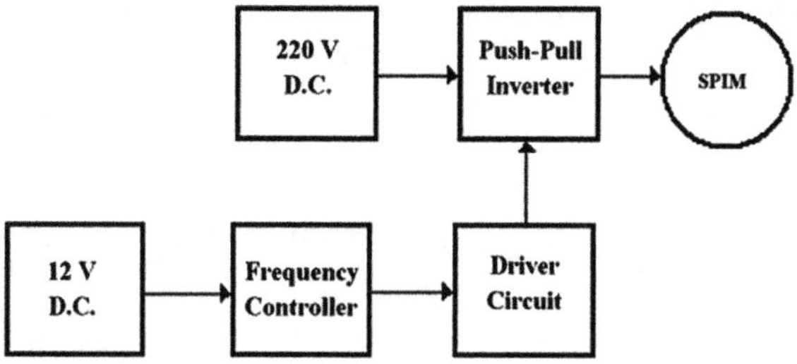

Fig. 12.3 Speed control of SPIM using V–F control

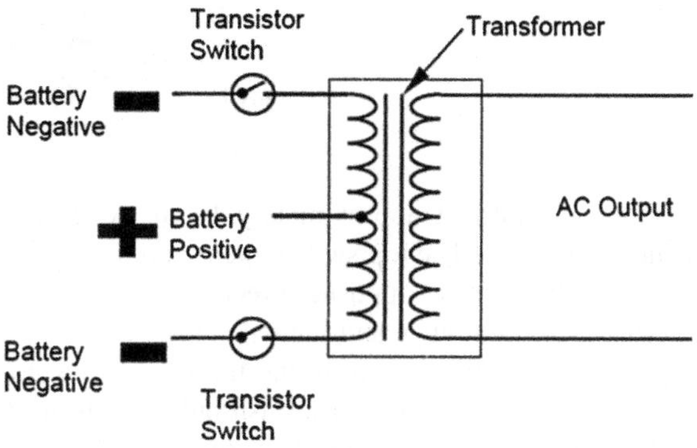

Fig. 12.4 Push-pull topology

At base speed and below, the volts-per-hertz ratio can be adjusted lower to minimize motor current when the motor is lightly loaded. This adjustment, which lowers the voltage to the motor, will reduce the magnetizing current to the motor. Consequently, the motor will produce less torque which is tolerable. This control is the most popular in industries and is popularly known as the constant V/f control [5].

The VFD is a system made up of active/passive power electronics devices. Figure 12.3 shows electronic speed control of the motor supply frequency. The basic concept of these drives is that a rectifier converts the fixed frequency supply to DC (which converts commercial power into a direct current). A DC link stage smoothes the rectified output to a stable DC voltage (or current). This DC is then inverted to provide a synthesized AC waveform at the motor terminals. The frequency and power of the AC supply delivered to the motor are controlled by inverter (Fig. 12.4).

12.3.2.5 Inverter Using Push-Pull

The basic theory of operation behind a push-pull design is as follows:

The top transistor switch closes and causes current to flow from the battery negative through the transformer primary to the battery positive. This induces a voltage in the secondary side of the transformer that is equal to the battery voltage times the turn's ratio of the transformer. Note: Only one switch at a time is closed (Fig. 12.5).

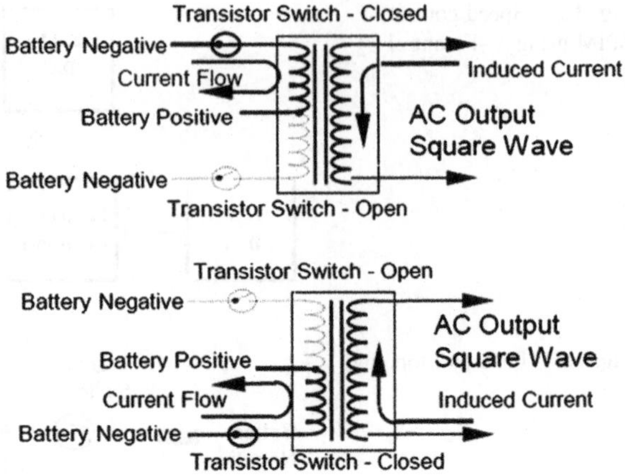

Fig. 12.5 Push-pull design

After a period of approximately 10 ms (one-half of a 50 Hz AC cycle), the switches flip-flop. The top switch opens and then the bottom switch closes allowing current to flow in the opposite direction.

This cycle continues and higher voltage AC power is the result. The push-pull approach is that the current in the transformer has to suddenly reverse directions. The transformer required for a push-pull design must have two primaries. Due to waveform characteristics, and lack of voltage regulation, the peak voltage of the output pulse is directly related to battery voltage. Since the transformer ratio is fixed, any change in battery voltage will affect the peak output voltage. For a square wave, RMS voltage is equal to peak voltage, and as a result, power output is dependent on battery voltage. Finally, most square wave inverters have mediocre efficiency (typically about 80 %), and the idle power draw is relatively high. To avoid rectangular waveform generated from inverter, a high value of capacitor is used which converts the square wave in sinusoidal waveform. This is done using charging and discharging property of capacitor.

12.3.3 Temperature to Voltage Converter

The temperature of refrigerator is sensed with thermistor as a sensor. The property of thermistor is given below.

A thermistor is a type of resistor whose resistance strongly depends on temperature. The word thermistor is combination of words "thermal" and "resistance." A thermistor is a temperature sensing element composed of sintered semiconductor material and sometime mixture of metallic oxide such as Mn, Ni, Co, Cu, and Fe, which exhibits a large change in resistance properties to a small change in temperature. Pure metals alloys nearly equal zero temperature coefficients of

This project negative temperature coefficient (NTC) thermistor is used. The variable resistance output terminal fed to input of IC SG 3524; its PWM IC varies the frequency by varying resistance of thermistor. This variable frequency fed to the input of LM 331 its convert that variable frequency in variable voltage proportional to frequency, adjusted by gain control.

12.3.4 Battery Charging System

There are two energy sources used in the project viz supply mains and photovoltaic cell (P-V Cell). A charging system with constant current constant voltage control (CCCV) is also used in the project.

12.3.5 Refrigerator Compressor

As the "heart" of the refrigeration system, the compressor is the workhorse behind the stable operation of the refrigerator or freezer. Most home-use refrigerators and freezers today utilize single speed compressors which are either "ON" or "OFF" and can only operate at one speed. This type of performance does not allow flexible operation to adapt to the different usage conditions, and such appliances experience during the day, nor does it efficiently utilize electric power once the unit is at a steady state.

The compressor having single-phase induction motor. The ratings of single-phase induction motor are 230 V AC. 500 W at peak load (Figs. 12.6 and 12.7)

Fig. 12.6 Model of inverter-based driver

Fig. 12.7 Complete circuitry of driver with measuring apparatus

12.4 Observations and Result

The Graph 12.1 reveals the progressive nature of frequency. At certain temperature T_1, i.e., 15 °C, the refrigerator works in non-inverting mode with constant frequency. When the temperature lowers down, it acts in inverting mode. When the temperature reduces below desired value, it turns OFF. The temperature is maintained at desired value. Note that during its ON/OFF cycle, the value of frequency as well as voltage, speed, and power consumption by compressor motor is less (Figs. 12.8, 12.9 and 12.10; Graphs 12.2, 12.3 and 12.4).

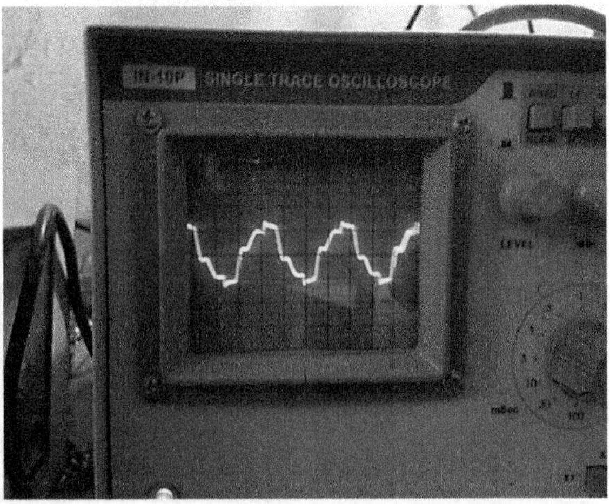

Fig. 12.8 Sine wave with distortion

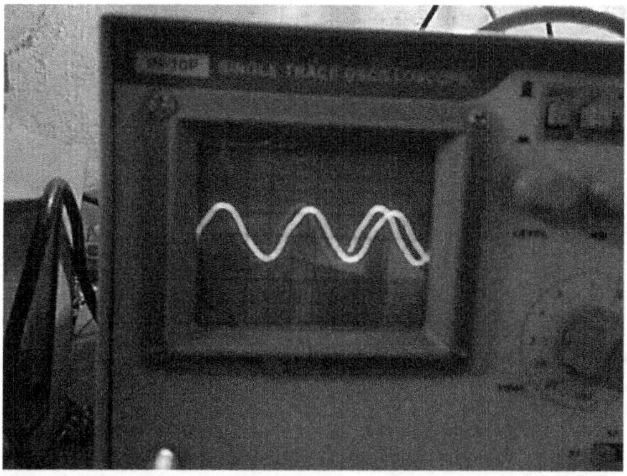

Fig. 12.9 Accurate voltage–frequency ratio

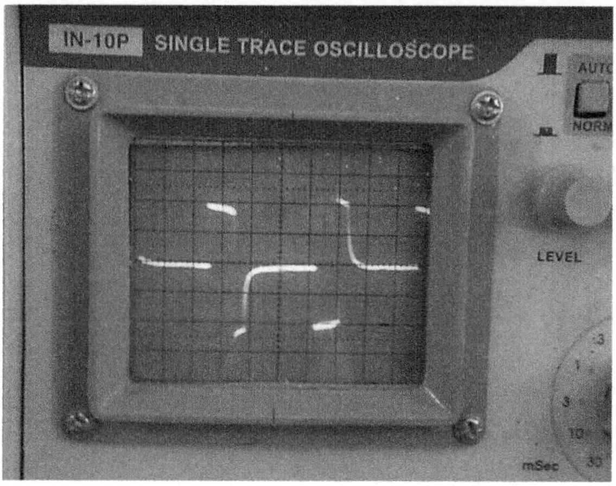

Fig. 12.10 Output of PWM

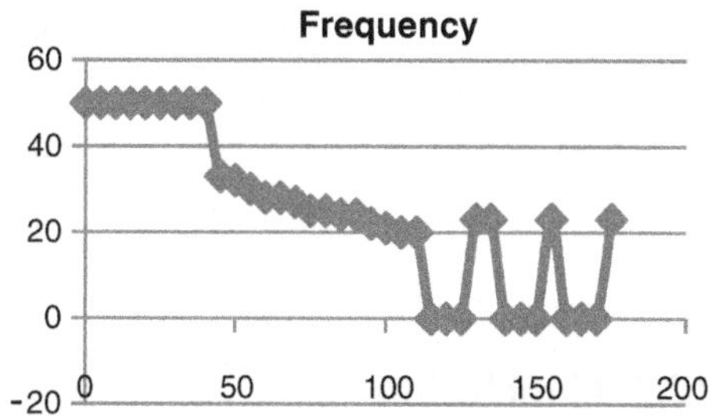

Graph 12.1 Variations in frequency

As shown in Graph 12.5, the temperature gradually decreases until required value is achieved, and then, it maintains that value. We can see from the Graph 12.5 that there is quite less variation in the temperature of refrigerator as compared to conventional one.

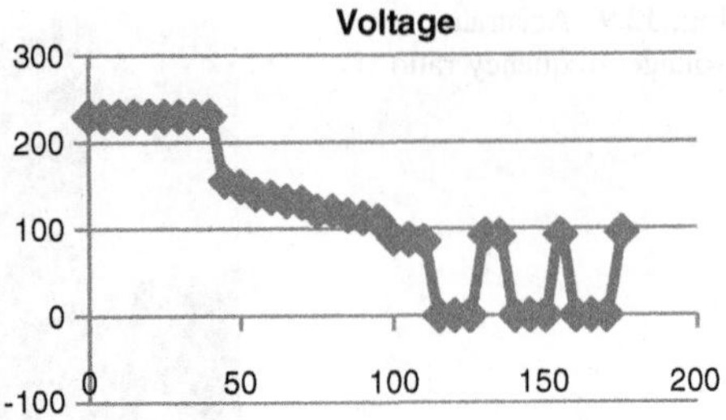

Graph 12.2 Variations in voltage

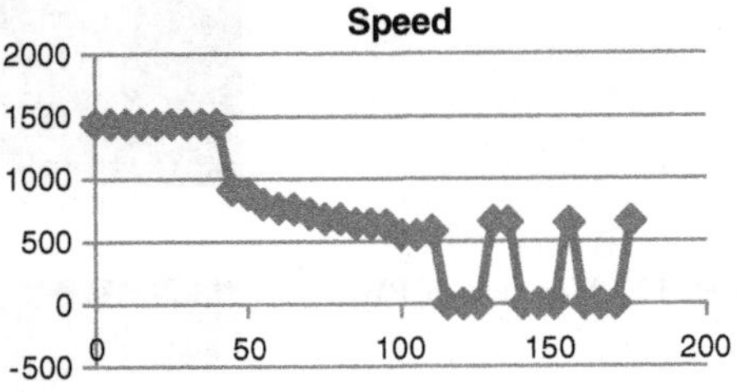

Graph 12.3 Variations in speed

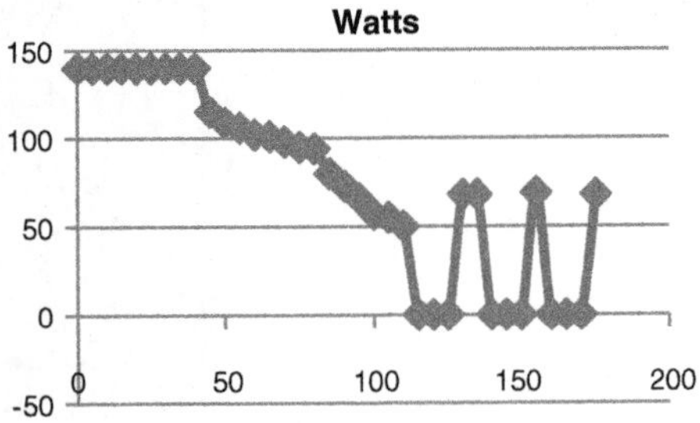

Graph 12.4 Variations in watts

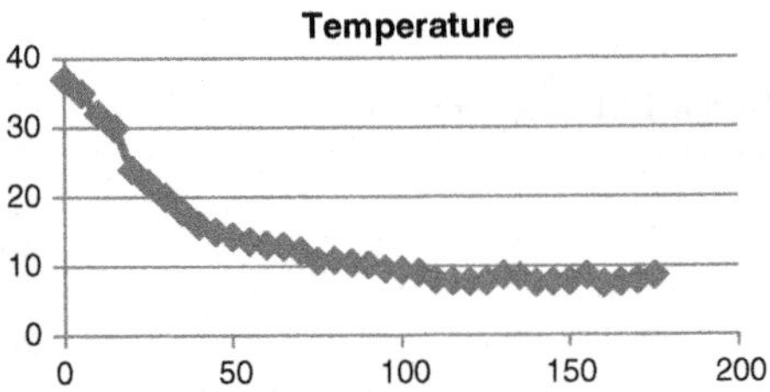

Graph 12.5 Variations in temperatures

12.5 Conclusion

In conventional refrigerator, the compressor works at full load whenever there is a slight change in temperature from the set value. Due to this, for a very small increment in temperature from the set value since the compressor is running at full load, it overcools the refrigerator which is the wastage of power.

The starting current of compressor in conventional refrigerators is very high which is avoided by controlling the voltage and frequency applied to the compressor. The controlling of voltage and frequency is based on the amount of temperature change from the set value, i.e., if the temperature change is more, the voltage and frequency applied will be more, and if the temperature change is less, the applied voltage and frequency will be less. Due to the small value of starting current, the harmonics generated are less which preserves the sinusoidal nature of the supply voltage and hence provides linearity. At the starting when the compressor is in inverter mode since the voltage, frequency, current, and harmonics are less, the losses in compressor such as copper loss and core loss are less. This reduces the power consumption about 50–55 % of the power consumed in conventional refrigerators. The project improves the performance of refrigerator by making it energy efficiency.

This project has vivid impact on the society especially in village and healthcare sectors, and also, it can be used commercially in an industry which involves the use of refrigeration such as chemical industries, ice manufacturing, and food processing industries.

In villages due to the problems of load shading, refrigerators cannot be used for domestic and commercial purposes. The healthcare services become critical in rural and remote areas where the people are either not having access to the electricity or suffering from heavy load shading. The problems still become more critical for the medicine such as vaccines and insulin which get spoiled if not stored out at lower temperatures. This project provides inelegant and efficient solution for such problems.

Apart from all these advantages, the use of solar panel increases the cost of the project which is not desirable especially for the implementation of project in rural areas. But this issue can be resolved by taking the aids and subsidies provided by the government for rural area such as Ministry of Natural Renewable Energy Government of India (MNRE) which provides the subsidies of about 30 % for purchase and installation of solar system and appliances.

References

1. Kanaoka S, Noda T, Hamano H, Kubota Y High efficiency refrigerator cooling system using an inverter compressor
2. Majhi B (2010) Analysis of single-phase SPWM inverter. In: Maswood AI, Al-Ammar E (eds.) Analysis of a PWM voltage source inverter with PI controller under non-ideal conditions. International power engineering conference (IPEC 2010)
3. Latt AZ, Win NN (2009) Variable speed drive of single phase induction motor using frequency control method. In: Proceedings of international conference on education technology and computer
4. Henderson DS Variable speed electric drives—characteristics and applications adjustable frequency control (inverters) fundamentals application consideration. Bulletin C870A
5. Yedamale P (2005) Bidirectional v/f control of single phase induction motor. Panasonic Corporation Refrigeration Devices Business Unit, Technical bulletin no. D-B16Ef12-E, 25 Dec 2009

Chapter 13
Power Quality Assessment of Grid-Connected Photovoltaic Plant

Anurag and Ravindra Arora

Abstract In the twenty-first century, the cost of production of electrical power from solar and wind energy has become highly competitive to the power from conventional sources of energy. Besides the stand-alone and comparatively smaller photovoltaic systems, large photovoltaic plants for the generation of electrical power have become quite common. These large sources integrate the electrical power with existing power grid. In this work, an assessment of the quality of electrical power was made by the measurement supplied by two 50-kW grid-connected photovoltaic plants at IIT Jodhpur. The requirements of power quality by International Electrotechnical Commission (IEC) have been taken into consideration. Parameters, for example, total harmonic distortion (THD) and crest factor (CF), and AC indices such as apparent power, real power, and power factor are analyzed with the help of power analyzer. The quality of DC power delivered by PV module and measured at DC/DC boost converter was also analyzed. It is a great concern for the operation of DC/DC microgrid. It was found that the quality of power supplied to the grid was highly affected by inverter design. The maximum THD in the voltage supplied to the grid was measured to be within the limits. The voltage swells and sags were found to be absent at both the plants. The ripple in DC power delivered by the PV modules was measured to be greater than the permissible limits.

Keywords Power quality · IEC · Grid-connected solar plant · THD

Anurag (✉) · R. Arora
Center of Excellence in Energy, Indian Institute of Technology Jodhpur, Jodhpur, India
e-mail: anurageng@iitj.ac.in

R. Arora
e-mail: rarora@iitj.ac.in

V. Vijay et al. (eds.), *Systems Thinking Approach for Social Problems*, Lecture Notes in Electrical Engineering 327,
DOI 10.1007/978-81-322-2141-8_13

13.1 Introduction

The upcoming generation of electrical energy trusts on renewable energy sources. Among all renewable energy sources, solar energy is the robust and cheap. Large photovoltaic power plants having an installed capacity of 50 MW and above are being installed. Hybridization of different sources of electrical power is going on. The stand-alone PV plants are useful for small power application and home electrification purposes. For large power application, grid-connected photovoltaic plants are playing an important role. Grid-connected PV plants supply electricity to power grid. When synchronization of solar plants with grid takes place, problems arise at point of common coupling (PCC) and load side. There is a need to evaluate the problems.

A PV plant contains PV modules and strings, power conditioning unit (PCU), a distribution box, energy meter, and grid-interfacing distribution box. The DC output of PV modules and strings goes through PCU, which converts the DC output to AC, synchronized to the existing AC power grid. An energy meter is connected at this junction to measure the power supplied to the grid. The power supplied by the inverters is disconnected from the grid in case of abnormal grid conditions in terms of voltage and frequency [1]. Only, International Electrotechnical Commission (IEC) 61727 states that the PV inverter will have an average lagging power factor greater than 0.9 when the output power of the plant is greater than 50 % [1]. According to IEC 62301, the total harmonic distortion (THD) in voltage is specified to be less than or equal to 2 % and THD in current is specified to be less than or equal to 5 %.

The aim of this work was to assess the power quality supplied by grid-connected PV plants. It is aimed to analyze the parameters influencing the quality and their probable sources. The results could improve the quality of power supplied by the solar PV panels.

13.2 Basic Concepts of Photovoltaic

A device made out of semiconductor material which converts solar radiation into electrical energy is called photovoltaic device, and the effect is called photovoltaic effect. Different types of materials are used, and each material has its own properties and advantages, but commonly used material is silicon. When photons fall on a PN junction, electron–hole pairs are generated and collected at the collector plates, giving rise to current generated by solar cell. The energy of falling photon should be greater than the band gap of semiconductor material. The simplest equivalent circuit of a solar cell consists of a current source driven by sunlight in parallel with a real diode. Figure 13.1 shows the equivalent circuit of a solar cell.

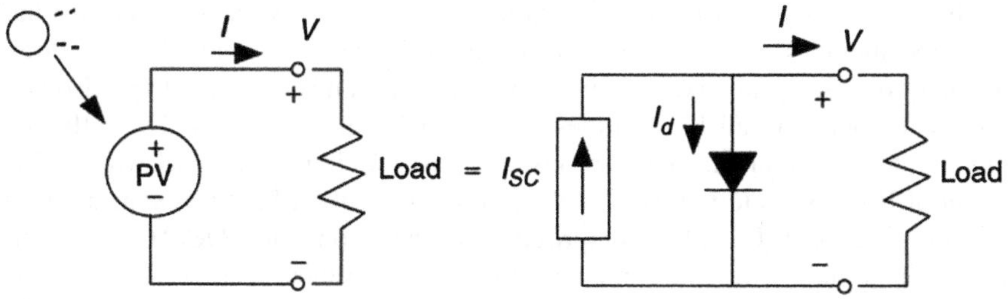

Fig. 13.1 Equivalent circuit of a PV cell [2]

Two basic parameters of a solar cell are I_{sc} and V_{oc}. I_{sc} is the short-circuit current, and V_{oc} is the open-circuit voltage. The calculations for these parameters are as follows:

Applying KCL in the circuit shown in Fig. 13.1,

$$I_{sc} = I + I_d$$

$$I = I_{sc} - I_d$$

$$I = I_{sc} - I_0\left(e^{qV/kT} - 1\right)$$

where I_0 is the dark current when there is no light, k Boltzmann's constant, and T the temperature in K. When leads of the cell are left open, $I = 0$ and $V = V_{oc}$. Solving for V_{oc}, the following results are obtained:

$$V_{OC} = \frac{kT}{q}\ln\left(\frac{I_{sc}}{I_0} + 1\right)$$

At room temperature, the equation becomes

$$I = I_{sc} - I_0(e^{38.9V_d} - 1),$$

where V_d is the voltage across diode.

13.3 IIT Jodhpur Photovoltaic Plant Structure

There are two academic blocks, block-I and block-II, installed with solar PV panels on their roof. Rooftop of block-I has PV panels made with thin-film amorphous silicon structure, and rooftop of block-II has crystalline silicon PV panels. The capacity of block-I is 43 kWp, and the capacity of block-II is 58 kWp, making the total installed capacity of two plants together approximately 100 kWp. The total number of modules is 114, and the total number of strings is 38. Thus, each string has three modules, which are connected in series. The total numbers of array junction box (AJB) are six, and similarly, the total numbers of PCU are also six. The DC output from all PV modules goes to PCUs through AJBs. The AJBs are optional in any PV plant. These are installed just for simplification of wiring. The output of PCUs goes to the AC distribution box (ACDB) where it is synchronized with the grid. A PCU unit is comprised of DC/DC converter, DC/AC converter, filters, maximum power point tracking (MPPT) algorithms, and some protection circuitry. Figure 13.2 shows the structure of the plant.

13.4 Parameters and Power Quality Issues

The power quality parameters are defined as follows:

1. Harmonics: Multiple of fundamental frequency, currents, and voltages may exist in a system, which distorts the fundamental frequency waveform. If a system has an operating frequency f, then harmonic contents may have frequencies f, $2f$, $3f$, $4f$ … nth order. It may contain both even and odd harmonics. For the measurement, the basic index is given as THD. THD includes the role of all individual contents of harmonics. THD in voltage is defined as follows:

$$\mathrm{THD} = \frac{\sqrt{V_2^2 + V_3^2 + V_4^2 + \cdots + V_n^2}}{V_1}$$

   where V_1 is the fundamental and rest are respective other components.

2. Voltage swells and sags: If the voltage magnitude is found to acquire greater than or equal to 110 % of nominal magnitude, then the voltage swells are said to be present. If lower voltage is observed in comparison with the nominal voltage for the duration of 1 min or lesser, then this phenomenon is called voltage sag or voltage dip [3]. The magnitude for sag is defined as 90 % of nominal voltage or lesser.
3. Crest factor (CF): The CF is the ratio of peak value to the root mean square (rms) value.

The measured THD in voltage, measured CF of current and voltage at noon time in the month of march is shown in Figs. 13.3 and 13.4 respectively.

For an ideal sinusoidal waveform, the CF is $\sqrt{2}$ or 1.414. The harmonic content distorts the ideal sinusoidal waveform, hence the CF. It is shown in Fig. 13.4 that the CF in voltage is around 1.432, which is nearly equal to $\sqrt{2}$. But in case of current, it deviates from $\sqrt{2}$ considerably, which means that current waveform is non-sinusoidal. It could be because of variation in the load. The other parameters measured at the block-II are listed in Table 13.1.

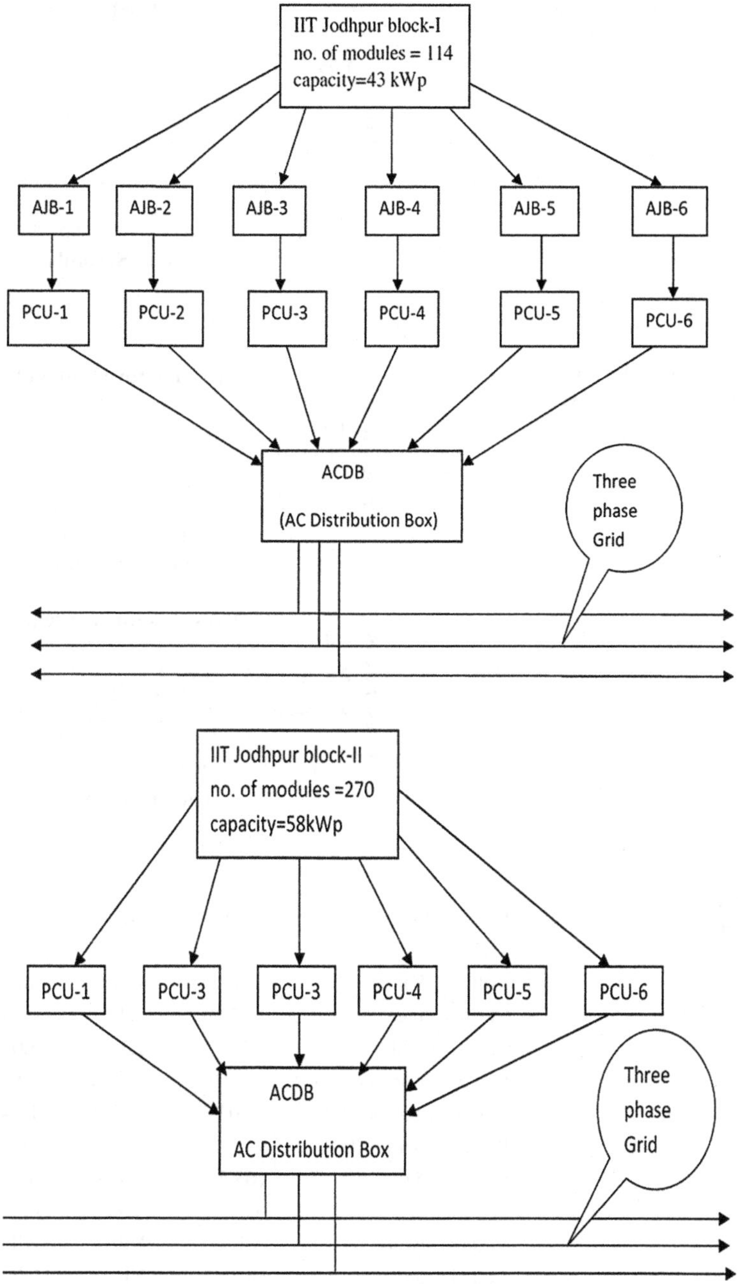

Fig. 13.2 IIT Jodhpur PV plant structure

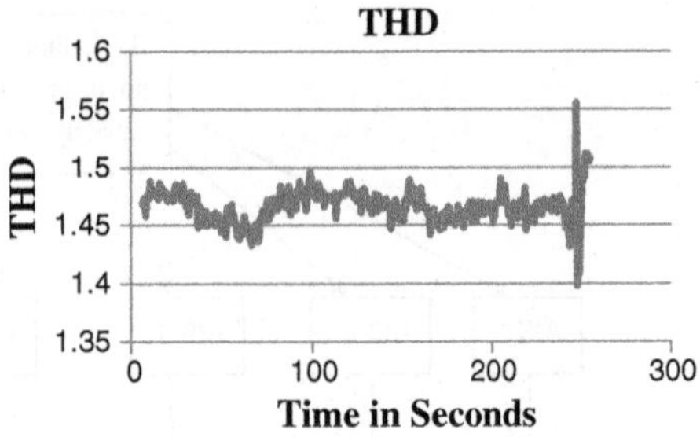

Fig. 13.3 Measured THD in voltage

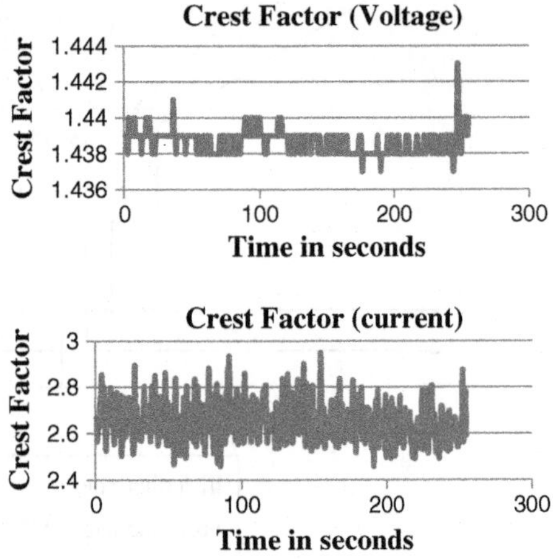

Fig. 13.4 Measured crest factors of voltage and current

Table 13.1 All AC parameters measured

Parameters	Morning	Noon	Evening
THD %	1.44	1.74	1.68
Voltage (V)	410	406	407
CF in voltage	1.43	1.45	1.44
CF in current	1.5	1.7	3.0
Maximum power (W)	57,163	44,810	690
Apparent power (VA)	67,077	53,010	1,244
Power variation (%)	91	77	3
Current (A)	163.3	130.8	3.0
PF	0.6	0.9	0.5
Voltage swells	Nil	Nil	Nil
Voltage sags	Nil	Nil	Nil

The nominal voltage is 440 V. For swell, voltage level should not be greater than 110 % of 440 V, which is 484 V. It can be observed that 484 V was never measured at block-II; rather, 425 V was the maximum value measured. It can be concluded that there is no voltage swell at block-II. For the voltage sag to occur, voltage level should not be less than 90 % of 440 V, which is 393 V. According to the measured values, voltage below 400 V was never measured. A minimum voltage of 400 V was measured.

13.5 AC Power Quality Assessment and Measurements by Power Analyzer

Yokogawa power analyzer was used for entire measurements. The three phases from ACDB were connected to the rear-side panel of the analyzer using delta connection method. The three phases A, B, and C were connected in such a fashion that they made a delta connection like A–B, B–C, and C–A. For the current measurement, clamp-type current tong tester was mounted on red, yellow, and blue phases at the ACDB box. The current from these sensors was fed to the rear panel of the analyzer. For analyzing the data on computer, a RS-232 interface was used. Figure 13.5 shows the complete setup on rooftop, just below the panel.

To get the trends of measurement for short duration, time was chosen to be 5 min. The power consumption measuring system (PCMS) had three different standards of IEC62301 Ed1, IEC 62301 Ed2 (auto), and IEC 62301 Ed2 (manual). Internationally recognized standard for the measurement technique and measurement accuracy for standby power is IEC62301. The IEC is a worldwide organization for standardization, comprising all national electrotechnical committees (IEC National Committees).

The internal structure of ACDB box is shown in Fig. 13.6.

Fig. 13.5 Experimental setup for AC power

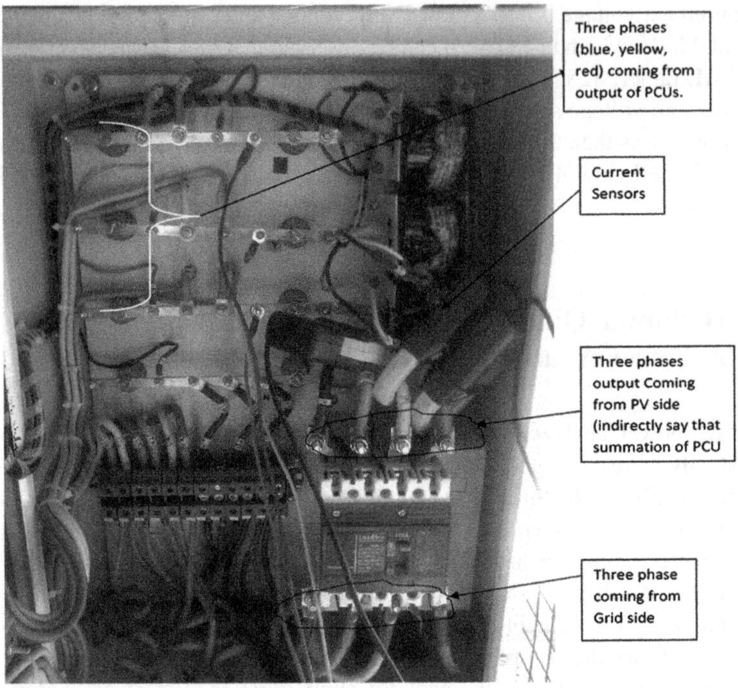

Fig. 13.6 Internal view of ACDB

13.6 Assessment of DC Power Quality

Considerations on power quality involve commonly AC. The power quality concerns are not only with AC but also with DC. The power delivered by solar panels is called raw DC, which is fluctuating, and hence, it needs to be taken care. For DC appliances, there is a need for constant and ideal power supply. For obtaining ideal DC, a DC/DC converter is used in the system. The complete system contains the following parts: a DC supply, function generator, oscilloscope, DC/DC boost converter, Yokogawa power analyzer, and DC motor as a load. The power obtained from the output of DC/DC boost converter is considered as a source, which may not have an ideal DC waveform as it contains ripple. The DC voltage provided by PV strings is 160 V, which can be converted into 260 V or any other voltage level as per requirement with the help of DC/DC converter. The DC power supply to boost converter was set at 18 V and the switching frequency at 25 kHz. Figure 13.7 shows the complete DC setup (Table 13.2).

The following measurements were taken at the output of DC/DC boost converter.

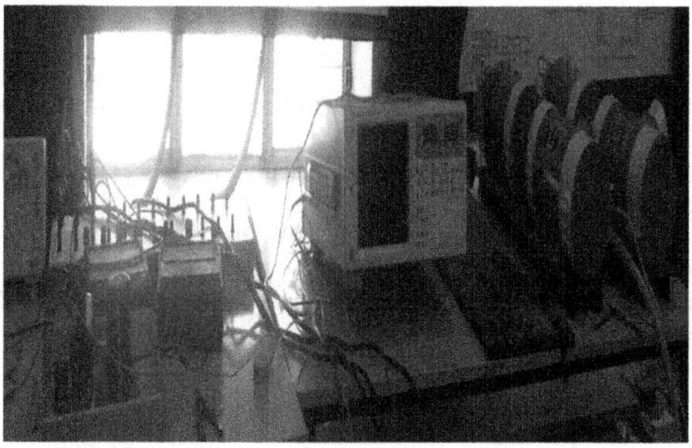

Fig. 13.7 Experimental setup for the measurement of DC power at DC/DC boost converter

Table 13.2 All DC parameters measured

No.	Function	Element	Order	Data	Units	Max	Min
1	Urms	3	–	0.16062 k	V	0.16063 k	0.16059 k
2	Irms	3	–	0.8958	A	0.8958	0.8924
3	P	3	–	0.1430 k	W	0.1430 k	0.1424 k

The 'ripple' is a kind of an AC waveform superimposed on DC output voltages. The amplitude of AC component could be the order of millivolts [4]. The occurrence of 'ripple' depends upon the operating frequency of DC/DC converter. The DC ripple was measured to be 9 %, which is higher than the IEC permissible limit of 4 %.

13.7 Conclusion

It can be concluded that there is no voltage sag at block-II and block-I at any hour. The THD is within limits as prescribed by IEC. However, the seventh harmonic component was measured to be 4.77 %. It appears that the design of the inverter did not provide filter for seventh- and higher-order harmonics, which can improve the AC power output quality further. The ripple in DC was measured to be 9 %, whereas the IEC requirement limit is within 4 %. The ripple can be minimized by using higher value smoothing capacitor. A smoothing capacitor is used to generate ripple-free DC.

References

1. Teodorescu R, Liserre M, Rodriguez P (2011) Grid converters photovoltaic and wind power systems. IEEE and Wiley Publication, New York
2. Masters GM (2004) Renewable and efficient electric power systems. Wiley Interscience Publication, New York
3. Dugan RC, McGranghan MF, Santoso S, Beaty HW (2002) Electrical power system quality, 2nd edn. McGraw-Hill, New York
4. Ericsson AB (2010) Output ripple and noise measurement methods for Ericsson power modules. Power modules, SE-16480 Stockholm, Sweden

Chapter 14
Online Frequency Domain Identification Method for Stable Systems

Priti Kshatriy and Utkal Mehta

Abstract An online estimation method for single-input single-output (SISO)-type stable systems is discussed based on frequency transformation technique. Reported method based on fast Fourier transform (FFT) is an off-line identification method means the controller is required to remove from the closed loop at the time of autotuning test. So the modified method is suggested for online identification and has been tested on several systems to show the effectiveness of the method. A relay with hysteresis in parallel with proportional–integral (PI) controller induces stationary oscillation cycle whose frequency and amplitude are used for system identification. We consider the development of a non-iterative approach with less computational efforts and a reasonable amount of data. A simulation study is given to illustrate the potential advantage of the presented method.

Keywords SISO · FFT · System identification · Relay experiment

14.1 Introduction

It is desirable to know the system before it is manipulated for control purposes. To estimate the system behavior, a relay experiment is probably most successfully used in the process industry [1]. Various methodologies on the relay test are reported and summarized in [1–3]. Some of the distinct advantages of the relay tuning are as follows: (i) It identifies system information around the important frequency, the ultimate frequency (the frequency where the phase angle is π), (ii) it is a closed-loop

P. Kshatriy (✉)
SAL Institute of Technology and Engineering Research, Gujarat Technical University, Gujarat, India
e-mail: priti.ec2009@gmail.com

U. Mehta
School of Engineering and Physics, The University of the South Pacific, Suva, Fiji
e-mail: utkal.mehta@usp.ac.fj

V. Vijay et al. (eds.), *Systems Thinking Approach for Social Problems*,
Lecture Notes in Electrical Engineering 327,
DOI 10.1007/978-81-322-2141-8_14

test; therefore, the system will not drift away from the nominal operating point, and (iii) for systems with a long time constant, it is a more time-efficient method than conventional step or pulse testing. The experimental time is roughly equal to two to four times the ultimate period. Despite this method has been subject of much interest in recent years and also has been field tested in a wide range of applications, basically, this technique is an off-line testing; i.e., some information is extracted after removing the controller from the loop. It has been noted [2, 4] that off-line testing may affect the operational process regulation which may not be acceptable for certain critical applications. Indeed, in certain key process, control areas such as vacuum control and environment control may be too expensive or dangerous for the control loop to be broken for tuning purposes.

A recent survey shows that the ratio of applications of proportional–integral–-derivative (PID) control, conventional advanced control (feed forward, override, valve position control, gain-scheduled PID, etc.), and model predictive control is about 100:10:1 [5]. An important feature of this controller is that it does not require a precise analytical model of the system that is being controlled. For this reason, PID controllers have been widely used in robotics, automation, system control, manufacturing, transportation, and interestingly in real-time multitasking applications [6]. However, the parameters of controller must be tuned according to the nature of the system. In practice, it has been shown that the PID gains are often tuned using experiences or trial and error methods. Again, due to varying nature of the system and environment, the performance of the closed-loop system is always deteriorated. It is important to re-tune the controller to regain the desired performance.

One of the simplest and most robust autotuning techniques for system controllers is a relay autotuning test. Many modified methods [2, 7–13] based on relay experiments are reported to rectify demerits in the original conventional method of relay feedback test. Most methods are discussed to estimate the system dynamics online, and then, based on estimated data, the new controller setting is suggested to improve the closed-loop performances. Again, some refinements to the original relay feedback method have been undertaken in identifying multiple points on the system frequency response. Based on the frequency domain describing function approach, many methods have been reported [1, 3] with improved accuracy to estimate process transfer function models. However, these approximate DF methods are basically iterative and also required some suitable initial guesses. A systematic time domain analysis was presented in [11] for identifying process dynamics with first-order model. In this approach, a relay is connected in series with a controller to tune the controller online. Other identification method has been reported, using fast Fourier transform (FFT), is very useful for process response identification [4, 14]. In this method, process input and output responses are obtained first from a single relay feedback test. The logged limit cycle oscillation is decomposed into the transient parts and the stationary cycle parts. Then, these parts are transformed to their frequency responses using the FFT and digital integration, respectively, to estimate the process frequency response. This method is basically an off-line since the controller is removed from the main line at the time of autotuning.

The paper is concerned with identification of single-input single-output (SISO) system based on FFT algorithm. The off-line method has some limitations as removing controller from main loop may not acceptable for some applications. To overcome this problem, an online system identification technique has been considered. A relay with hysteresis in parallel with PI controller induces stationary oscillation cycle whose frequency and amplitude are used for system identification. The system input and output are decomposed into their transient part and steady-state part, respectively. The system frequency response can be obtained by transforming transient part and steady-state part into their frequency response using FFT and digital integration, respectively. A number of examples are given to illustrate simplicity, effectiveness, and potential advantage of the online FFT-based technique.

14.2 Revisited FFT-Relay Method

Departing from the conventional relay test where the controller is replaced by the relay, the presented test is to carry out online without breaking the closed-loop control by placing the relay in parallel with the controller. Figure 14.1 shows the tuning scheme in which the relay height is increased from zero to some acceptable value when re-tuning is necessary.

Aiming to estimate the system frequency response $G(j\omega)$ accurately without disconnecting controller from the main line. Let us take the controller of type PI at the time of relay test as

$$C(s) = K_p + \frac{K_i}{s} \tag{14.1}$$

where K_p and K_i are its controller gains. This structure is simple with only two parameters yet it is one of the most common and adequate ones used, especially in the process control industries. When a relay is invoked with amplitude $\pm h$ and hysteresis $\pm \varepsilon$ in parallel with the controller, a stable oscillation will result if the system has a phase lag of at least π radians.

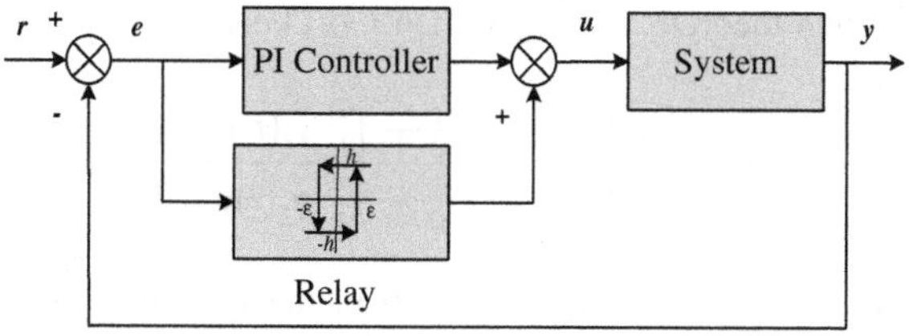

Fig. 14.1 Relay feedback structure

We want to extract the dynamic information on the given process, such as frequency response, from the measured values of $y(t)$ and $u(t)$. In the following, the method is given to compute system frequency response using the FFT [14].

Transient part and steady-state part of the system input $u(t)$ and output $y(t)$ are extracted from its time response data. By replacing transient part by steady-state part, one can obtain steady-state part of system input [14]. As we know,

$$u(t) = u_s(t) + \Delta u(t) \tag{14.2}$$

$$y(t) = y_s(t) + \Delta y(t) \tag{14.3}$$

By subtracting steady-state part u_s from system input u, transient part Δu can be obtained. Same method is applied to get steady-state part y_s and transient part Δy of system output.

For a system $G(s) = Y(s)/U(s)$, from (14.2) and (14.3)

$$G(s) = \frac{\Delta Y(s) + Y_s(s)}{\Delta U(s) + U_s(s)} \tag{14.4}$$

where

$\Delta Y(s)$ Laplace transform of transient part of $y(t)$
$\Delta U(s)$ Laplace transform of transient part of $u(t)$
$Y_s(s)$ Laplace transform of steady-state part of $y(t)$
$U_s(s)$ Laplace transform of steady-state part of $u(t)$

For the periodic part of time response, the following Lemma [15] holds. The periodic function $f(t)$ can be described as

$$f(t) = \begin{cases} f(t+T_c), & t \in [0, +\infty) \\ 0, & t \in (-\infty, 0) \end{cases} \tag{14.5}$$

where T_c is the time period of function $f(t)$.

Assume that £$\{f(t)\} = F(s)$ exists. Then, the Laplace transform of $f(t)$ is given by

$$F(s) = \frac{1}{1 - e^{-sT_c}} \int\limits_0^{T_c} f(t) e^{-sT_c} dt \tag{14.6}$$

If we apply above theorem to (14.4), then G(s) becomes

$$G(s) = \frac{\Delta Y(s) + \frac{1}{1 - e^{-sTc}} \int_0^{T_c} y_s(t) e^{-st} dt}{\Delta U(s) + \frac{1}{1 - e^{-sTc}} \int_0^{T_c} u_s(t) e^{-st} dt} \tag{14.7}$$

where T_c is the period of the steady-state part of the system output obtained from the relay feedback test. If we put $s = j\omega$, (14.7) becomes

$$G(j\omega) = \frac{\Delta Y(j\omega) + \frac{1}{1-e^{-j\omega T_c}} \int_0^{T_c} y_s(t)e^{-j\omega t}dt}{\Delta U(j\omega) + \frac{1}{1-e^{-j\omega T_c}} \int_0^{T_c} u_s(t)e^{-j\omega t}dt} \tag{14.8}$$

Suppose that $t = T_f$ is transient period for u and y and after $t = T_f$, both Δu and Δy are approximately zero. Then, the Fourier transform of Δy is given by

$$\begin{aligned} \Delta Y(j\omega) &= \int_0^{\infty} \Delta y(t)e^{-j\omega t}dt \\ &\approx \int_0^{T_f} \Delta y(t)e^{-j\omega t}dt \end{aligned} \tag{14.9}$$

Equation (14.4) can be computed at discrete frequencies with the standard FFT algorithm, which is an efficient and reliable way for calculating DFT more quickly. Suppose that $y(kT)$, $k = 1, 2, 3, \ldots, N - 1$ are samples of $y(t)$ where T is the sampling interval and $(N - 1)T = T_f$ are formed from (14.6) and we have its FFT as

$$\begin{aligned} \text{FFT}(\Delta y(kT)) &= T\sum_{k=0}^{N-1} \Delta y(kT)e^{-j\omega_l kT} \\ &\approx \Delta Y(j\omega_l) \quad l = 1, 2, 3, \ldots, M \end{aligned} \tag{14.10}$$

where M is the number of frequency response points to be identified on Nyquist plot and $\omega_l = 2\pi l/(NT)$.

$Y_s(j\omega)$ in (14.8) at discrete frequency ω_l are computed using digital integral as

$$Y_s(j\omega_l) = \frac{1}{1 - e^{-j\omega_l Tc}} \sum_{k=0}^{Nc} y_s(kT)e^{-j\omega_l kT}T \quad l = 1, 2, 3, \ldots, M \tag{14.11}$$

where ω_l and M are defined as in (14.9) and $N_c = (T_c - T)/T$. $\Delta U(j\omega_l)$ and $U_s(j\omega_l)$ can be calculated in the same way. Consequently, the system frequency response is obtained as

$$G(j\omega_l) = \frac{\Delta Y(j\omega_l) + Y_s(j\omega_l)}{\Delta U(j\omega_l) + U_s(j\omega_l)} \quad l = 1, 2, 3, \ldots, M \tag{14.12}$$

To calculate FFT of $\Delta y(kT)$ given by (14.10), we have to give value to time period T_f which can be obtained either from time response data or by $T_f = (N - 1)$ T. It shows that T_f is related to frequency response points to be identified between zero frequency to phase cross over frequency ω_c on Nyquist plot. It observed from (14.11) that the frequency response points to be identified by the FFT algorithm are at the discrete frequencies 0, $\Delta\omega$, $2\Delta\omega$, $3\Delta\omega$, ..., $(M-1)\Delta\omega$, where $\Delta\omega = \omega_{l+1} - \omega_l = 2\pi/NT$. The definition of M means that $\omega_c \approx (M-1)\Delta\omega$, thus

$$\omega_c \approx (M-1)\frac{2\pi}{NT} \tag{14.13}$$

The oscillation period can also be measured online at the time of relay test and can be estimated as

$$\omega_c \approx \frac{2\pi}{T_c} \tag{14.14}$$

From Eqs. (14.13) and (14.14)

$$N \approx (M-1)\frac{T_c}{T} \tag{14.15}$$

where M should be specified by user.

This FFT-based method gives a high accuracy to identify multiple points on frequency response from a single relay test. However, it has been found that the input and output must be recorded from the initial time to calculate the transient parts Δy and Δu accurately. Here, the initial time is defined when the relay feedback is performed to the system. Therefore, it is required for the system at the steady state before a relay feedback is applied at $t = 0$. The transient parts $\Delta y(\Delta u)$ and steady-state parts $y_s(u_s)$ are required to decompose first to determine FFT from its time response data accurately using some computational efforts.

14.3 Examples

In this section, proposed method has been applied on several systems to show robustness and accuracy of the method. The amplitude level should be sufficient to prevent false switching caused by noisy signals. It is important to keep the output oscillation amplitude in the prescribed limit as per the tolerable system variable swing and decide values for the relay heights that produce a limit cycle with acceptable amplitude level. To overcome the undesirable relay chattering caused by noisy signals, the width of the hysteresis of the relay is set to twice the standard deviation of the noise. For ease in simulation study, the relay with $h = \pm 0.1$ and $\varepsilon = \pm 0.01$ is taken although fairly low values of relay height could be used.

Example 1 Consider second-order plus dead time system

$$G(s) = \frac{1}{(s+1)(3s+1)}e^{-s} \tag{14.16}$$

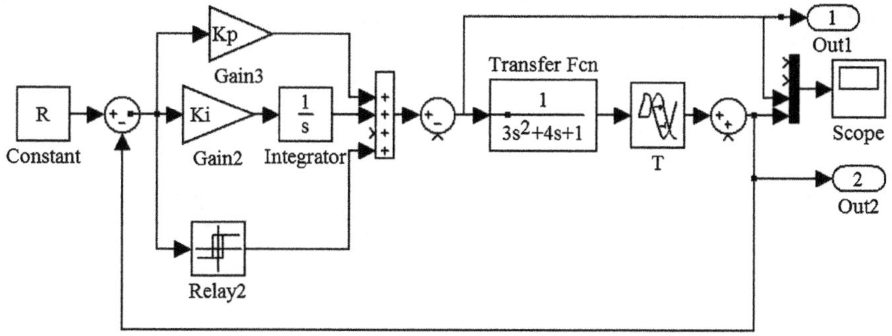

Fig. 14.2 Simulink block diagram of relay feedback applied to system

Fig. 14.3 Transient part of system input

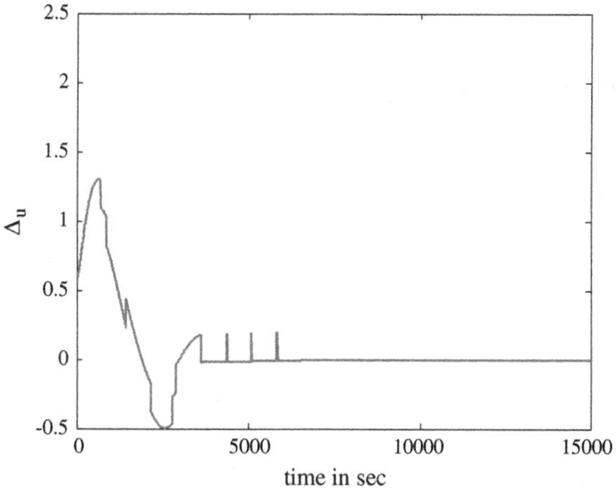

For this system, $K_p = 0.35$, $K_i = 0.35$, $h = 0.1$, and $\varepsilon = 0.01$ have been taken. Relay feedback has been applied as shown in Fig. 14.2 from which the system input and output parameter has been extracted. The system input and output are then decomposed into their transient part (Figs. 14.3 and 14.6) and steady-state part (Figs. 14.4 and 14.5), respectively. FFT and digital integration have been applied to transient parts and steady-state parts, respectively, to obtain their frequency response. These two parts are then combined to calculate system frequency response and compared with actual frequency response. The signals measured from the simulation are plotted in Fig. 14.7. The method gives an accurate result which is shown from the Nyquist plot in Fig. 14.8.

Example 2 Consider second-order plus dead time system

$$G(s) = \frac{1}{(s+1)^2} e^{-s} \tag{14.17}$$

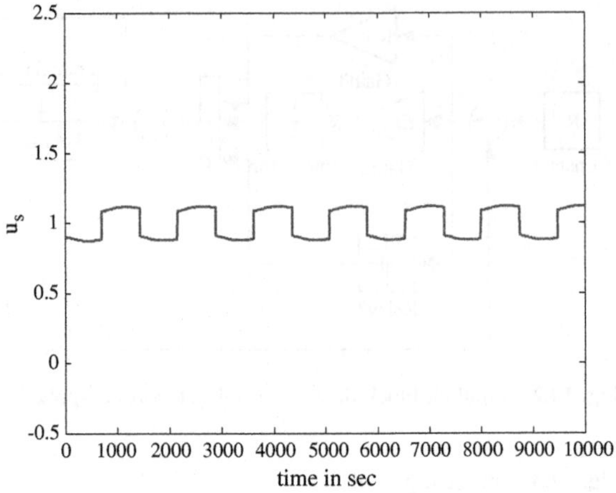

Fig. 14.4 Steady-state part of system input signal

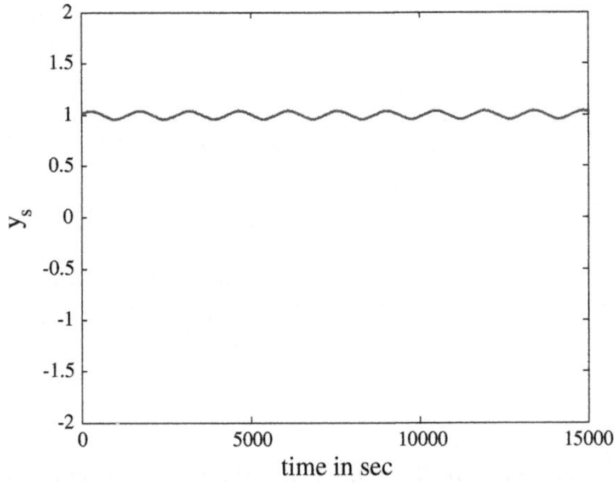

Fig. 14.5 Steady-state part of system output signal

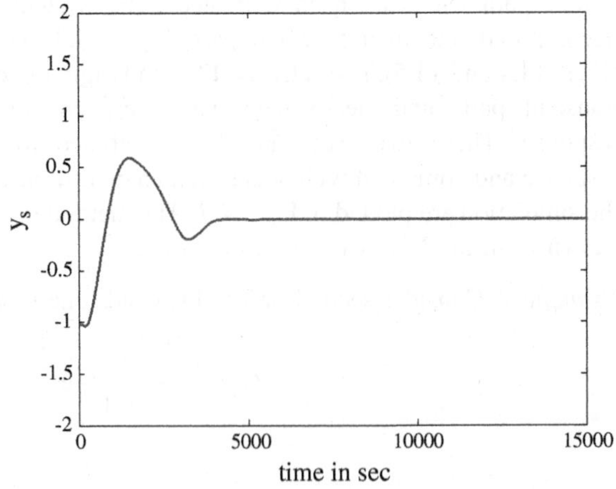

Fig. 14.6 Transient part of system output

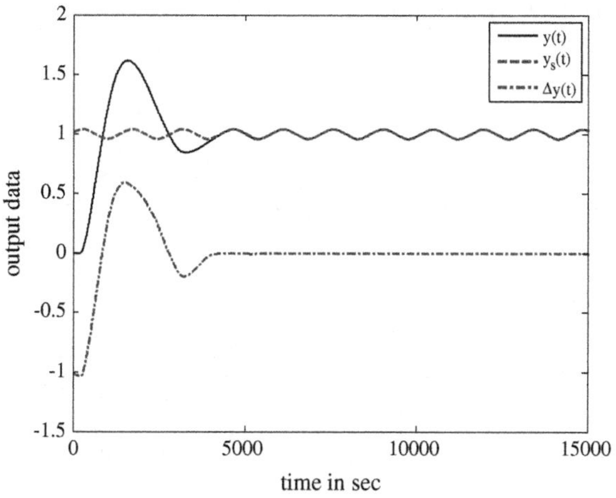

Fig. 14.7 System output signals

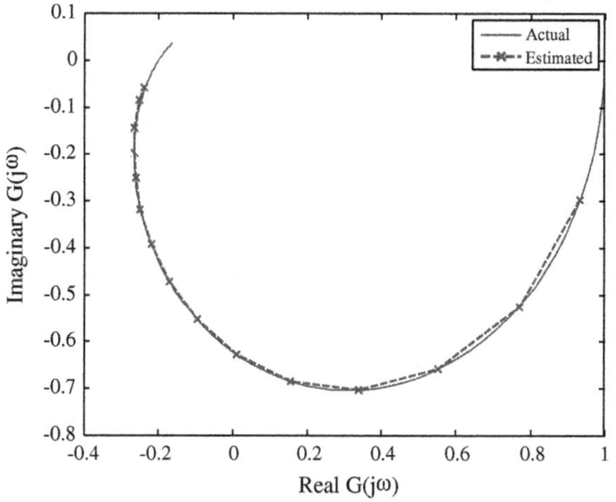

Fig. 14.8 System Nyquist plots

For this plant, Mehta and Majhi [13] have proposed controller parameter, $K_p = 0.35$, $K_i = 0.35$, $h = 0.1$, and $\varepsilon = 0.001$. With $M = 15$ and $T = 0.005$, time period $T_c = 5.41$ has been calculated manually. System input (output) is decomposed into its transient part and steady-state part as shown in Fig. 14.9 (Fig. 14.10). Using proposed methodology, multiple points on frequency response curve have been identified accurately as in Fig. 14.11.

Example 3 Consider non-minimum phase system

$$G(s) = \frac{-1.5s + 1}{(s+1)^3} \tag{14.18}$$

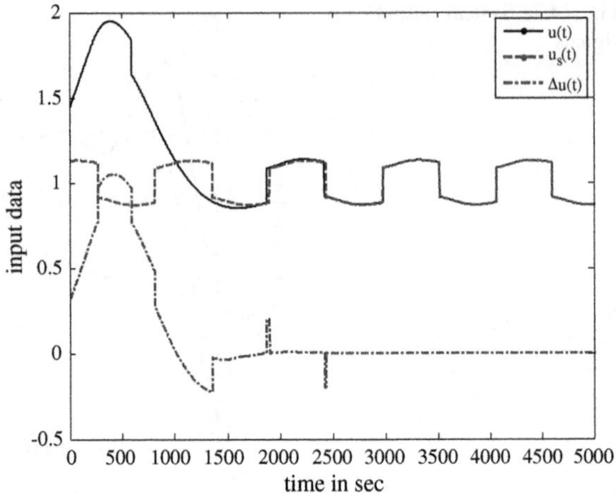

Fig. 14.9 System input signals

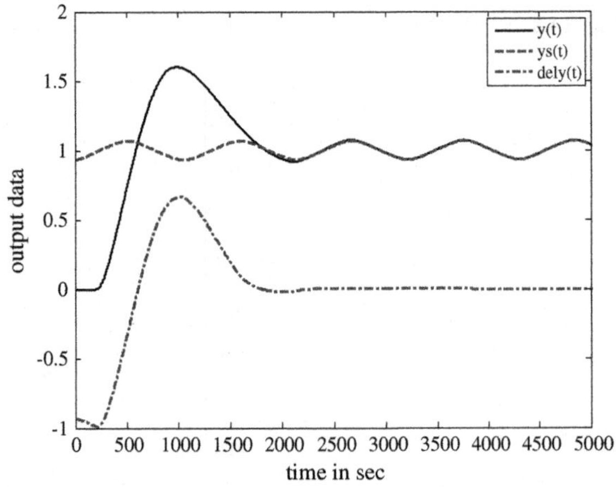

Fig. 14.10 System output signals

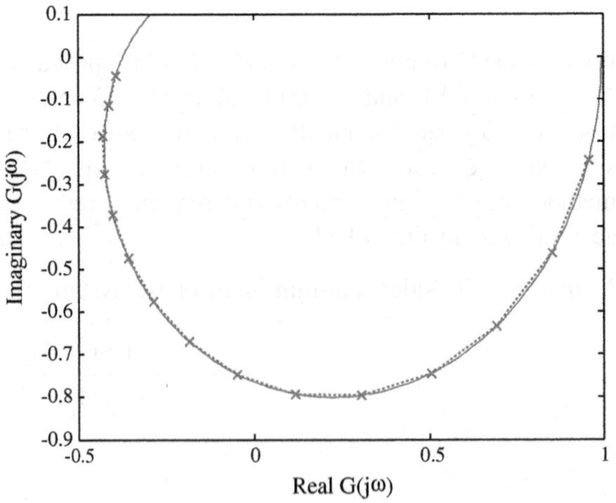

Fig. 14.11 System Nyquist plots

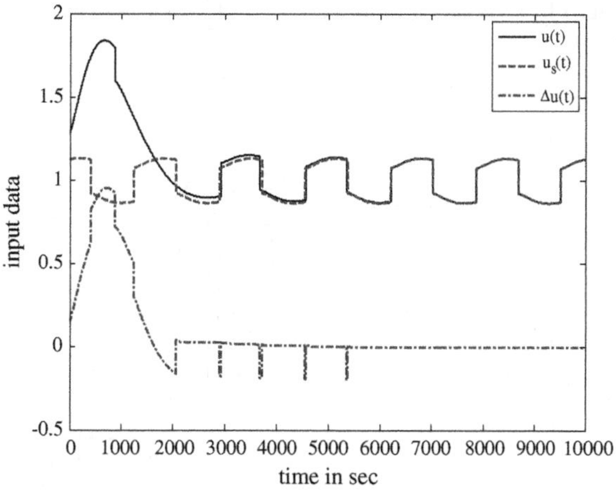

Fig. 14.12 System input signals

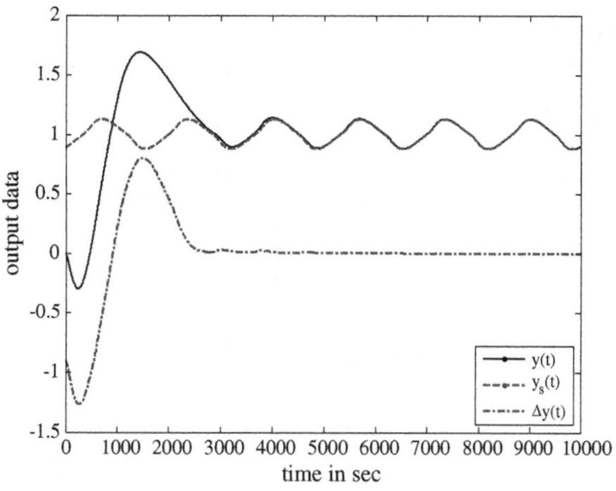

Fig. 14.13 System output signals

For this plant, Mehta and Majhi [13] have proposed controller parameter, $K_p = 0.172$, $K_i = 0.181$, $h = 0.1$, and $\varepsilon = 0.001$. With $M = 15$ and $T = 0.005$, time period $T_c = 8.31$ has been obtained. The transient part and steady-state part of system input are as shown in Fig. 14.12. Same for system output is as shown in Fig. 14.13. Estimated frequency response points in important frequency range from zero to critical frequency (i.e., $[0,\omega_c]$) is as shown in Fig. 14.14.

Example 4 Consider the very high-order system

$$G(s) = \frac{1}{(s+1)^{20}} \tag{14.19}$$

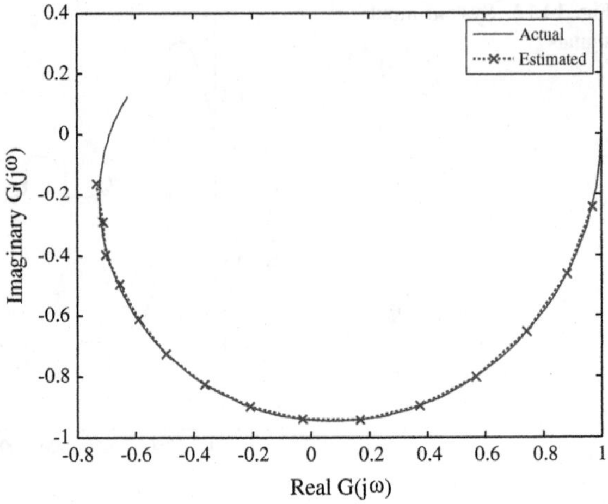

Fig. 14.14 System Nyquist plots

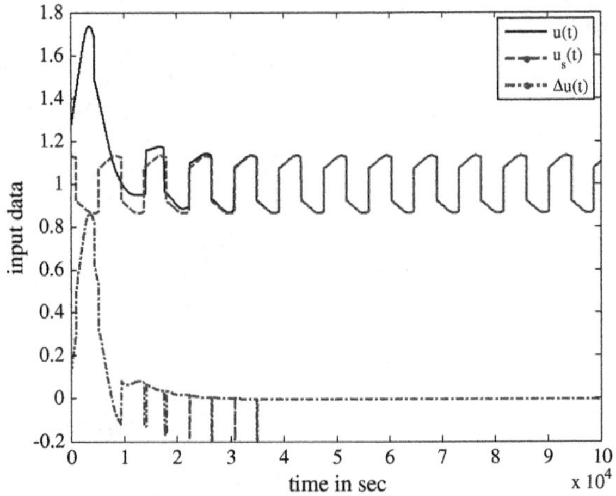

Fig. 14.15 System input signals

$K_p = 0.172$, $K_i = 0.181$, $h = 0.1$, and $\varepsilon = 0.001$ has been proposed by Majhi [11]. With $M = 15$ and $T = 0.005$, time period $T_c = 8.31$ has been calculated. The system input (output) is decomposed into transient part and steady-state part as shown in Fig. 14.15 (Fig. 14.16). DFT-based algorithm gives accurate result as shown in Fig. 14.17.

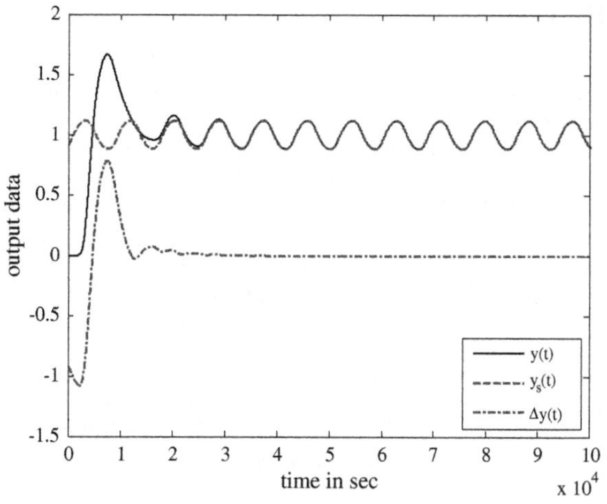

Fig. 14.16 System output signals

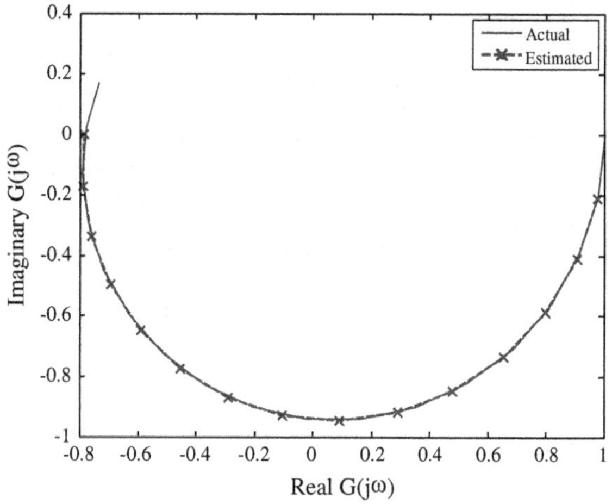

Fig. 14.17 System Nyquist plots

14.4 Conclusions

An online technique is presented for system frequency response identification in context of the relay experiment. The method has several features. First, it can obtain multiple points on system frequency response simultaneously without breaking the closed-loop control and this increases applicability for certain critical systems. Second, this approach is robust as all measurements are made nearby setpoint value without removing the controller from the main line. Third, it can be used in the presence of a static load disturbance since the PI action is always present during the test. The estimated system frequency response is useful generally for controller design.

References

1. Atherton DP (2006) Relay autotuning: an overview and alternative approach. Ind Eng Chem Res 45:4075–4080
2. Tan KK, Lee TH, Jiang X (2000) Robust on-line relay automatic tuning of PID control system. ISA Trans 39(2):219–232
3. Wang QG, Lee TH, Lin C (2003) Relay feedback: analysis, identification and control. Springer, London
4. Wang QG, Hang CC, Bi Q (1997) A technique for frequency response identification from relay feedback. IEEE Trans Control Syst Technol 7(1):122–128
5. Kano M, Ogawa M (2009) The state of the art in advanced chemical process control in Japan. In: IFAC symposium ADCHEM2009
6. Udaykumar YR, Sreenivasappa VB (2009) Design and implementation of FPGA based low power digital PID controllers. In: Fourth international conference on industrial and information systems, ICIIS 2009, 28–31 Dec 2009
7. Schei TS (1992) A method for closed loop automatic tuning of PID controllers. Automatica 28 (3)
8. Tan KK, Ferdous R, Huang S (2002) Closed-loop automatic tuning of PID controller for nonlinear systems. Chem Eng Sci 57:3005–3011
9. Ho WK, Honga Y, Hanssonb A, Hjalmarssonc H, Denga JW (2003) Relay auto-tuning of PID controllers using iterative feedback tuning. Automatica 39:149–157
10. Arruda GHM, Barros PR (2003) Relay-based gain and phase margins PI controller design. IEEE Trans Instrum Measur 52(5):1548–1553
11. Majhi S (2005) On-line PI control of stable processes. J Process Control 15:859–867
12. Tsay TS (2009) On-line computing of PI/lead compensators for industry processes with gain and phase specifications. Comp Chem Eng 33:1468–1474
13. Mehta U, Majhi S (2012) On-line relay test for automatic tuning of PI controllers for stable processes. Trans Inst Measur Control 34(7):903–913
14. Wang QG, Bi Q, Zou B (1997) Use of FFT in relay feedback systems. IET Elect Lett 33 (12):1099–1100
15. Kuhfittig PKF (1978) Introduction to the Laplace transform. Plenum, New York

Chapter 15
High-Temperature Solar Selective Coating

Belal Usmani and S. Harinipriya

Abstract Solar energy has maximum potential in comparison with other renewable energy sources. Solar thermal conversion is one of the direct methods for harnessing solar energy using solar selective absorbers. This review article summarizes the recent research progress on the high-temperature solar selective coatings, methodology, and process involved in coating, computer modeling, as well as designs of coatings, optical, compositional, and structural properties of selective coatings. The major bottleneck in developing a solar selective coating is its stability in air at temperature higher than 450 °C that possess thermal and structural stability in both individual and combined layers. New possibilities to overcome the above-mentioned problems on the performance of solar selective coatings are discussed.

Keywords High-temperature solar selective coating · Cermet · Nanocermet · Solar thermal energy · Absorptance · Emittance

15.1 Introduction

The current global energy problem can be attributed to insufficient fossil fuel supplies and excessive greenhouse gas emissions resulting from increasing fossil fuel consumption. Energy supply has arguably become one of the most important problems facing humanity [1]. The energy crisis is further exacerbated by major concerns about global warming from greenhouse gas emissions due to increasing fossil fuel consumption [2–4]. At this large scale, solar energy seems to be the most viable choice to meet our clean energy demand. The sun continuously delivers to the earth

B. Usmani (✉)
Center of Excellence in Energy, Indian Institute of Technology Jodhpur,
Jodhpur 342011, Rajasthan, India
e-mail: busmani@iitj.ac.in

S. Harinipriya
Center of Excellence in Systems Science, Indian Institute of Technology Jodhpur,
Jodhpur 342011, Rajasthan, India

V. Vijay et al. (eds.), *Systems Thinking Approach for Social Problems*,
Lecture Notes in Electrical Engineering 327,
DOI 10.1007/978-81-322-2141-8_15

120,000 TW of energy, which dramatically exceeds our current rate of energy needs (13 TW) [5]. The direct system of harnessing solar energy can be broadly classified into two categories: (a) thermal systems that convert solar energy into thermal energy and (b) photovoltaic (PV) systems which convert solar energy directly into electrical energy. There are four main systems of solar thermal electricity: solar towers, dishes, linear fresnel, and the parabolic troughs. These systems require high-absorptance, low-emittance solar selective absorber coatings stable at high temperatures and in the wavelength range 0.3 μm < λ < 3 μm as well as high reflectance at 3 μm < λ < 50 μm in order to achieve maximum efficiency [6, 7]. Different types of solar selective coatings, including multilayer gradient coating [8], optical interference coating [9], and cermet composite coating [10], have been used in solar thermal power generation.

This review summarizes the recent progress on high-temperature solar selective structures based on (i) metal–dielectric composites, known as cermet (ceramic–-metal), which is used for high-temperature applications (stable up to 650 °C), (ii) noble metal nanoparticles and refractory oxide coatings, and (iii) transition metal nitride host matrices as ideal high-temperature coatings with good thermal stability and high degree of spectral selectivity and improved oxidation resistance [11, 12].

15.2 Solar Spectral Selective Mechanisms

Solar (spectral) selective absorber surfaces for solar collectors were pioneered by Tabor in 1995 and later carried out by several other researchers [13–16]. Tabor proposed two different concepts for developing solar selective surfaces. The first type constitutes a low-emissivity metal base covered by a thin surface layer such as black chrome, black nickel, copper oxide, etc., which is non-transparent in the visible region. The second type consists of metallic systems with low emissivity, which exhibit high absorptivity in the visible spectrum due to color or a finely divided structure [16].

Improving the properties of the selective coating on the receiver and operating temperature of solar thermal system from 400 to >450 °C can increase the overall solar-to-electricity efficiency and reduce the cost of electricity from solar thermal system plants [17]. Different types of cermet coatings have been developed using Pt, W, Mo, Cu, Ag, Co, Ag, SS, Cr, Ti, and Au as metal and Al_2O_3, AlN, Si_2O, MgO, AlON, Cr_2O_3, and AlON as the dielectric material and are discussed below.

15.3 Cermet as a Solar Selective Coating

Cermet materials as solar absorbers have attracted interest since the 1950s, when Tobar [13], Gier and Dunkle [15] have studied practically useful cermet films for solar selective absorption. Several cermet selective absorbers have been commercialized.

Zhang et al. [18] designed a numerical model and experimentally deposited high-efficient Mo–Al_2O_3 cermet solar absorber and also calculated a photothermal conversion efficiency as high as 0.914 at 350 °C with a concentration factor of 26 for the film structure consisting of a double cermet layer on a Mo metal thermal reflector with an Al_2O_3 anti-reflecting coating. Sohon and Bucher [19] modeled solar thermal performance of metal/dielectric multilayer coatings of various material combinations and also calculated maximum solar absorptance $\alpha = 0.94$ with low emittance $\varepsilon = 0.16$ ($T = 1{,}100$ K) for Al_2O_3/noble metal (Rh, Ir) coatings. Lux-Steiner et al. [20] developed a program to calculate the optical properties of metal/insulator-multilayer systems from complex refractive indices of corresponding bulk materials. Sohon et al. [21] optimized and prepared Al_2O_3/Pt and $Al_2O_3/MoSi_2$ multilayer films by computer simulation and RF magnetron sputtering and also calculated the solar absorptance up to $\alpha = 0.95$, emittance $\varepsilon = 0.08$ to $\varepsilon = 0.2$ for Pt/Al_2O_3 and $\alpha = 0.92$, $\varepsilon = 0.14$ for tetragonal $MoSi_2/Al_2O_3$. Zhang et al. [22] developed a model to calculate the reflectivity of germanium implanted with high-dose oxygen and also calculated the reflectivity spectra for a series of special layer structures. Niklasson and Granqvist [23] investigated the relation between microstructure and spectral properties of cermet selective surfaces. Niklasson and Granqvist [24] presented an extensive analysis of the optical properties of co-evaporated Co–Al_2O_3 composite films. Nejati et al. [25] investigated the solar absorptance and thermal emittance of various configurations of solar selective coatings. Farooq and Lee [26] addressed the optical performance of several metal/dielectric composites like Sm, Ru, Tm, Ti, Re, W, V, Tb, and Er, in alumina or quartz on the basis of their refractive indices and also described the effect of surface roughness and porosity in the antireflection layer. Yue et al. [27] prepared an Al_xO_y–AlN_x–Al selective absorber surface by DC magnetron reactive sputtering with aluminum alloy in air and argon and also studied the high-temperature (400–600 °C) optical properties and stability of the coatings. Liu et al. [28] prepared a new solar selective absorbing coating of NbTiON/SiON on Cu substrate using reactive magnetron sputtering method, and also exhibited a high absorptance (α) of 0.95 and a low emittance (ε) of 0.07. Du et al. [29] investigated the phase structure, microstructure, surface roughness, and chemical composition of $Ti_{1-x}Al_xN$ ($0.25 \le x \le 0.75$) coatings, deposited by reactive magnetron cosputtering. Niklasson and Granqvist [30] investigated the complex dielectric function of well-characterized Co–Al_2O_3 cermet films, produced by coevaporation, over the 0.3 μm < λ < 40 μm wavelength range. Zhang and Mills [31] presented a new cermet film structure of solar thermal absorber and calculated the results for low- and high-temperature performance. Zhang and Shen [32] designed a high solar performance W–AlN cermet solar coating and deposited experimentally, and also calculated the dielectric function and the complex reflective index of W–AlON cermet materials using Sheng's approximation. Zhang [33] presented data on the properties of solar selective surfaces incorporating a SS–AlN cermet, which is deposited by DC sputtering, and also achieved a solar absorptance of 0.93–0.96 and emittance of 0.03–0.04 at room temperature. Niklasson [34] presented a detailed investigation on the structure of Co–Al_2O_3 composite coatings. Barshilia et al. [35]

deposited $Cr_xO_y/Cr/Cr_2O_3$ multilayer absorber coating and studied the structural, chemical, and optical properties of these coatings along with the thermal stability in vacuum and air. Barshilia et al. [36] prepared Ag–Al_2O_3 nanocermet-based spectrally selective solar absorber coating and optimized the coating process to exhibit high absorptance ($\alpha = 0.93$) in the visible region and low emittance ($\varepsilon = 0.04$–0.05 at 82 °C) in the infrared region of solar spectrum. Farooq and Raja [37] studied the performance of selective coatings to have maximum solar absorptance and mininum thermal emittance, in relation to base layer that should have acceptable thermal emittance and minimum diffusion into the composite coating. Farooq and Hutchins [38] studied the choice of material for optically solar selective coatings on the basis of their optical constants and observed that higher values of both n and k of the material are more suitable in solar selective coatings. Zhang et al. [39] developed a commercial-scale cylindrical direct current (dc) magnetron sputter coater for the deposition of metal–aluminum nitride (M–AlN) cermet solar selective coatings onto batches of tube and described the construction and operation of this sputter coater. Antonania et al. [40] investigated the stability of solar multilayer coating based on W–Al_2O_3 graded cermet layer before and after annealing processes at high temperature. Esposito et al. [41] optimized the solar absorptance higher than 0.94 and hemispherical emittance at 580 °C lower than 0.13 of double cermet layer. Erben and Tlhanyl [42] reported two coating techniques: galvanizing and chemical vapor deposition (CVD) of developing solar selective absorber systems for temperatures above 300 °C. Teixeira et al. [43] presented a numerical model that allows us to correlate the selectivity of the produced absorbers to the collector efficiency and deposited cermet coatings of Cr–Cr_2O_3 by reactive DC magnetron sputtering and also described the structural and optical properties of multilayer films. Jaworske and Shumway [44] reported a high-temperature durability of solar selective coatings composed of nickel and alumina oxide, titanium and alumina oxide, and platinum and alumina oxide. Zhang [45] reviewed the two recent key developments in solar selective coatings, double cermet layer film structures and M–AlN cermet solar coating deposited by two targets DC sputtering technology and also described the application of solar collector tubes to solar thermal electricity. Zhang and Mills [46] presented the calculated results for Cu–SiO_2 cermet in detail and also experimental results. Zhang et al. [47] developed a physical model which explains the experimental results for germanium implanted with high-dose oxygen. Fan and Zavracky [48] investigate the thermal stability of MgO/Au cermet films. McKenzie [49] prepared cermet solar selective surfaces of absorptance over 0.90 and emittance less than 0.05 by vacuum coevaporation of alumina and spinel with gold, silver, chromium, and copper. Nyberg and Buhrman [50] reported a new physical property of coevaporated metal–dielectric films: the drastic increase in the surface roughness on Mo/Al_2O_3 when the deposition temperature exceeds a certain minimum temperature. Rebouta et al. [51] reported the characterization of TiAlN/TiAlON/SiO_2 tandem absorber; the first two layers were deposited by magnetron sputtering, and third layer was prepared by plasma-enhanced chemical vapor deposition (PECVD) and determined the optical constants of individual layers by first measuring spectral transmittance and reflectance of the individual layers. Du et al. [52] designed a new solar selective coating, $Ti_{0.5}Al_{0.5}N$ and $Ti_{0.25}Al_{0.75}N$ coatings were chosen as absorber layers, and AlN coating was chosen as anti-reflecting layer and calculated the absorptance 0.945 and low emittance 0.04 (82 °C) of the solar selective coating. Xinkang et al. [53] prepared a series of Mo–Al_2O_3 films by direct current (DC, for metal Mo) and radio frequency (RF, for dielectric Al_2O_3) magnetron sputtering techniques and evaluated thermal stability at different temperatures and also investigated the variation of microstructure, diffusion of the component, and influences on the spectral performances. Vien et al. [54] reported features of Pt–-Al_2O_3 cermet coatings which is deposited by RF cosputtering on metallic substrate and obtained absorptivity α = 0.92, emissivity ε = 0.14 (at 300 °C), and also obtained the stability of selective absorbers at a temperature of up to 400 °C when cermet coating is deposited on stainless steel substrate and over 600 °C on superalloy substrate. Segaud et al. [55] prepared a thin film of Pt–Al_2O_3 cermets by RF cosputtering on glass and metallic substrate and investigated the wide ranges of composition and thickness. Craighead et al. [56] described an ultrahigh absorptivity selective absorber based on a Pt–Al_2O_3 composite with a graded refractive index depth profile. Nuru et al. [57] reported the optimization and further selectivity improvement of the radio frequency sputtered graded Pt–Al_2O_3 deposited on Mo base layer exhibiting a high solar absorption and significant thermal stability up to 650 °C in the air with and without antireflecting layers.

15.4 Others Solar Selective Coatings

Moller and Honicke [58] presented a new electrolytic process to grow solar selective layers on aluminum and also investigated the optical properties of the layers. Kunic et al. [59] studied the spectral selectivity and degradation properties of a novel thickness-sensitive spectrally selective (TSSS) coating based on organic polymer resin binders and trisilanol polyhedral oligomeric silsesquioxane (POSS) dispersant and also demostrated clearly that coating can be used in solar thermal system. Kozelj et al. [60] studied the formation of a protective layer from (3-mercaptopropyl) trimethoxysilane (MPTMS) on commercial sunselect, cermet-based spectrally selective coating by non-electrochemical, electrochemical cyclic voltammetry (CV) in the presence of a redox probe (Cd^{+2}), and potentiodynamic (PD) techniques. Orel et al. [61] studied to make paints of variety of colors and whose spectral selectivity would be independent of the thickness of deposited layer of paint. Xiao et al. [62] prepared a copper oxide thin film as solar selective absorbers by one-step chemical conversion method and characterized the composition, structure, and optical properties of thin films and indicated that the composition, structure, and optical properties of thin films were greatly influenced by reaction temperature and time concentration of NaOH. Cindrella [63] analyzed the efficiency of the solar selective coatings with various combinations of the optical properties and their impact on the performance of solar thermal systems of different concentration ratio (Crs). Tharamani and Mayanna [64] reported the development of the Cu–Ni alloy coating as a selective surface for solar energy. The coatings were deposited by using Hull cell and optimized deposition parameter to achieve high solar absorptance $\alpha = 0.94$ and low emittance $\varepsilon = 0.08$.

15.5 Future

In order to make solar thermal system more efficient, in addition to improvement in the solar thermal system design and associated support structure for solar fields, solar absorber coatings with improved optical properties and thermal stability need to be developed. New, more efficient selective coatings will be needed that have both high solar absorptance and low thermal emittance at elevated (>400 °C) temperatures. The coating need to be stable in air at elevated temperatures in case the vacuum is breached. Current coatings do not have the stability and performance desire for moving to higher operating temperatures. For efficient photo–thermal conversion, solar absorber surfaces must have low reflectance at wavelength $\lambda \leq 2$ μm and a high reflectance at $\lambda \geq 2$ μm; an improved spectrally selective surface should be thermally stable above 450 °C, ideally in air, and have solar absorptance (α) greater than about 0.96 and thermal emittance (ε) below about 0.07 at 400 °C. Still, none of the existing coatings used commercially have proven to be stable in air at 400 °C. Designing and fabrication of a solar selective coating that is stable in air at temperatures greater than 450 °C requires (i) high thermal and structural stabilities for both the combined and individual layers, (ii) excellent adhesion between the substrate and adjacent layers, (iii) suitable texture to drive the nucleation and subsequent growth of layers with desire morphology, (iv) enhanced resistance to thermal and mechanical stresses, and (v) acceptable thermal and electrical conductivities. Materials should have low diffusion coefficient at high temperature and should be stable with respect to chemical interactions with any oxidation products, including any secondary phase present, over long period of exposure time at elevated temperatures. Selecting materials with elevated melting points and large negative free energies of formation may meet this objective. Nanocermets are likely to be potential candidates for solar energy conversion because they exhibit strong absorption in the visible region, which occurs because of surface plasmon resonance phenomenon, also known as quantum confinement effect [65]. The optical properties of nanocermet strongly dependent on the particle shape, size, and concentration of the particle in the matrix, particle distribution and local dielectric environment of the dielectric matrix [66]. Stable nanocrystalline are the most desirable for diffusion barrier application and also need to develop nanostructured materials for solar energy conversion as special physical effects related to the nanometer scale do rise to interesting microscopic properties [67].

15.6 Summary and Conclusions

Renewable energy sources are important sources of energy for meeting the increasing energy demand of the world; solar thermal energy system is one of the expected systems to harness maximum energy from solar energy which have maximum potential rather than other renewable energy sources. In this article, we have reviewed the recent progress of high-temperature solar selective coatings for solar thermal applications.

For high-temperature applications, a large number of solar selective coatings such as Pt–Al_2O_3, Mo–Al_2O_3, W–Al_2O_3, W–AlON, SS–AlN, Co–Al_2O_3, W–AlN, Ag–Al_2O_3, Cr–Cr_2O_3, Cu–SiO_2, Mo–SiO_2, etc., have been developed. Researchers have studied Pt–Al_2O_3 coatings for high-temperature applications because of their high absorptance and low emittance at high operating temperatures, but it could not be commercialized due to high cost of platinum. Researchers have developed Mo–Al_2O_3, W–Al_2O_3, Co–Al_2O_3, and cermet coatings for high-temperature applications. The Mo–Al_2O_3 cermet coatings have been used in the Luz receiver tubes which were used for solar thermal power plants. Although these coatings have good thermal stability in vacuum, they have low thermal stability (≤300 °C) in air. Siemens, Germany have modified the Mo–Al_2O_3 cermet coating and developed a novel exhibit higher thermal stability. W–Al_2O_3-based cermet coatings are also being successfully produced commercially, which are reported to be stable up to 500 °C in vacuum. Mo–SiO_2 and SS–AlN cermet coatings have been successfully commercialized for receiver tubes by ENEA and TurboSun. Recently, researchers have focused to develop new cermet coatings based on transition metal nitrides/oxynitrides/oxides and silicides.

References

1. Smalley RE (2005) MRS Bull 30:412
2. Keeling CD, Whorf TP, Wahlen M, Vanderplicht J (1995) Nature 375:666
3. Mann ME, Bradley RS, Hughes MK, Jones PD (1998) Science 280:2029
4. DOE Argonne National Laboratory (2003) Basic research needs for the hydrogen economy. Report of DOE BES workshop on hydrogen production, storage, and use, 13–15 may 2003
5. Crabtree GW, Lewis NS (2007) Phys Today 60:37
6. Hahn RE, Seraphin BO (1978) In: Hass G, Francombe MH (eds) Physics of thin films, vol 10. Academic Press, New York, p 1
7. Seraphin BO (1979) Solar energy conversion: solid state physics aspects. In: Seraphin BO (ed) Topics in applied physics, vol 31. Springer, Berlin, p 5
8. Harding GL (1979) Alternative grading profile for sputtered solar selective surfaces. J Vac Sci Technol 16:2111–2113
9. Cao Y, Tian J, HU X (2000) Ni-Cr selective surface based on polyamide substrate. Thin Solid Films 365:49–52
10. Zhang QC, Mills DR (1992) New cermet film structures with much improved selectivity for solar thermal application. Appl Phys Lett 60:545–547
11. Blickensderfer R, Deardoff DK, Lincoln RL (1977) Sol Energy 19:429

12. Rebouta L, Vaz F, Andritschky M, da Silva MF (1995) Surf Coat Technol 76–77:70
13. Tabor H (1956) Selective radiation: I. Wavelength discrimination, II. wave front discrimination. Bull Res Counc Isr 5A(2):119–134
14. Shaffer LH (1958) Wavelength-dependent (selective) processes for the utilization of solar energy. Sol Energy 2:21
15. Gier JT, Dunkle RV (1958) Selective spectral characteristics as an important factor in the efficiency of solar collectors. In: Transactions of the conference on the use of solar energy, vol 2. University of Arizona Press, Tucson, p 41
16. Tabor H (1961) Solar collectors, selective surfaces and heat engines. Proc Natl Acad Sci USA 47:1271–1278
17. Kennedy CE (2008) Symposium on 14th Biennial CSP solar PACES (solar power and chemical energy systems), Las Vegas, Nevada, 4–7 Mar 2008
18. Zhang Q-C, Yin Y, Mills DR (1996) Sol Energy Mater Sol Cells 40:43–53
19. Sohon JH, Bucher E (1996) Sol Energy Mater Sol Cells 43:59–65
20. Lux-steiner MCh, Kirchner R, Liebemann E, Bucher E (1994) Sol Energy Mater Sol Cells 33:453–464
21. Sohon JH, Binder G, Bucher E (1994) Sol Energy Mater Sol Cells 33:403–65
22. Zhang Q-C, Kelly JC, Mills DR (1991) Appl Opt 30:13
23. Niklasson GA, Granqvist CG (1981) Sol Energy Mater 59:173–180
24. Niklasson GA, Granqvist CG (1984) J Appl Phys 55:9
25. Nejati MR, Fathollahi V, Asadi MK (2005) Sol Energy 78:235–241
26. Farooq M, Lee ZH (2003) Renew Energy 28:1421–1431
27. Yue S, Yueyan S, Fengchun W (2003) Sol Energy Mater Sol Cells 77:393–403
28. Liu Y, Wang C, Xue Y (2012) Sol Energy Mater Sol Cells 96:131–136
29. Du M, Hao L, Liu X, Jiang L, Wang S, Lv F, Li Z, Mi J (2011) Phys Procedia 18:222–226
30. Niklasson GA, Granqvist CG (1982) Appl Phys Lett 41:8
31. Zhang Q-C, Mills DR (1992) Appl Phys Lett 60(5):545–547
32. Zhang Q-C, Shen YG (2004) Sol Energy Mater Sol Cells 81:25–37
33. Zhang Q-C (1998) Sol Energy Mater Sol Cells 52:95–106
34. Niklasson GA (1988) Sol Energy Mater 17:217–226
35. Barshilia HC, Selvakumar N, Rajam KS (2008) J Appl Phys 103:023507
36. Barshilia HC, Kumar P, Rajam KS, Biswas A (2011) Sol Energy Mater Sol Cells 95:1707–1715
37. Farooq M, Raja IA (2008) Renew Energy 33:1275–1285
38. Farooq M, Hutchins MG (2002) Sol Energy Mater Sol Cells 71:73–83
39. Zhang QC, Zhao K, Zhang B-C, Wang L-F, Shen Z-L, Zhou Z-J, Lu D-Q, Xie D-L, Li B-F (1998) Sol Energy 64(1–3):109–114
40. Antonania A, Castadlo A, Addonizio ML, Esposito S (2010) Sol Energy Mater Sol Cells 94:1604–1611
41. Esposito S, Antonaia A, Addonozio ML, Aprea S (2009) Thin Solid Film 517:6000–6006
42. Erben E, Tlhanyl BA (1984) Ind Eng Chem Prod Res Dev 23:659–661
43. Teixeira V, Sousa E, Costa MF, Nunes C, Rosa L, Carvalho MJ, Collares-Pereira M, Roman E, Gago J (2002) Vacuum 64:299–305
44. Jaworske DA, Shumway DA (2003) Solar selective coatings for high temperature applications. In: CP654 Space technology and applications international forum, STAIF
45. Zhang QC (2000) Sol Energy Mater Sol Cells 62:63–74
46. Zhang QC, Mills DR (1992) Sol Energy Mater Sol Cells 27:273–290
47. Zhang QC, Kelly JC, Mills DR (1990) J Appl Phys 68(9):4788–4794
48. Fan JCC, Zavracky PM (1976) Appl Phys Lett 29(8):478–480
49. Mckenzie DR (1979) Appl Phys Lett 34(1):25–28
50. Nyberg GA, Buhrman RA (1982) Appl Phys Lett 40(2):129–131
51. Rebouta L, Pitaes A, Andritschky M, Capela P, Cerqueira MF, Matilainen A, Pischow K (2011) Surf Coat Technol 211:41–44

52. Du M, Hao L, Mi J, Lv F, Liu X, Jiang L, Wang S (2011) Sol Energy Mater Sol Cells 95:1193–1196
53. Xinkang D, Cong W, Tianmin W, Long Z, Buliang C, Ning R (2008) Thin Solid Films 516:3917–3977
54. Vien TK, Sella C, Lafait J, Berthier S (1985) Thin Solid Film 126:17–22
55. Segaud JP, Drevillon B, Pascual E, Ossikovski R, Monnier G, Rimbourg L (1993) Thin Solid Films 234:503–507
56. Craighead HG, Howard RE, Sweeney JE (1981) Appl Phys Lett 39(1)
57. Nuru ZY, Arendse CJ, Nemutudi R, Nemraoui O, Maaza M (2011) Phys B
58. Moller T, Honicke D (1998) Sol Energy Mater Sol Cells 54:397–403
59. Kunic R, Mihelcic M, Orel B, Perse LS, Bizjak B, Kovac J, Brunold S (2011) Sol Energy Mater Sol Cells 95:2965–2975
60. Kozelj M, Vuk AS, Jerman I, Orel B (2009) Sol Energy Mater Sol Cells 93:1733–1742
61. Orel B, Spreizer H, Perse LS, Fir M, Vuk AS, Merlini D, Vodlan M, Kohl M (2007) Sol Energy Mater Sol Cells 91:93–107
62. Xiao X, Miao L, Xu G, Lu L, Su Z, Wang N, Tanemura S (2011) Appl Surf Sci 257:10729–10736
63. Cindrella L (2007) Sol Energy Mater Sol Cells 91:1898–1901
64. Tharamani CN, Mayanna SM (2007) Sol Energy Mater Sol Cells 91:664–669
65. Barshilia HC, Kumar P, Rajam KS, Biswas A (2011) Structure and optical properties of Ag–Al_2O_3 nanocermet solar selective coatings prepared using unbalanced magnetron sputtering. Sol Energy Mater Sol Cells 95:1707–1715
66. Yin Y (2007) Nanocomposite thin films for solar energy conversion. In: Zhang S (ed) Nanocomposite thin films and coatings. Imperial College Press, London, pp 381–414
67. Oelhafen P, Schuler A (2005) Nanostructured materials of solar energy conversion. Sol Energy 79:110–121

Chapter 16
Performance Analysis of IOCL Rawara Photovoltaic Plant and Interpretation

Anurag, Piyush Kapoor, Belal Usmani and S. Harinipriya

Abstract Solar Energy intensity varies geographically in India, but western Rajasthan receives the highest annual radiation energy. The present work will describe the performance parameters of IOCL Rawara Plant which have capacity of 5 MW. The technology used is polycrystalline silicon. The generated power is fed to the 32 kV grid. The data have been recorded from October 2012 to May 2013. The AC power generated by different blocks is plotted against hours of a day, and also, wind speed and module temperature are taken into account. The accumulated energy of all operating month has been recorded.

Keywords Solar energy · AC power · Performance

16.1 Introduction

In order to discuss the potential for solar in western Rajasthan it is desirable to examine the climate condition of that region. The variables of importance from the point of view of generating solar generating solar electricity are solar radiation, sunshine hours, and ambient temperature. Western Rajasthan experiences a hot and dry climate with a sever summer, relatively clear skies, a short monsoon period and a cold winter [1]. A photovoltaic (PV) plant is made of number of PV modules, power inverters, energy meters and distribution boxes. The DC output form PV modules goes through Power Conditioning Unit (PCU) where DC convert into AC. The generated AC power can be used for appliances and sometimes fed to the grid.

Anurag · B. Usmani
Center of Excellence in Energy, Indian Institute of Technology,
Jodhpur 342011, Rajasthan, India

P. Kapoor · S. Harinipriya (✉)
Center of Excellence in System Science, Indian Institute of Technology,
Jodhpur 342011, Rajasthan, India
e-mail: shpriya@iitj.ac.in

V. Vijay et al. (eds.), *Systems Thinking Approach for Social Problems*,
Lecture Notes in Electrical Engineering 327,
DOI 10.1007/978-81-322-2141-8_16

For selecting a photovoltaic plant location, initially parameters are defined and analyzed. The essential parameter being solar irradiation, and other may be wind speed, atmospheric temperature etc. Here study of a 5 MW crystalline PV plant of IOCL had been done.

16.2 Analysis of Data

The main parameters which were analyzed are total AC generation from different blocks, module temperature, ambient temperature, wind speed, and accumulated energy. The major part of AC generation was found in peak hours or sunny hours (9.30 a.m. to 3.30 p.m.). The following curves show the variation in AC power for different months (Figs. 16.1 and 16.2).

The maximum AC generation is found in noon and minimum generation found in morning and evening hours. More fluctuations are seen in the DC generation of April month. It could be attributed to the dust storm and pre-monsoon activities. The other parameters such as wind speed, module temperature, ambient temperature, and solar insolation were also plotted for different months. It is observed that

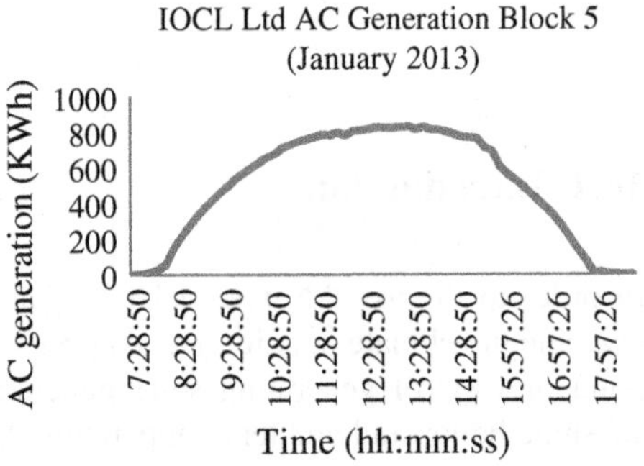

Fig. 16.1 AC generation of January

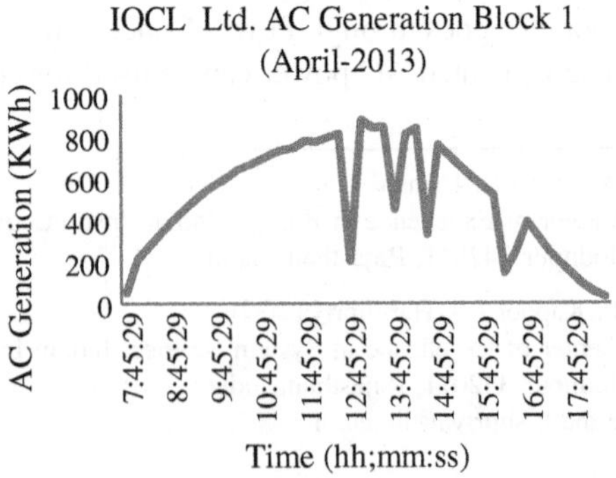

Fig. 16.2 AC generation of April

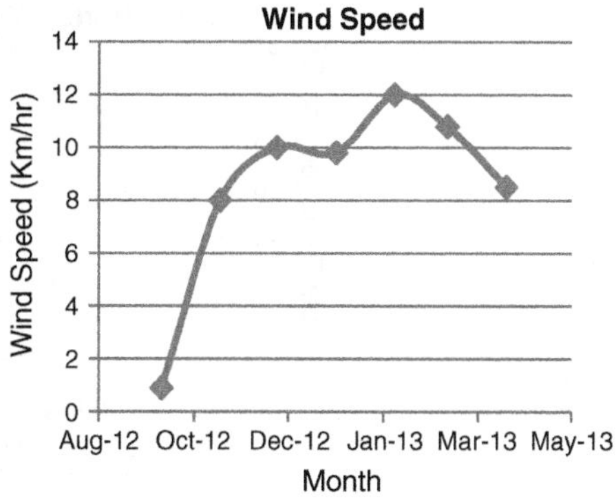

Fig. 16.3 Wind speed for different months

higher the wind speed, lower the module temperature. The wind speed profile is shown in Fig. 16.3.

Module temperature depends upon factors such as the module encapsulating material, thermal dissipation, and absorption properties of working point of the module, and atmospheric parameters such as irradiance level, ambient temperature, wind speed, and particular installing conditions [2]. The module temperature is recorded for a particular module for a complete month and then averaged. The trajectories of both module and ambient temperature curves are nearly same except for deviation of 2–5 °C. The minimum temperature is recorded during December and January. The maximum temperature is obtained in the month of April. The profile of ambient temperature and module temperature is shown in Figs. 16.4 and 16.5, respectively. The solar irradiance is measured to be maximum in the month of December. This high irradiance in winter is mainly attributed to the fact that Rajasthan state receives with 300–330 clear sunny days and average daily solar incidence of 5–7 kWh/m^2 [3].

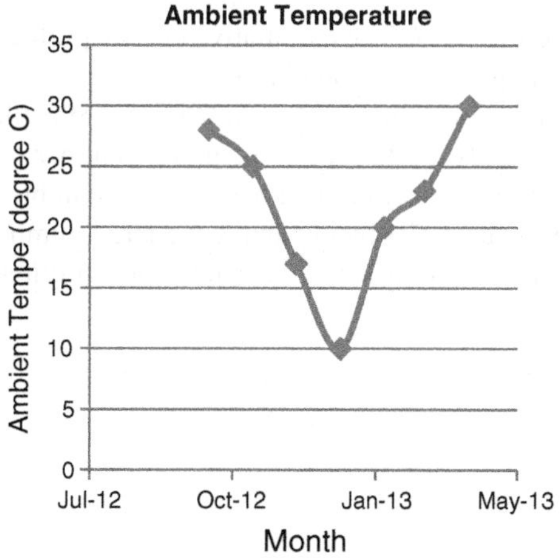

Fig. 16.4 Ambient temperature for different months

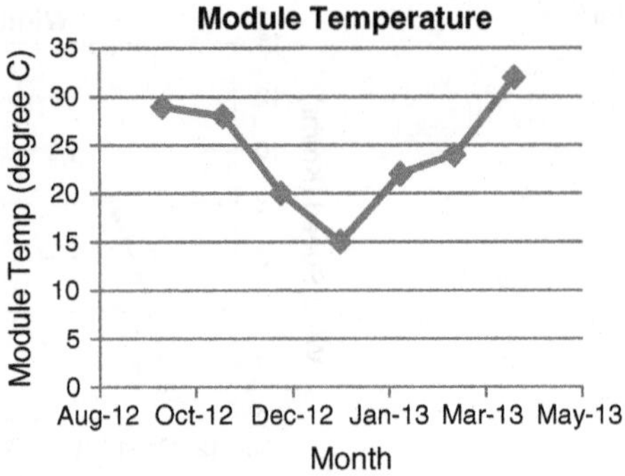

Fig. 16.5 Module temperature for different months

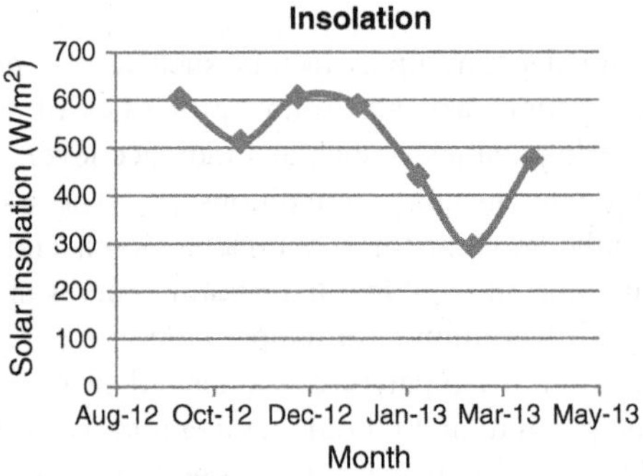

Fig. 16.6 Solar insolation for different months

The insolation curve is shown in Fig. 16.6. The two highest points are seen in curve in the month of December and April, which shows irradiance 607 and 490 W/ m^2, respectively. The monthly comparison of AC generation of all blocks is shown in Fig. 16.7. The cumulative energy profile is plotted against time in Fig. 16.8. The energy curve is found linear, as energy is the time multiplication of power. The total energy reaches up to 25,824,795 kWh till the end of May. The AC generation depends on the size of plant say number of modules installed in a specific area. In current scenario, plant has five different blocks, each block with different number of modules. The blocks that have large number of modules generate large AC power. The variation of AC generation from different blocks is shown in Fig. 16.7.

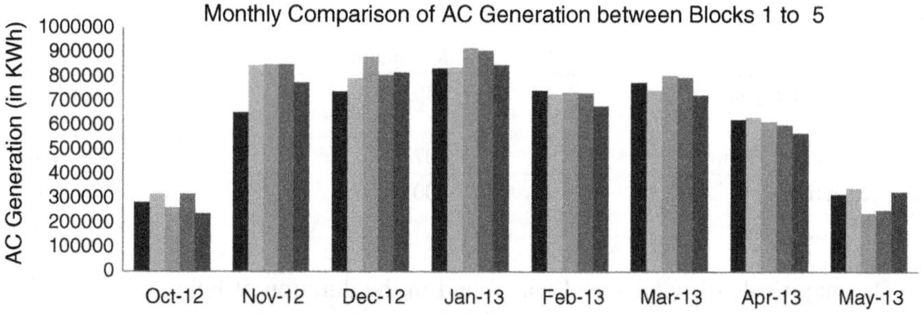

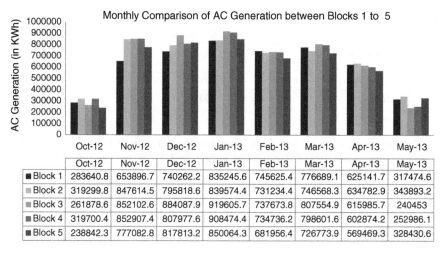

	Oct-12	Nov-12	Dec-12	Jan-13	Feb-13	Mar-13	Apr-13	May-13
Block 1	283640.8	653896.7	740262.2	835245.6	745625.4	776689.1	625141.7	317474.6
Block 2	319299.8	847614.5	795818.6	839574.4	731234.4	746568.3	634782.9	343893.2
Block 3	261878.6	852102.6	884087.9	919605.7	737673.8	807554.9	615985.7	240453
Block 4	319700.4	852907.4	807977.6	908474.4	734736.2	798601.6	602874.2	252986.1
Block 5	238842.3	777082.8	817813.2	850064.3	681956.4	726773.9	569469.3	328430.6

Fig. 16.7 AC generation of different blocks for different months

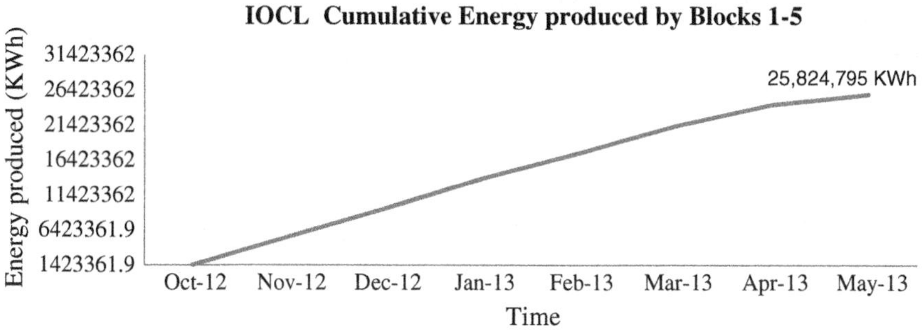

Fig. 16.8 Total energy produced for different months

16.3 Conclusion

The observations are made by data and graphs. The module temperature is slightly higher than ambient temperature. The maximum wind speed was 12 km/h in the month of February. Other parameters are listed in the table below.

Parameter	Maximum	Month
Wind speed (km/h)	12	February
Module temperature (°C)	32	April

(continued)

(continued)

Parameter	Maximum	Month
Ambient temperature (°C)	30	April
Solar insolation (W/m^2)	607	December
AC generation (MWh)	900	January

The major role of wind speed was found in the duration of February–June. The average wind speed in the duration of February–June is greater than rest of the months.

References

1. Sukhatme SP, Nayak JK (1997) Solar energy in western rajastnan. Currt sci 72(1)
2. Alonso Garci'a MC, Balenzategui JL (2004) Estimation of phoroviltaic module yearly temperature and performance based on nominal operation cell temperature calculations. Science Direct publication
3. http://eng.riico.co.in/upload/Solar-Energy.pdf

Chapter 17
Wearable Cardiac Detector

R. Harshitha, Manasa Manohar, P. Dhanya, P. Manoj, S. Swathi, M. Amogh, Viswanath Talasila, H.S. Jamadagni and B.S. Nanda Kumar

Abstract India has the highest incidence of heart-related diseases in the world. The early detection of various heart conditions can be a significant aid for immediate medical intervention. In rural communities, access to healthcare facilities is limited, and early detection of diseases is difficult. For the aged in urban areas, it is desirable to have portable devices to monitor and diagnose such diseases. Here, we propose the development of a wearable cardiac detector (WCD) which is designed to detect four kinds of arrhythmia—ventricular tachycardia, ventricular bradycardia, myocardial infarction, and hypertrophy. The detection includes a signal processing system based on Pan–Tompkins algorithm which detects the QRS complexes of electrocardiogram (ECG) signals, as well as standard methods to filter noise present in the ECG signals. The proposed WCD is portable, cheap, and lightweight unlike the widely used Holter monitor which is bulky, expensive, and cannot be worn all the time. It is specially designed for patients who have had a surgery and are at potential risk of relapse. It allows heart risk patients not to be restricted to the hospital. The device sends alarm messages about the location of the patient and stores the ECG signal data for further analysis. The proposed WCD is part of a long-term initiative to enhance the medical facilities to people from all walks of life and hopeful to be beneficial to the society. The effort is a joint collaboration along

R. Harshitha (✉) · P. Manoj · S. Swathi · M. Amogh · V. Talasila
Department of Telecommunication Engineering, MS Ramaiah Institute of Technology, Bangalore, India
e-mail: harshitha.ramamurthy@gmail.com

M. Manohar
Department of Instrumentation Technology, MS Ramaiah Institute of Technology, Bangalore, India

P. Dhanya
Department of Computer Science Engineering, MS Ramaiah Institute of Technology, Bangalore, India

H.S. Jamadagni
Department of Electronic Systems Engineering, IISC, Bangalore, India

B.S.N. Kumar
Community Medicine, MS Ramaiah Medical College, Bangalore, India

V. Vijay et al. (eds.), *Systems Thinking Approach for Social Problems*,
Lecture Notes in Electrical Engineering 327,
DOI 10.1007/978-81-322-2141-8_17

with the Community Medicine and Cardiac Specialty departments of MS Ramaiah Medical College.

Keywords Wearable arrhythmia detector · Portable cardiac monitor and detector · Ventricular tachycardia · Ventricular bradycardia · Myocardial infraction · Hypertrophy · Pan–Tompkins algorithm · Einthoven's triangle · QRS complex of ECG

17.1 Introduction

It has been estimated by a leading daily, the Indian Express that by 2020, 40 % of the deaths in India will be caused by cardiovascular diseases. With over 3 million deaths owing to cardiovascular diseases every year, India is set to be the 'heart disease capital of the world' in few years, said doctors on the eve of World Heart Day (September 29).

Owing to this, there is a need for round-the-clock monitoring and storage of the ECG signal. The components of a typical ECG signal are shown in Fig. 17.1.

It corresponds to the depolarization of the right and left ventricles of the human heart. In adults, it normally lasts 0.06–0.10 s; in children and during physical activity, it may be shorter. Typically, an ECG has five deflections, arbitrarily named 'P' to 'T' waves. The 'Q,' 'R,' and 'S' waves occur in rapid succession, do not all appear in all leads, and reflect a single event and thus are usually considered together. A 'Q' wave is any downward deflection after the 'P' wave. An 'R' wave

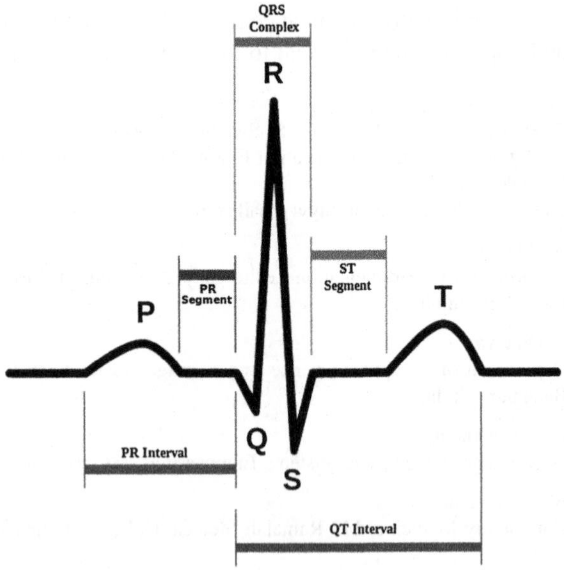

Fig. 17.1 Components of a typical ECG signal

follows as an upward deflection, and the 'S' wave is any downward deflection after the 'R' wave. The 'T' wave follows the 'S' wave, and in some cases, an additional 'U' wave follows the 'T' wave. The ECG data used in this project are obtained from MIT-BIH Arrhythmia Database [1]. The MIT-BIH Arrhythmia Database has been obtained from 47 subjects at Boston's Beth Israel Hospital (now the Beth Israel Deaconess Medical Center) studied by the BIH Arrhythmia Laboratory between 1975 and 1979. The recordings here are digitized at 360 samples per second per channel with 11-bit resolution over a 10 mV range. Doctors all over the world use this as their reference. We have also verified the data obtained from MIT-BIH with the doctors from Bangalore.

17.2 Objectives

We aim to classify the ECG signals based on the widely acclaimed Pan–Tompkins algorithm which provides a success rate of 99.3 % [2]. For the purpose of software simulation, data from MIT-BIH are used. The device we intend to build will monitor the ECG signals round the clock. Its lightweight, portable, and robust characteristics make it more user-friendly. This device is different from other ECG monitoring devices in that it goes one step ahead from automatically detecting arrhythmia to drawing attention to the patients' distress immediately via the communication module. This ensures that the patient does not lose valuable minute.

17.3 Synopsis

This project deals with the study and analysis of ECG signals by utilizing MATLAB software effectively. ECG signal is acquired from the electrodes placed at strategic points of the body explained in the subsequent paragraphs. ECG signal analysis includes acquisition of real-time ECG data, noise reduction, and signal processing which includes feature extraction from ECG signals, calculating the heart rate based on which the arrhythmias are detected. All the while, ECG data are stored in the device for future analysis by the cardiologist.

On detection of an arrhythmia, a message is sent to the nearest point of contact to ensure that medical attention reaches the patient within a matter of minutes. This is achieved by means of the communication system based on a global system for mobile (GSM) module.

17.4 Block Diagram

Figure 17.2.

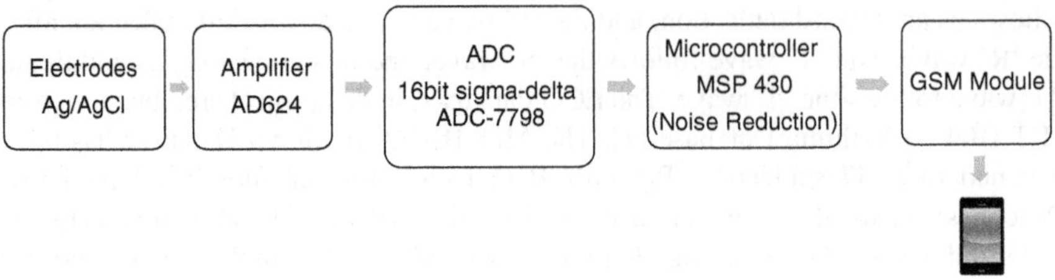

Fig. 17.2 Block diagram

17.5 Block Breakdown

17.5.1 Electrodes

We intend to use silver–silver chloride (Ag–AgCl) electrode [3], and it consists of base lining material, conductive gel, and electrode buckle. In order to record the ECG signal, a transducer capable of converting the ionic potentials generated within the body into electrical potential which can be measured using electronic instruments is required.

Ag–AgCl electrodes most closely approximate as a non-polarizable electrodes which behave as resistor. Ag–AgCl electrode is preferred over the electrodes for two reasons. Firstly, when bought in contact with an electrolyte, Ag–AgCl establishes a stable contact potential. It cannot form a compound on metal surface consisting of metal ions and electrolyte anions. Secondly, the electron exchange reaction in Ag–AgCl can continue indefinitely, until the AgCl is depleted, and hence, it is referred as non-polarizable. Since QRS signals are of the order of mV, these electrodes have a contact potential which will have sufficient time to stabilize the negligible current ($<10^{-12}$ A).

Hence, this is best suited for the measurement of biopotentials. The placement of these electrodes is explained below:

Lead I Positive electrode is on the left arm, and negative electrode is on the right. It measures the potential difference across the chest between two arms.

Lead II Positive electrode is on the left leg, while the negative electrode is on the right arm.

Lead III Positive electrode is on the left leg, and the negative electrode is on the left arm.

This positioning is known as 'Einthoven's triangle' [4]. Information gathered between these leads is known as bipolar. When the accuracy of 3-lead positioning is compared to the 12-lead positioning, the former leads to 97 % accuracy, whereas the latter results in 91 % accuracy. Since our product is aimed at ambulatory patients, 3-lead positioning is generally preferred to the laborious 12-lead attachment (Fig. 17.3).

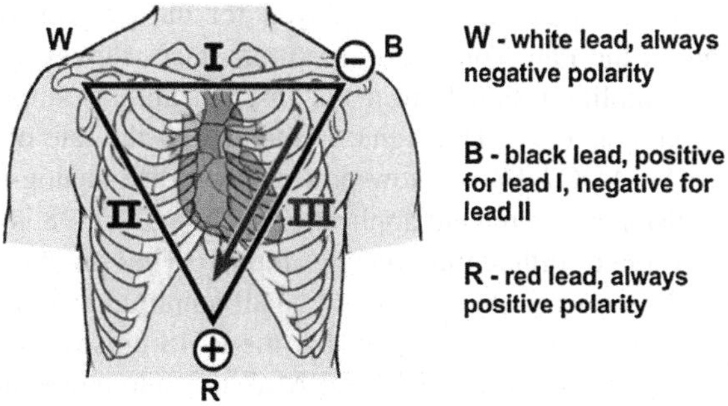

Fig. 17.3 Einthoven's triangle

17.5.2 Instrumentation Amplifier

ECG signals are generally weak signals of order 1–4 mV, which makes it difficult for the analysis by signal conditioning circuit. In addition to this, the 3-lead system we use in this device will provide us with a signal which will be distorted with various types of noise. The following characteristics have to be fulfilled by the amplifier we use:

- *Gain*: As mentioned earlier, it is required that the signal is amplified to an order of 1 V for signal analysis for which an amplifier of gain 1000 is needed.
- *Common-mode rejection ratio (CMRR)*: The signal from the electrodes will be distorted by a large ac common-mode component (up to 1.5 V) and a large variable dc component (300 mV). As per the Association for the Advancement of Medical Instrumentation (AAMI), 89 dB is the minimum CMRR required for standard ECG.
- *Input impedance*: To achieve low-power input and a good frequency response, we require high input impedance.

The AD624, which is a high-precision and low-noise amplifier, satisfies all of the above requirements. Input noise is typically less than 4 nV/$\sqrt{\text{Hz}}$ at 1 kHz [5].

17.5.3 Filtering

For ECG recordings concurrent with acceptable medical standards, signal acquisition must be noise free. The signal acquisition is susceptible to the interference from other biological and environmental sources.

The main sources of noise in ECG are as follows:

1. Baseline wander (low-frequency noise)
2. Power line interference (50 Hz or 60 Hz noise from power lines)
3. Muscle noise (This noise is very difficult to remove as it is in the same region as the actual signal. It is usually corrected in software.)
4. Other interference (i.e., radio frequency noise from other equipment)

A band-pass filter is employed for the purpose of noise reduction. We have chosen the Pan–Tompkins algorithm since it gives a success rate of 99.3 % [PT]. The amplified signal from the instrumentation amplifier is fed to an analog-to-digital convertor. The signal is sampled at the rate of 200 samples/s.

The ADC7798 is a low-power, low-noise analog-to-digital convertor for high-precision measurement applications. The ADC7798 is a 16/24-bit sigma-delta ADC with three differential analog inputs. The on-chip, low-noise instrumentation amplifier means that signals of small amplitude can be interfaced directly with the ADC. With a gain setting of 64, the RMS noise is 40 nV for the ADC7798 when the update rate equals 4.17 Hz. Also, the output data rate is software programmable and can be varied from 4.17 to 470 Hz [6].

The QRS detection is done in three stages. Stage 1: Noise Reduction

An integer coefficient band-pass filter is implemented by cascade of low-pass and high-pass filters whose pass band should be between 5 and 15 Hz. The filter is a fast, recursive filter in which poles are located to cancel zeros in the unit circle in the z-plane. This approach gives us the filter coefficients of the filter which can be implemented by means of a microprocessor. For practical purposes, a filter with a 3-dB pass band of 5–12 Hz is realized.

The transfer function for the second-order low-pass filter is as follows:

$$|H_a(j\Omega)|^2 = 1/1 + (j\Omega/j\Omega_c)^{2N} \quad (17.1)$$

The transfer function for the high-pass filter is as follows:

$$|H(k)|^2 = 1/1 + (\Omega_c/\Omega)^{2N} \quad (17.2)$$

To obtain the QRS complex slope information, a five-point derivative is used whose transfer function is as follows:

$$H(z) = (1/8T)\left(-z^{-2} - 2z^{-1} + 2z^{1} + z^{2}\right) \quad (17.3)$$

This is followed by the squaring of the signal, point by point, using the equation:

$$y(nT) = [x(nT)]^2 \quad (17.4)$$

In the end, we use a moving window integrator filter whose window width is 30 samples to determine the R slope information. The width of the window should be such that it should not be too wide or too narrow to avoid missing important information about the signal.

$$y(nT) = (1/N)[x(nT - (N-1)T) + x(nT - (N-2)T) + \cdots + x(nT)] \quad (17.5)$$

The output of the integrator block will have the noise-free signal which can be used for signal analysis.

17.5.4 MSP430F2013 Microcontroller

A very low-power, cheap, reliable, and relatively feature full microcontroller unit (MCU) was required. After much analysis, the MSP430F2013 MCU by Texas Instruments was selected to perform the core functions. MSP430 ensures that application chooses appropriate clock and peripherals that are needed to perform a particular task. There are several features that make MSP430 suitable for low-power and portable applications which are hardly present in 8051 microcontroller. It has six operating modes including the active working mode. Whenever a particular peripheral is not functional, it shifts to low-power mode and facilitates power consumption. The CPU is disabled, but the peripherals that are required remain active. No change in RAM content will be allowed. This feature is absent in the other microcontrollers. MSP430 makes use of reduced instruction set computing (RISC) processor and provides higher performance and quicker execution than the 8051 counterparts. The major reason for choosing MSP is the presence of an inbuilt ADC and DAC, which is not present in 8051. It converts the analog ECG signal into digital format for comparison with the threshold values. It has options of 10/12-bit successive approximation register or 16/24-bit sigma-delta ADC. We utilize sigma-delta feature as it employs the error feedback mechanism which measures the difference between two signals and results in better conversion with high precision and high fidelity. The output of the microcontroller needs to be interfaced with the GSM module. The microcontroller has serial peripheral interface (SPI), interintegrated circuit (I2C), and universal serial communication interface (USCI) components on chip [7].

MSP430 adopts four-wire JTAG and Spy Bi wire interface as opposed to the aging RS-232 present in the 8051. It is imperative to utilize a foolproof controller which has an inbuilt watchdog timer which notifies a hardware fault or software error and prevents system from collapsing. MSP430 has an inbuilt watchdog timer and chosen as an ideal controller for our project for all the above-mentioned features.

17.5.5 Signal Analysis

1. *For calculating heart rate*: This follows the Pan–Tompkins algorithm.
   QRS wave is isolated, and the signal is passed to the thresholding phase. Two sets of thresholds are used to detect QRS complexes. The first set thresholds the filtered ECG, while the second threshold is used on the output of the moving window integrator. This increases the reliability of detection. The algorithm uses a dual-threshold technique to find missed beats and thereby reduces false negatives. There are two separate threshold levels in each of the two sets of thresholds. One level is half of the other. The thresholds continuously adapt to the characteristics of the signal since they are based upon the most recent signal and noise peaks that are detected in the ongoing processed signals. If the

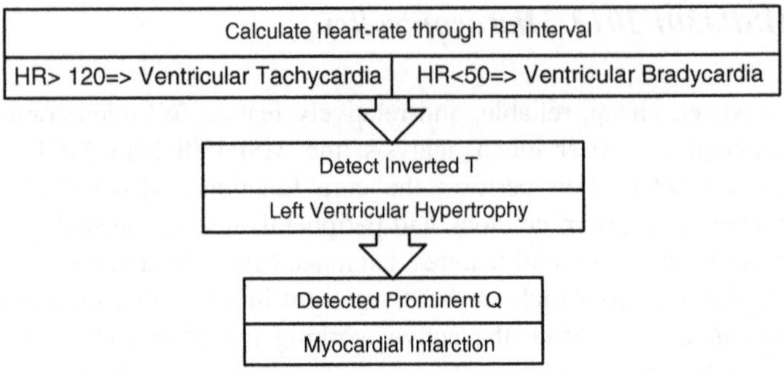

Fig. 17.4 Detection of arrhythmia

program does not find a QRS complex in the time interval corresponding to 166 % of the current average RR interval, the maximal peak detected in that time interval that lies between these two thresholds is considered to be a possible QRS complex, and the lower of the two thresholds is applied.

2. *For inverted T and Prominent Q detection*: For the purpose of diagnosis, we often need to extract various features from the preprocessed ECG data, including QRS intervals, QRS amplitudes, PR intervals, and ST intervals. Inverted T indicates left ventricular hypertrophy, while prominent Q denotes myocardial infarction (heart attack) (Fig. 17.4).

17.5.6 GSM Module

The GSM module SIM 900 is interfaced with MSP430 to send an SMS to an emergency service or ambulance. Connections are made using the universal asynchronous receiver transmitter (UART) protocol. The microcontroller is coded to send AT commands over the UART port. Predefined number for the SMS is coded previously. A GSM phone can also be used instead of a dedicated GSM module. GSM was chosen owing to its features such as reliability and user adaptability (Fig. 17.5).

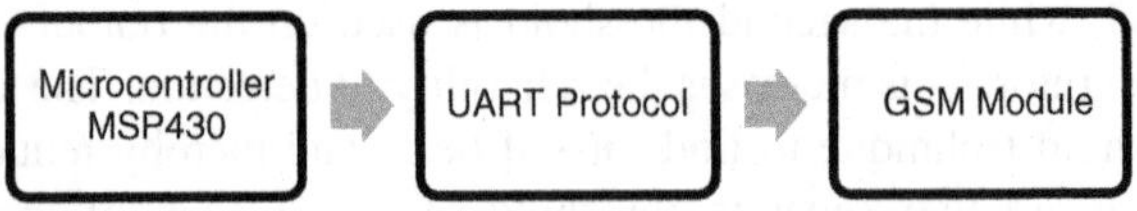

Fig. 17.5 GSM module

17.5.7 Additional Feature

On occasion, the arrhythmia may be such that the patient may require blood transfusion within a stipulated time period. It is useful to have the device enhanced with the ability to contact a blood bank or suitable donors as well. To this end, the contact numbers of a blood bank and a few matching donors can also be stored. On purchase of the device, the blood group of the user is determined and appropriate donor numbers are loaded. A button is added on the device which when pressed will send messages to the blood bank and the donors. Thus, both and patient and the donor will reach the hospital together, thereby saving precious time.

17.6 Feasibility

- *Cost effective*: Our product is 10 times cheaper than the existing Holter monitor
- *Communication*: Send message to call for help, detect the location of the patient, and proper medical intervention is received.
- *Portable*: Device can be worn beneath the clothes, while the patient is on the move.
- *Easily accessible*: Components have been used, and all components used are safe for the patients.
- *User-friendly*: Easy usage for patients of all ages.

17.7 Results

Figures 17.6 and 17.7

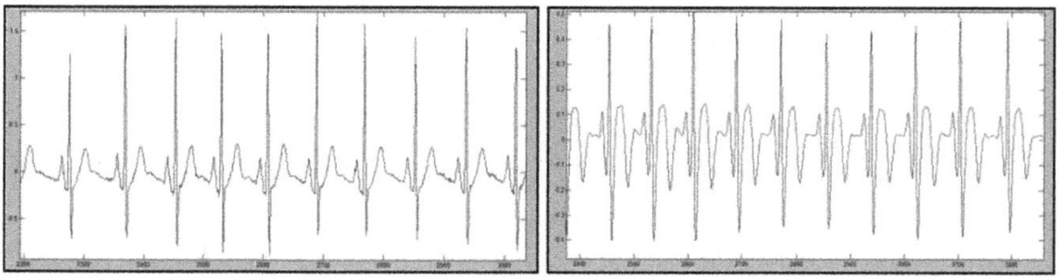

Fig. 17.6 Original ECG signal and filtered ECG signal

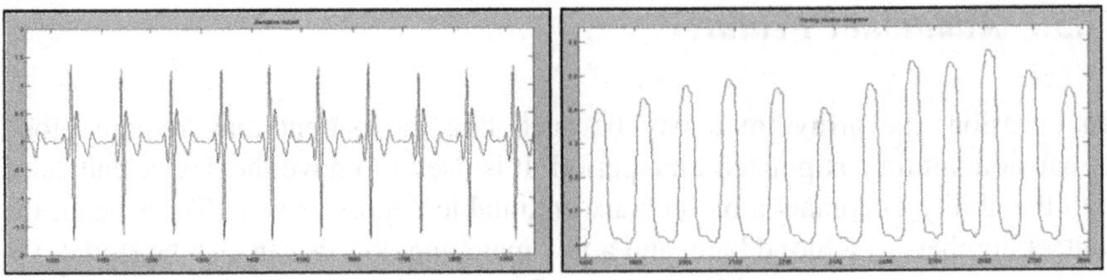

Fig. 17.7 Derivative output and moving window integral output

References

1. MIT BIH Database
2. Pan J, Tompkins WJ (1985) A Real-time QRS detection algorithm. IEEE Trans Biomed Eng 32 (3):230
3. Townsend N (2001) Medical electronics. IEEE Eng Med Biol 20(3):45–50 (Michaelmas tern). http://www.robots.ox.ac.uk/~neil/teaching/lectures/med_elec/lecture2.pdf
4. Brugada P, Brugada J, Mont L, Smeets J, Andries EW (1991) A new approach to the differential diagnosis of a regular tachycardia with a wide QRS complex. Circulation 8:1649 (Brugada criteria)
5. Singh S, Netaji GN (2010) Pattern analysis of different ECG signals using Pan–Tomkins Algorithm. Int J Comput Sci Eng 2:2502
6. http://www.analog.com/static/imported-files/data_sheets/AD7798_7799.pdf
7. Chavan MS, Agarwala RA, Uplane MD (2008) Suppression of baseline wander and powerline interference in ECG using digital IIR filters. Int J Circuits Syst Signal Process 2:356

Chapter 18
LQR-Based TS-Fuzzy Logic Controller Design for Inverted Pendulum-Coupled Cart System

Bharat Sharma and Barjeev Tyagi

Abstract In this paper, an effective design technique for heuristic Takagi-Sugeno fuzzy logic controller (TS-FLC) for nonlinear inverted pendulum (IP) and cart system has been proposed. IP is linearized around distinct combinations of localized points and their respective linear quadratic regulator (LQR) gains are obtained. Set of these localized points are used to decide the range of input fuzzy membership function, and the LQR controller gains are used to obtain basic TS rule base for nonlinear model. Angle and angular velocity are used to design the controller for upright stabilization of pendulum. Cart position and cart velocity are the inputs for cart control. The main aim is to control pendulum in upright unstable equilibrium point and cart position at desired value simultaneously. Physical constraints of the system such as cart track length and controller output are considered in the designing of FLC. The results obtained by FLC are compared with LQR. The results show that FLC is better than LQR because it can be further tuned to satisfy the constraints. Simulation results show the effectiveness and robustness of proposed TS-FLC over LQR controller.

Keywords TS-FLC · LQR · Inverted pendulum

18.1 Introduction

Inverted pendulum (IP) is naturally unstable, nonlinear, strong coupling, and high-order single input multiple output system. The International Federation of Automatic Control considered it as "benchmark control problem" for design,

B. Sharma (✉) · B. Tyagi
Electrical Engineering Department, Indian Institute of Technology Roorkee, Roorkee, India
e-mail: bharatsharma20@hotmail.com

B. Tyagi
e-mail: btyagfee@iitr.ernet.in

V. Vijay et al. (eds.), *Systems Thinking Approach for Social Problems*, Lecture Notes in Electrical Engineering 327,
DOI 10.1007/978-81-322-2141-8_18

implementation, and comparison of novel control techniques and theories [1]. Physical system based upon concepts of IP are pedubots, seagways, missile guidance, space rocket, earthquake tolerant structures, gait pattern generation of bipedal dynamic walking, aircraft stabilization in turbulent airflow and its landing system, etc [2]. Inverted pendulum-coupled cart system (IPCS) is a subclass of this domain. Additionally, constraints on track length and actuator output constraint further increase the complexity of controller implementation.

Many control techniques are proposed so far for IPCS such as bang-bang control, neural network control, adaptive PID control, energy-based control, sliding mode control, predictive control, and many other are reported in [3]. These techniques are not novel but not in use previously.

FLC is among popular control methodologies. Because it is simple, flexible and grant assertive low cost conclusion from vague, imprecise and ambiguous input information. Originally, Zadeh [4] introduced the concept of fuzzy logic theory. Later, Mamdani and Assilian [5] successfully implemented FLC for control of steam engine. There is no need of plant dynamic equations and modeling because Mamdani FLC's (MFLC) consequent part is also fuzzified and rule base is established according to expert's knowledge and experience. Hence, intuitive MFLC is cost effective, simple, and easy solution to complex systems whose conventional analytical solution is very difficult to obtain if not impossible. But the reliability of MFLC is often controversial because critics have been dubious over its ability to foresee the closed-loop instabilities and limit cycles especially in critical processes such as power industries, which involved risk to human life and/or to environment. Takagi and Sugeno [6] suggested impressive amalgamation of available control approaches to FLC in order to make it more adaptive, robust to nonlinearities, and disturbances. It is more compact and computationally efficient than MFLC and assured continuity of control surface. But precise and substantial design procedure is not suggested.

Various hybrid TS-FLCs are available in literature for IPCS such as sliding mode control TS-FLC switching for construction of linear aggregated system [7, 8], embodiment of GA for optimization and tuning of parameters for automatic design of TS-FLC [9–12], feedback linearization using TS-FLC [13], PID FLC [14], and Fuzzy linear quadratic regulator (LQR) [15]. But these techniques have not considered physical constraints of system.

This paper proposed systematic method to obtain nonlinear compensator by interweaving the locally partitioned LQRs along with consideration of constraints such as track length and actuator output limit. LQR is obtained by linearizing the model around desired points.

The paper is organized as follows: Section 18.2 deals with the mathematical dynamic models of the system for the controller design and simulation evaluation. Section 18.3 gives brief overview of FLC. Section 18.4 goes through the main steps in the design of the control algorithms. In this work, LQR gain for each rule in FLC is calculated and then incorporated in output membership function. Section 18.5 presents some simulation, its comparative survey thus outlining any discrepancies. Finally, some conclusions are drawn in Sect. 18.6.

18.2 Modeling

The pendulum is mounted over the cart as shown in Fig. 18.1. Cart is used to balance the pendulum upright unstable equilibrium point. The movement of cart is controlled by DC motors attached at the end of tracks pulling through belt. Pendulum rotates about its pivoted point in *x*–*y* direction while cart can move only in horizontal direction. Hence, system has two degree of freedom and can be represented in *x*–*y* coordinates.

Initially, Euler-Lagrange's approach is used to derive nonlinear dynamics of system [16]. Then Lyapunov's method of linearization is applied to obtain generalized linear state space across different points [17]. The considered parameters are tabulated in Table 18.1.

Taking, θ = pendulum angle from vertical position; x = Displacement of Cart from center of track.

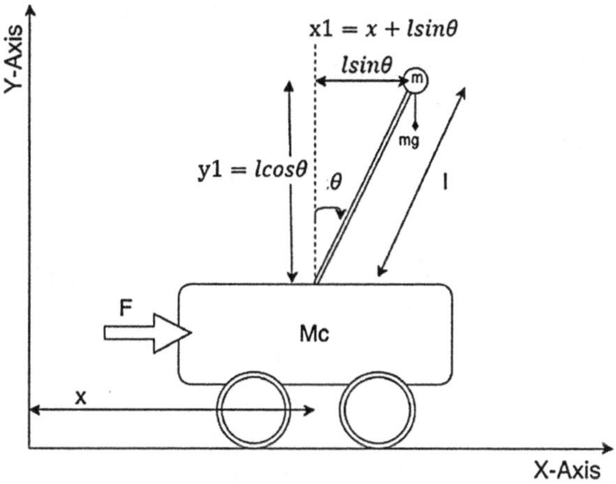

Fig. 18.1 Inverted pendulum system

Table 18.1 Parameters of inverted pendulum

Parameters	Value
g—Gravity	9.81 m/s^2
l—Pole length	0.36 m
M—Cart mass	2.4 kg
m—Pole mass	0.23 kg
I—Moment of inertia of the pole	0.099 kg m^2
b—Cart friction coefficient	0.05 Ns/m
d—Pendulum damping coefficient	0.005 Nm s/rad
x—Track length	±0.5 m
F—Maximum force	±20 N
u—Maximum control voltage	±2.5 V

Net velocity of pendulum is given as:

$$V_1^2 = \dot{x}_1^2 + \dot{y}_1^2 \tag{18.2.1}$$

The kinetic energy of pendulum is given as:

$$K_1 = \frac{1}{2}V_1^2 + \frac{1}{2}I\dot{\theta}^2 \tag{18.2.2}$$

The kinetic energy of cart is given as:

$$K_2 = \frac{1}{2}M\dot{x}^2 \tag{18.2.3}$$

The net kinetic energy of system is given as:

$$K = K_1 + K_2 \tag{18.2.4}$$

Potential of pendulum/system is given as:

$$P = mgl\cos\theta \tag{18.2.5}$$

Net Lagrangian equation:

$$L = K - P \tag{18.2.6}$$

$$L = \frac{1}{2}\left(M\dot{x}^2 + m\dot{x}^2 + 2ml\cos\theta\dot{x} + ml^2\dot{\theta}^2 + I\dot{\theta}^2\right) - mgl\cos\theta \tag{18.2.7}$$

Euler-Langrage's equation for the cart and pendulum is given as:

$$\frac{\mathrm{d}}{\mathrm{d}t}\left(\frac{\partial L}{\partial \dot{x}}\right) - \frac{\partial L}{\partial x} + b\dot{x} = F \tag{18.2.8}$$

$$\frac{\mathrm{d}}{\mathrm{d}t}\left(\frac{\partial L}{\partial \dot{\theta}}\right) - \frac{\partial L}{\partial \theta} + d\dot{x} = 0 \tag{18.2.9}$$

Substituting Eq. (18.2.7) in above functions, we get sum of all moments and forces acting upon system:

$$ml\cos\theta \times \ddot{\theta} + (m + M) \times \ddot{x} + b \times \dot{x} - ml\sin\theta \times \dot{\theta}^2 = F \tag{18.2.10}$$

$$\left(I + ml^2\right) \times \ddot{\theta} + ml\cos\theta \times \ddot{x} + d \times \dot{\theta} - mgl\sin\theta = 0 \tag{18.2.11}$$

$$\Rightarrow \begin{pmatrix} ml\cos\theta & (m+M) \\ (I + ml^2) & ml\cos\theta \end{pmatrix}\begin{pmatrix} \ddot{\theta} \\ \ddot{x} \end{pmatrix} = \begin{pmatrix} -b \times \dot{x} + ml\sin\theta \times \dot{\theta}^2 + F \\ -d \times \dot{\theta} + mgl\sin\theta \end{pmatrix} \tag{18.2.12}$$

$$\Rightarrow \begin{pmatrix} \ddot{\theta} \\ \ddot{x} \end{pmatrix} = \frac{1}{\Delta}\begin{pmatrix} -ml\cos\theta \times \alpha + (I + ml^2) \times \beta \\ (m+M) \times \alpha - ml\cos\theta \times \beta \end{pmatrix} + \frac{1}{\Delta}\begin{pmatrix} -ml\cos\theta \\ m + M \end{pmatrix}F \tag{18.2.13}$$

where

$$\Delta = \left(I(m+M) + m^2l^2\sin^2\theta\right)$$
$$\alpha = \left(-b\dot{x} + ml\sin\theta\dot{\theta}^2\right)$$
$$\beta = \left(-d\times\dot{\theta} + mgl\sin\theta\right)$$

$$\Rightarrow \begin{pmatrix}\ddot{\theta}\\ \ddot{x}\end{pmatrix} = \begin{pmatrix} f_1\left(\theta,\dot{\theta},\dot{x}\right) + g_1(\theta)\\ f_2\left(\theta,\dot{\theta},\dot{x}\right) + g_2(\theta)\end{pmatrix} \quad (18.2.14)$$

Using Lyapunov's linearization method for obtaining linear model of above nonlinear Eq. (18.2.14) for states- $\left(\theta \quad \dot{\theta} \quad x \quad \dot{x}\right)$:

$$\begin{pmatrix}\frac{\partial\ddot{\theta}}{\partial\theta}\\ \frac{\partial\ddot{x}}{\partial\theta}\end{pmatrix} = \begin{pmatrix}a21\\ a41\end{pmatrix} = \frac{1}{\Delta}\begin{pmatrix}\gamma + m^2l^2\sin 2\theta\times f_1\left(\theta,\dot{\theta},\dot{x}\right)\\ \sigma + m^2l^2\sin 2\theta\times f_2\left(\theta,\dot{\theta},\dot{x}\right)\end{pmatrix} \quad (18.2.15)$$

where

$$\gamma = ml\sin\theta\times\beta - m^2l^2g\cos^2\theta + ml\left(I + m^2l^2\right)\cos\theta\dot{\theta}^2$$
$$\sigma = m^2l^2\cos^2\theta\dot{\theta}^2 + ml\sin\theta\times\alpha + m(m+M)gl\cos\theta$$

$$\begin{pmatrix}\frac{\partial\ddot{\theta}}{\partial\dot{\theta}}\\ \frac{\partial\ddot{x}}{\partial\dot{\theta}}\end{pmatrix} = \begin{pmatrix}a22\\ a42\end{pmatrix} = \frac{1}{\Delta}\begin{pmatrix}-d(m+M) - m^2l^2\sin 2\theta\dot{\theta}\\ mld\cos\theta + 2ml\left(I + ml^2\sin\theta\dot{\theta}\right)\end{pmatrix} \quad (18.2.16)$$

$$\begin{pmatrix}\frac{\partial\ddot{\theta}}{\partial F}\\ \frac{\partial\ddot{x}}{\partial F}\end{pmatrix} = \begin{pmatrix}b2\\ b4\end{pmatrix} = \frac{1}{\Delta}\begin{pmatrix}ml\cos\theta\\ (I + ml^2)\end{pmatrix} \quad (18.2.17)$$

Other coefficients are zero. Therefore, state matrix, input matrix, and output matrix are:

$$A = \begin{pmatrix}0 & 1 & 0 & 0\\ a21 & a22 & 0 & a24\\ 0 & 0 & 0 & 1\\ a41 & a42 & 0 & a4\end{pmatrix}; \quad B = \begin{pmatrix}0\\ b2\\ 0\\ b4\end{pmatrix}$$

$$C = \begin{pmatrix}1 & 0 & 0 & 0\\ 0 & 0 & 1 & 0\end{pmatrix}$$

To design the LQR, (*A*, *B*) pair should be controllable at different values of localized points.

18.3 Takagi-Sugeno Fuzzy Logic Controller

FLC has three stages as shown in Fig. 18.2 [19]:

1. Input stage performs fuzzification, i.e., converting crisp input into linguistic fuzzy variables. Membership functions, their grade, shape, number, width, and overlap should be for each variable.
2. Processing stage includes rule base and inference system. Rule base division incorporates control decision based upon different combination of fuzzified input sets. Inference mechanism division evaluates firing strength of individual rule depending upon the given inputs to the controller from which controlled output is obtained.
3. Output stage performs defuzzification, i.e., reconverts linguistic output results from inference division to crisp values so that it is used in actuators.

The TS-FLC is also known as functional fuzzy system [18]. Its premise part is same to that of MFLC, which consisted conjunction of different fuzzy sets, while consequent part is functional. Taking consequent part as linear function of state variables, ith rule is defined as:

$$\text{If } z_1(t) \text{ is } M_{i1} \quad \text{and...and} \quad z_j(t) \text{ is } M_{ij},$$

$$\text{Then } x^i(t) = A_i x(t) + B_i u(t), \quad i = 1, 2 \ldots r$$

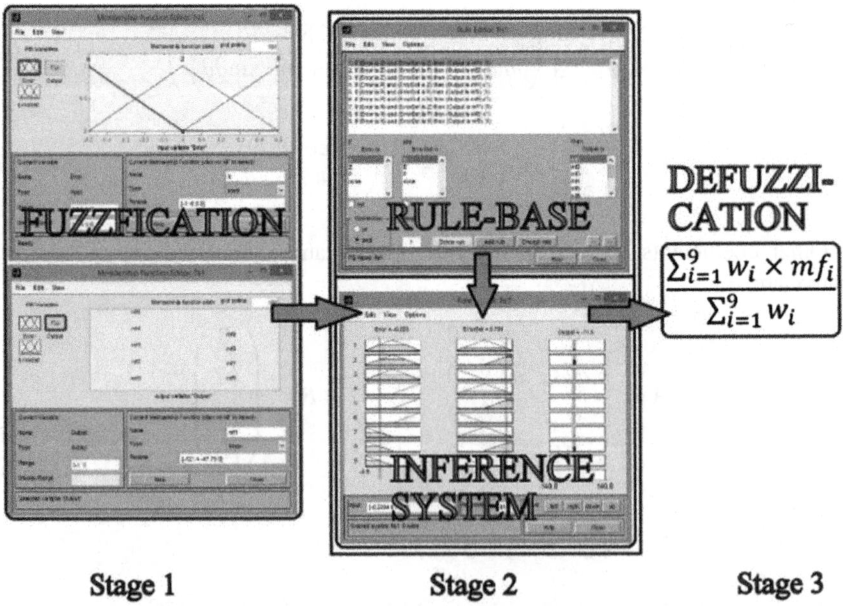

Fig. 18.2 Design of FLC

where M_{ij}, ($i = 1, 2 \ldots r$, $j = 1, 2 \ldots g$) are fuzzy sets and r is number of rules. $x(t) \in \mathbb{R}^n$, $u(t) \in \mathbb{R}^m$, and $z(t) \in \mathbb{R}^g$ are state vector, input vector, and input variable, respectively. A_i and B_i are state matrix and input matrix, respectively, of appropriate order. The overall state equation in the form of weighted average is given as:

$$\dot{x}(t) = \frac{\sum_{i=1}^{r} \mu_i(z(t))\{A_i x(t) + B_i u(t)\}}{\sum_{i=1}^{r} \mu_i(z(t))} \tag{18.3.1}$$

If $r = 1$, we get typical linear system. For $r > 1$, certain rules will turn and together contribute to output.

$$\omega_i = \frac{\mu_i(z(t))}{\sum_{i=1}^{r} \mu_i(z(t))}$$

$$\text{such that} \quad \omega_i > 0 \quad \text{and} \quad \sum_{i=1}^{r} \omega_i = 1 \tag{18.3.2}$$

Then, we can modify Eq. (18.3.1) as:

$$\dot{x}(t) = \sum_{i=1}^{r} \omega_i\{A_i x(t) + B_i u(t)\} \tag{18.3.4}$$

18.4 Controller Design

Basic idea is to obtain different linearized model across domain of interest and derive their feedback gains through LQR technique. And then, these pieces have been incorporated together in TS-FLC, and LQR gains have been taken as rule-based gains.

Architecture of control system is shown in Fig. 18.3.

Separate TS-FLC controllers are constructed to save time, memory, and rule-base complexity. Net output is the sum of their individual outputs. In this work, "prod," "max," and "weighted average" are selected as AND, OR, and Defuzzification method after choosing TS-FLC-type structure from Fuzzy Logic Toolbox in MATLAB [19]. Linear relationship between control voltage and force is considered. Sometimes, additional asymmetric voltage signal is required to compensate static friction effect.

Steps for constructing LQR TS-FLC:

1. Select input–output variable for FLC and their universe of discourse. It is better to choose symmetric universe of discourse. Since, angle above 0.5 rad is

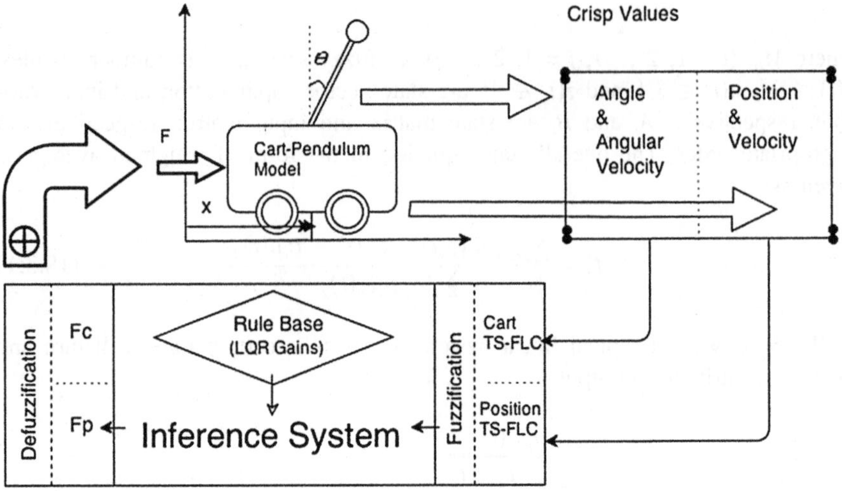

Fig. 18.3 Architecture of control system

generally considered under swing-up control and therefore [−0.5 0.5] is considered as range of angle variable for stabilization. Also, the maximum range of cart is 0.5 m, so the gain of linear controller across this point should be considered as maximum attainable gain. Thus, [−0.5 0.5] is considered as range of cart position. Universe of discourse is normalized, i.e., [−1 1].

2. Add membership functions (MF) such that it attains maximum truth value at considered linearization points. Triangular and trapezoidal MF can be used if simplicity is required. On the other side, Gaussian and Bell MF have advantage of being smooth and continuous at every point. Sigmoidal MF is open right/left sided. The number of MFs and their placement is more crucial than their shape. Resolution improves with increasing MFs [18]. Here, only three triangular membership functions, i.e., Negative–Zero–Positive are considered for input variables angle, angular velocity, cart position, and cart velocity.
3. In this work, input variable is considered as error and rate of change of error. For every ith combination of error and error dot, corresponding linearized model is fabricated and thus its ith negative feedback gain "K_i" is obtained:

$$\begin{aligned} &\text{If } z_1(t) \text{ is } M_{i1} \quad \text{and...and} \quad z_j(t) \text{ is } M_{ij}, \\ &\text{Then } u(t) = -K_i x(t) \end{aligned} \tag{18.4.1}$$

These gains are nothing but parameters of linear equation of consequent part. Substituting above equation in Eq. (18.3.4) we get net closed-loop state-space:

$$\dot{x}(t) = \sum_{j=1}^{r} \omega_i \{A_i - B_i K_i\} x(t) \tag{18.4.2}$$

K_i is calculated exclusively for each antecedent part by employing LQR gains using "lqr(*A*, *B*, *C*, *D*)" command in MATLAB [20] satisfying Performance Index (PI) with infinite time horizon:

$$J(u(t)) = \frac{1}{2} * \int_{to}^{\infty} [x'(t)Q(t)x(t) + u'(t)R(t)u(t)]\mathrm{d}t \tag{18.4.3}$$

where $Q(t)$ is positive semi-definite state weight matrix and $R(t)$ is positive definite input weight matrix. These weighted matrices are symmetric in nature. Mostly, these are diagonal matrices. Due to quadratic nature of PI, LQR emphasizes on large terms compared to small terms. Since pendulum regularization is primary motive, therefore maximum weight is given to its coefficient in Q matrix. Large value is also given to cart position coefficient for achieving secondary aim of cart servo control. Henceforth, Q and R matrix taken as diag(1000000 500 5000 500) and 100, respectively, derived after trial and error. It is important to note that for infinite horizon problem, poles should be stable if not controllable otherwise PI will become infinite. Naidu [16] mentioned that LQR ensures phase margin is greater than or equal to 60° and infinite gain margin. And, hence smooth fuzzy scheduling between these local linear controllers with assured stability is obtained.

4. Due to complicated cross-effects and desire for further improvement in system constraint margins, preprocessing and postprocessing gains of FLC can be tuned. Passino and Yurkovich [18] mentioned that input gain has inverse effect on scaling of MFs. On increasing input gain, MFs are uniformly contracted which causes increment in resolution but at the cost of chattering. Tuning of these parameters also affects the performance of system similar to that of PD controller. On other hand, this effect is opposite between output gains and MFs, i.e., linguistic outcome will become stronger on increasing output gain. Increasing output gain will support proportional effect. FLC Toolbox GUI Rule Viewer and Surface Viewer further give better insight of controller.

Controllers are being tested upon nonlinear models along with nonlinearities (cart friction coefficient, pendulum damping coefficient friction, time delay of 10 ms at input–output interfacing, and saturator block at output of controller) which were neglected in its designing stage.

18.5 Simulation and Results

Simulation results are obtained for initial angle of 0.105 rad or approximately 6°. Other states are considered to be zero. Performance criteria chosen for comparison of control techniques are as follows: maximum deviation from reference point (Δ) and settling time (t_s) of cart and pendulum. Robustness against disturbance and

Table 18.2 Analysis of performance

Controller	Deviation		Settling time	
	Pendulum (rad)	Cart (m)	Pendulum (s)	Cart (s)
Performance of typical closed-loop system				
LQR	−0.052	0.433	1.6	3.6
TS-FLC	−0.032	0.395	1.4	5.6
Performance of tracking control				
TS-FLC	−0.066	0.71	1.56	3.56
LQR	−0.042	0.67	1.46	5.55
Performance under disturbance				
LQR	−0.084	0.433	3.73	5.7
TS-FLC	−0.038	0.395	1.37	5.65
Parameter variation: (i) half				
LQR	−0.076	0.567	1.04	3.65
TS-FLC	−0.049	0.485	0.9	5.45
(ii) Doubled				
LQR	−0.082	0.54	1.9	5.4
TS-FLC	−0.05	0.46	1.7	5.45

parameter variations is also considered. Simulation results are mentioned in Table 18.2.

Typical closed response is shown in Fig. 18.4, it is observed that apart from settling time of cart, all other performance parameters of TS-FLC are better than LQR.

Figure 18.5 shows the servo control of cart position for reference of −0.3 m. Both of the controllers track the desired position. But their maximum deviation of pendulum has been increased due to extra force required for tracking.

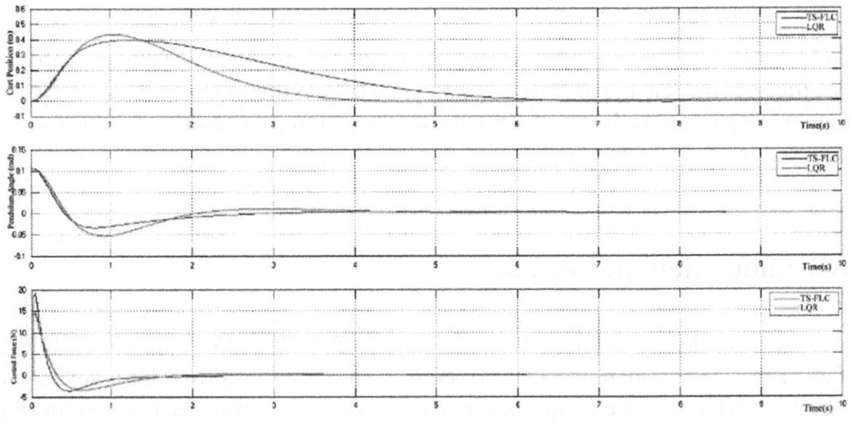

Fig. 18.4 Closed response of IPCS

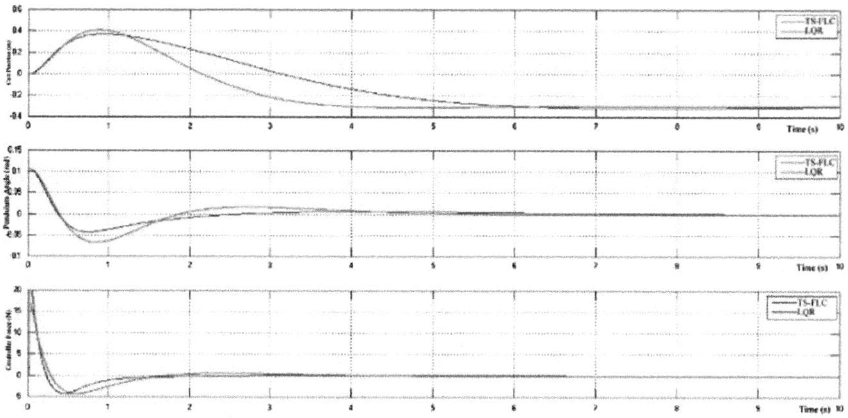

Fig. 18.5 Cart tracking control

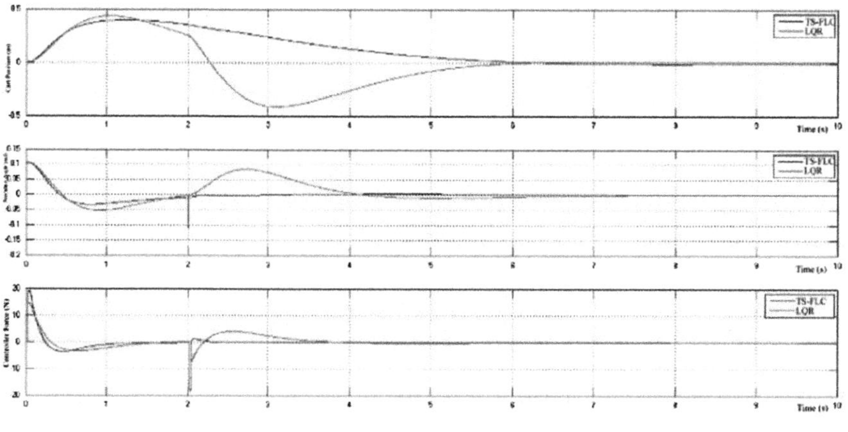

Fig. 18.6 Disturbance control

Disturbance of 0.1 rad at 2 s is given to the system. It can be interpreted from Fig. 18.6 that TS-FLC is almost unaffected by disturbance, while LQR may get hampered.

Various parameters of system have been halved and doubled to analyze the robustness. Settling time of pendulum and cart is decreased, respectively, when various parameters were halved and then doubled, but overall performance is aggravated as shown in Fig. 18.7.

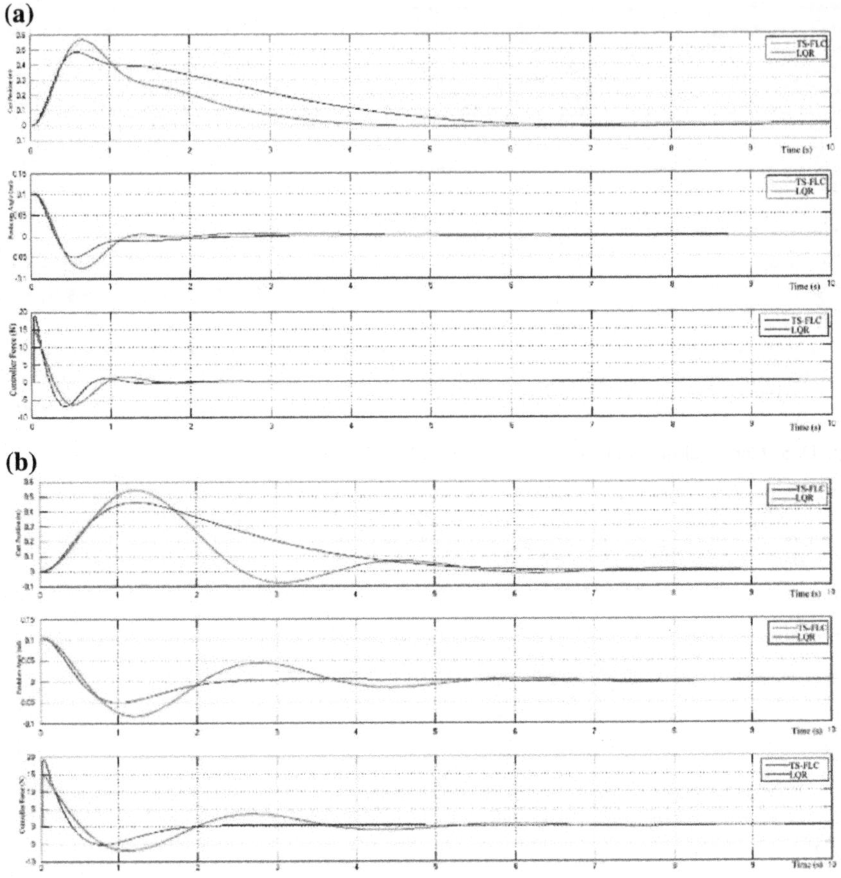

Fig. 18.7 Parameter variation. **a** Halved. **b** Doubled

18.6 Conclusion

In this paper, simple systematic TS-FLC design procedure in MATLAB environment is discussed. Chattering-free transition occurred between different sub-systems in TS-FLC. Pendulum regulation and cart servo are successfully achieved by both controllers under given constraints. However, suggested FLC is more effective in handling disturbances. Also, LQR is not able to control the cart under track constraint when there are parametric variations in system model. So, it can be admitted that TS-FLC is more robust. Thus, such extension and switching among different controllers through TS-FLC is superior to single controller.

References

1. Davison EJ (1990) Benchmark problems for control system design. International Federation of Automatic Control
2. Boubaker O (2012) The inverted pendulum benchmark in nonlinear control theory: a survey. Int J Adv Robot Syst 10:1–7
3. Boubaker O (2012) The inverted pendulum: a fundamental benchmark in control theory and robotics. J Electron Syst 2(4):154–164
4. Zadeh LA (1965) Fuzzy sets. Inf Control 8(3):338–353
5. Mamdani EH, Assilian S (1975) An experiment with in linguistic synthesis with a fuzzy logic controller. Int J Man Mach Stud 7:1–13
6. Takagi T, Sugeno M (1985) Fuzzy identification of systems and its applications to modeling and control. IEEE Trans Syst Man Cybern 15(1):116–132
7. Chen J-Y (1997) Expert SMC-based fuzzy control with genetic. J Franklin Inst 245 (3):589–610
8. Wu B, Yu X, Man Z (1998) Design of fuzzy sliding-mode control system. Fuzzy Sets Syst 95 (8):295–306
9. Lee MA, Takagi H (1993) Integrating design stages of fuzzy systems using genetic algorithms. In: Second IEEE international conference on fuzzy system, San Francisco
10. Shieh CY, Nair SS (1993) A new self tuning fuzzy controller design and experiments. In: Second proceedings of IEEE international conference on fuzzy systems, vol 2
11. Lam HK, Leung FH, Tam PKS (2003) Design and stability analysis of fuzzy model-based nonlinear controller for nonlinear systems using genetic algorithm. IEEE Trans Syst Man Cybern 33(2):250–257
12. Wong C-C, Her S-M (1999) A self-generating method for fuzzy systems design. Fuzzy Sets Syst 103:13–25
13. Olivier PD (2005) Feedback linearization using TSK fuzzy approximants. In: Thirty seventh south-eastern symposium on system theory, Mar 2005
14. Sun YL, Er MJ (2004) Hybrid fuzzy control of robotics systems. IEEE Trans Fuzzy Syst 12 (6):755–765
15. Concha J, Cipriano A (1997) A design method for stable fuzzy LQR controllers. In: Sixth IEEE international conference on fuzzy systems
16. Naidu DS (2003) Optimal control system. CRC Press, Boca Raton
17. Slotine JJE, Li WP (1991) Applied nonlinear control. Prentice Hall, Upper Saddle River
18. Passino KM, Yurkovich S (1998) Fuzzy control. Addison-Wesley, California
19. Zadeh LA (1995) Fuzzy logic toolbox. MathWorks, Massachusetts
20. Getting Started Guide (2001) Control system toolbox. MathWorks, Massachusetts

Chapter 19
An Effective Analysis of Healthcare Systems Using a Systems Theoretic Approach

Alok Trivedi and Shalini Rajawat

Abstract The use of accreditation and quality measurement and reporting to improve healthcare quality and patient safety has been widespread across many countries. A review of the literature reveals no association between the accreditation system and the quality measurement and reporting systems, even when hospital compliance with these systems is satisfactory. Improvement of healthcare outcomes needs to be based on an appreciation of the whole system that contributes to those outcomes. The research literature currently lacks an appropriate analysis and is fragmented among activities. This paper aims to propose an integrated research model of these two systems and to demonstrate the usefulness of the resulting model for strategic research planning. In this paper, we discuss how to improve the overall performance of quality in healthcare systems and, additionally, what methods a researcher needs to adopt for system effectiveness.

Keywords Health care · Hierarchical systems · Supply/input/process/output/key stakeholder (SIPOKS) · Systems theory

19.1 Introduction

The use of accreditation systems to improve healthcare quality and patient safety has been widespread across many countries [1–4]. Quality measurement incorporating clinical indicators and quality indicators and reporting systems have grown substantially as the more visible aspects of hospitals' quality improvement efforts [5–9]. Taken together, these systems comprise the health administration segment of the healthcare system, for convenience labelled the health administration system.

A. Trivedi (✉) · S. Rajawat
Department of Computer Science, Vivekananda Institute of Technology, Jaipur, India
e-mail: aloktrivedi@live.com

S. Rajawat
e-mail: shalinirajawat19@gmail.com

V. Vijay et al. (eds.), *Systems Thinking Approach for Social Problems*,
Lecture Notes in Electrical Engineering 327,
DOI 10.1007/978-81-322-2141-8_19

The health administration system is believed to influence quality outcomes, and considerable resources are spent by participating hospitals in this belief.

There is rich research literature on the association of the accreditation and measurement/reporting systems to quality in health care, but the results are unsatisfactory. The outcome of quality is not well correlated with accreditation requirements, even when hospital compliance with accreditation and measurement/reporting requirements is acceptable. In general, partial, inconsistent or conflicting results have been discovered [7, 10]. An important feature of this research is that it is concerned only with correlation, rather than the processes through which the impact of the systems occurs, and is fragmented: specialized to specific clinical or management perspectives or a system or subsystem taken in isolation. The fragmented research on the determinants of quality in health care reveals partial observation and ambiguous results.

Owing to these findings, some arguments have been made for 'a more systematic use of theories in planning and evaluating quality improvement activities in clinical practice' [11–13]. The idea is to use theories to describe the model lying behind a specific intervention and then design research to evaluate the model. The need for a theoretically driven approach to understanding complex social interventions and their effects has been strongly advocated [14] as the way to gain knowledge about the overall systemic effects of the health administration segment acting on the healthcare system, especially knowledge that will inform decisions about the use of healthcare resources to support the most valued processes. Yet, for all the interest in the use of the accreditation and measurement/reporting systems to improve healthcare quality and patient safety, the science of healthcare performance measurement and management is still relatively embryonic, and there remains a paucity of hard evidence to guide policy and research planning from a firm theoretical basis.

According to systems theory, patient safety and quality of health care is an emergent property of the entire healthcare system [14], and it follows that the improvement of healthcare outcomes needs to be based in a systematic appreciation of the whole system that contributes to those outcomes. Yet the research literature currently lacks an appropriate analysis of this sort. Therefore, the first aim of this paper is to use systems theoretic approach to develop an integrated research model of the accreditation and quality measurement/reporting systems, taken in relation to the hospital-level healthcare system. The second aim is to demonstrate the usefulness of the resulting model for strategic research planning. The paper provides an example of an adaptive control study derived from the proposed model. It demonstrates a template for more advanced strategic research planning of quality improvement throughout the healthcare systems.

19.2 Methods and Designs

To achieve these aims, the research was conducted through the combination of a theoretical based study and a literature-based empirical study. The theoretical based study combines the basic concepts of systems theory and a general systems flow with the supply/input/process/output/key stakeholder (SIPOKS) process model to form the systems theoretic approach. The approach is used to initially develop a basic high-level integrated systems model as a framework to guide the investigation of the effect of the accreditation and measurement/reporting systems on healthcare quality and then to examine, from a systems flow perspective in a lower level, the causal links within the system that impact on its effectiveness.

A literature-based empirical study used existing research or documentation as evidence with which to validate the proposed general relationships derived from the model. Australian experiences in accreditation and clinical performance data reporting, especially from the Australian Commission on Safety and Quality in Health Care and the Australian Council on Healthcare Standards (ACHS), are used as the major evidence base for evaluating the performance of these relationships and further as a design base to develop an example of an adaptive control study. In order to gain a better understanding of systems theory, healthcare systems hierarchy, general systems flow, the SIPOKS process model and the Australian research base are used to form the adaptive control study, and their use is elaborated below.

19.2.1 Systems Theory

The foundation of systems theory rests on two pairs of concepts: emergence and hierarchy, and control and communication [14, 15]. According to the first pair of systems theory concepts, a general model of complex systems can be expressed in terms of a hierarchy of levels of organization [16]. The safety and quality characteristics of complex systems are an emergent property of the system as a whole, not a property of individual system components. According to the second pair of basic system theory concepts, an open and dynamic complex system like the healthcare system is viewed as a suite of interrelated subsystems that are kept in a state of dynamic equilibrium by feedback loops of information and control [17]. Specifically, their relevant emergent properties are controlled by a set of safety and quality constraints related to the behaviour of the system components or subsystems [18]. Regulation to required standards is the common form that enforcement of safety constraints takes in complex systems and is expressed through hierarchical regulation relationships [19]. Since control implies the need for communication, reverse communication within the system hierarchy from controlled to controller is required to stimulate systems' behaviour towards the accepted standard of safety and quality [20].

In the study, systems theory is applied to construct a healthcare system hierarchy which consists of interacted systems linked with control and communication in different layers.

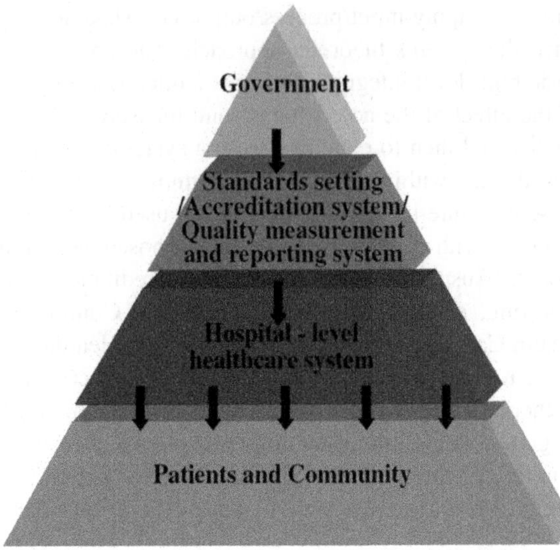

Fig. 1 Health systems hierarchy

19.2.2 Healthcare Systems Hierarchy

For the purposes of this paper, the overall healthcare system may be simplified to a 4-layer model shown in Fig. 19.1. Inter-layer relationships are characterized by vertical control and communication, but the full regulatory structure also includes significant horizontal interrelations as well as self-regulation. The proper functioning of all these relationships is important to the ultimate achievement of quality health care [19]. The focus of this paper is on the middle two layers, the health administration system, with its two subsystems—the accreditation system and the quality measurement and reporting systems—in relation to the hospital-level healthcare system. A holistic healthcare systems relationship model made of the two-layer system hierarchy is constructed for detailed analysis. The relationship of the two health administration subsystems to one another and to the hospital-level healthcare system 'below' then forms the focus of this study.

19.2.3 General Systems Flow and SIPOKS Process Model

A general systems flow is used where systems receive inputs and utilize, transform, and otherwise act on them to create outputs, whether to other systems or the external environment [21]. The SIPOKS process model partitions the overall systems flow into suppliers, input, process, output and key stakeholders for convenient analysis. In this model, the supplier provides the required inputs to a system process, including people, equipment, materials, working procedures and methods, and general working environment. The process then utilizes, transforms and otherwise acts on the inputs to produce a set of outputs that are used by key stakeholders, which may be supplied to other processes in the same or different systems. A stakeholder is defined as any group that is impacted or interested in the performance of the process, and the word 'key' denotes important stakeholders.

Applied to each of several interrelated subsystems, SIPOKS can, for example, usefully analyse an extended process into shorter phases and link the analysis of interacting processes from different hierarchical levels for a specific purpose [22]. In this way, the SIPOKS analysis can provide insight into cause–effect relationships within systems.

A literature-based empirical study used a wide range of existing research or documentation, from several countries, concerning the functioning of these systems and their interrelations as a basis for imputing general flow processes to them, of the hierarchical control and communication type which were of interest to the safety and quality of health care. The model processes obtained were thus broadly validated by their occurrence throughout developed healthcare systems.

19.3 Results

Using a systems theoretic approach, the holistic healthcare systems relationship model was developed, the overall effectiveness of the accreditation and measurement/reporting systems for providing quality of care was identified and system weaknesses from a system flow perspective were discovered. An example of an adaptive control study involving accreditation surveyors derived from this approach is developed.

19.3.1 The Holistic Healthcare Systems Relationship Model

In practice, the hospital-level healthcare system is typically impacted by numerous accreditation and clinical or quality performance reporting systems, typically clinically differentiated. Ultimately, these systems need to be discriminated and treated individually; however, in this study, only their overall, shared interrelations are considered. Combining the concept of healthcare systems hierarchy and an analysis of control and communication relationships, a basic holistic healthcare systems relationship model is designed. As both the horizontal and vertical control/communication relationships are potentially relevant to maintaining an acceptable level of quality within the healthcare systems hierarchy, there are in principle four model relationships of direct interest. These are labelled P1–P4 in Fig. 19.2. Of note, the fourth relationship (P4) has hitherto essentially gone unrecognized and its potential remains underdeveloped. It will figure prominently in the study design proposal.

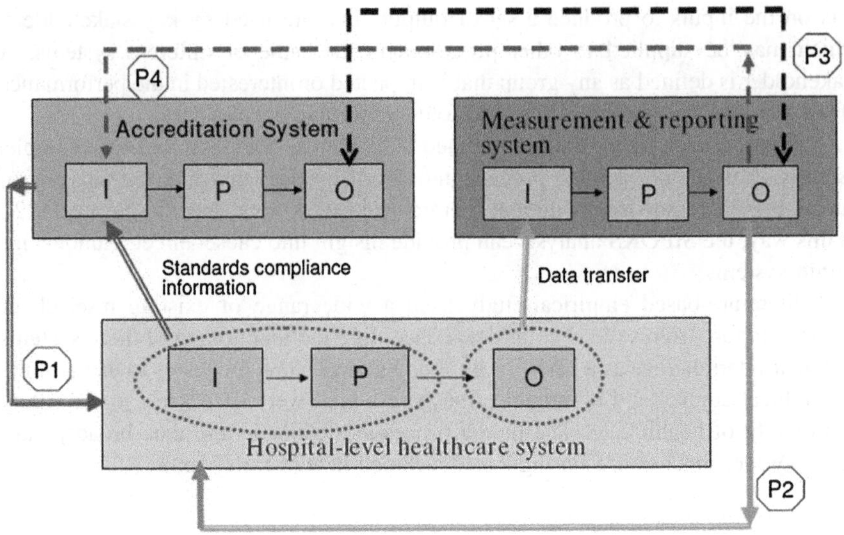

Fig. 2 The holistic healthcare systems relationship model

Two of these four relationships (P1 and P2) are vertical control and communication relationships, providing the outputs of the accreditation and the quality measurement/reporting systems, respectively, as inputs to the hospital-level healthcare system to improve quality of care. Relationship P1 is a control relationship that provides hierarchically determined practitioner standards, while the P2 communicates outcomes to the hospital-level healthcare system for its own internal control response. The remaining two relations, P3 and P4, are horizontal control and communication relationships within the health administration system, to improve the focus and impact of accreditation on quality of care. P3 communicates correlations in the outputs of the accreditation and the quality measurement/reporting systems, and P4 provides feedback from the output of the quality measurement/reporting system to the accreditation system input. While P3 is concerned with communication, P4 is intended as a control relationship. Strictly, P3 is at present a quasi-relationship because in practice there is no specific recipient of this research-generated information within the healthcare system. The information is in effect an indicator of system-wide coherence and in that sense relevant to all; however, no one has a mandate to respond to it. In a more systematic institutional design, this information might be fed to a government source (top layer of Fig. 19.1) or other system-wide responsible entity to initiate and focus healthcare improvement research of the kind proposed below. This is a potentially important relationship outside the scope of this paper, or it might also be directly utilized by P4 to focus its feedback as proposed here. In what follows it is treated as a potential communication relationship.

With respect to each of these relationships, the question arises as to what is known about its impact on quality of care. While considerable research has been directed towards understanding and improving P1 and P2 to produce improved health care, only P1 is well understood and supports practical reform. P2 has no satisfactory outcome as discussed below under 'effectiveness of quality measurement and reporting system'. As for P3 and P4, complex, ambivalent findings hold in P3 and it currently lacks an effective feedback role, while little or no research has been directed towards understanding and improving P4. These claims are reviewed below for the four relationships.

19.4 System Theoretic Thinking in Research Design

Different research designs or approaches have been proposed to resolve issues of the accreditation and the quality measurement and reporting system separately. Braithwaite et al. [23] apply multi-method approaches to focus on two central aims: to examine the relationships between accreditation status and processes, and the clinical performance and culture of healthcare organization; to examine the influence of accreditation surveyors and the effect of accreditation surveyors on their own health organizations. Joly et al. [24] propose a logic model approach which focuses on inputs, strategies, outputs and multiple level outcomes, with emphasis on accredited public health agencies as the input of interest. However, these approaches need to be complemented by research using systems theoretic approach to help address the weaknesses from a system flow perspective and to understand quality of care as an emergent whole-system property.

The quality measurement and reporting system are playing an increasingly important role in the healthcare systems. It has been utilized as a potentially complementary tool to accreditation for improving quality. However, the available evidence suggests that both the weakness of quality measurement and reporting system and the impetus from external systems to stimulate improvement can influence the effectiveness of quality of care. Therefore, improving the effectiveness of P2 requires an increase in the quality of feedback and the utilization of performance reports for data accuracy, validity, meaningfulness, timeliness and the right stimulation from external systems, like government. This invites the systems theoretic approach to design a series of systematic studies covering each weakness from the system flow perspective and cascading up and down external systems to establish adequate control/communication relationships between systems. If we facilitate two pairs of basic systems theory concepts and system flow perspective in our research, the quality measurement and reporting system can be enhanced to support the right information in the right time for quality improvement in health care.

Reform must focus on how to get competition right and how to put in place the enabling conditions, such as the right information, the right incentives and time horizons and the right mindsets [25]. The research approach for the effectiveness analysis of the middle two-layer subsystems in Fig. 19.1 can be used as an example for the effectiveness analysis of the whole healthcare systems hierarchy. Using systems theoretic model explores the relationships between systems and assesses their effectiveness in a clear system flow thinking. The overall analysis results of the approach provide an integrated conceptual cascade effect, which might bring possible adaptive control studies from the additional implicit relationships between systems, or potential feasible ways of achieving the complete and adequate control/communication to create value-based competition in healthcare systems could be realized more.

To improve the overall performance of quality in healthcare systems, researcher needs to adopt system theoretic thinking in research design and a holistic view of systems effectiveness. However, systems thinking can be challenging, especially taking a broad view of systems. This may limit the application of the systems theoretic approach.

19.5 Conclusion

As safety and quality is an emergent property of the healthcare system as a whole, not a property of individual system components or subsystems, the assessment of safety and quality from any perspective in one system or using any one tool is unlikely to give the complete picture.

A systems theoretic approach supported by research evidence provides the necessary holistic insight to understand the overall relationship and effectiveness among the three systems. A systems analysis reveals four intersystem relationships, the fourth hitherto unreported, along with the unsatisfactory vertical control and communication between the quality measurement/reporting system and hospital-level healthcare systems, and little or no concrete horizontal control and communication between the accreditation system and the measurement/reporting system. Overall, the health administration systems do not yet have significant positive impact on the quality of care. To help advance the science of safety and quality improvement in healthcare systems and to inform decisions on the use of healthcare resources for optimized results, the paper examines system issues using the system flow SIPOKS process model to give more supporting information on the system weaknesses. It provides a system thinking structure to assist the design of quality improvement strategies. An example of adaptive control study design is derived from the implicit P4 relationship between the health administration systems that can overcome the present fragmented state of communication and control relationships among the relevant systems.

The effectiveness of quality delivered by each subsystem in the healthcare systems hierarchy can be affected by other subsystems. However, this research develops a prototype of using a systems theoretic approach for the effectiveness analysis within only the middle two-layer subsystems. There are not enough attentions to the effects from other systems such as the organizational context in hospital-level healthcare systems, patients, community and the role of government. We believe that the basic two pairs' concepts of systems theory and the system flow model can be applied to other layers in the healthcare systems. It is hoped that this analysis will stimulate wider debate on the application of holistic systems analysis for improving the effectiveness of systems on quality and safety in health care. In this paper, we discussed how to improve the overall performance of quality in healthcare systems and, additionally, what methods a researcher needs to adopt for system effectiveness. This paper aimed to propose an integrated research model of the two systems and to demonstrate the usefulness of the resulting model for strategic research planning.

References

1. Scrivens E (1997) Assessing the value of accreditation systems. Eur J Public Health 7:4–8
2. Pawlson L, O'Kane ME (2002) Professionalism, regulation, and the market: impact on accountability for quality of care. Health Aff 21(3):200–207
3. Ovretveit J (2005) Which interventions are effective for improving patient safety? A review of research evidence Stockholm. Karolinska Institute, Medical Management Centre, Sweden
4. International Society for Quality in Health Care (2003) Global review of initiatives to improve quality in health care. World Health Organization, Geneva
5. Gibberd R, Hancock S, Howley P, Richards K (2004) Using indicators to quantify the potential to improve the quality of health care. Int J Qual Health Care 16(Suppl I):i37–i43
6. Williams SC, Schmaltz SP, Morton DJ, Koss RG, Loeb JM (2005) Quality of Care in US hospitals as reflected by standardized measures, 2002–2004. N Engl J Med 353(3):164–255
7. Fung CH, Lim YW, Mattke S, Damberg C, Shekelle PG (2008) Systematic review: the evidence that publishing patient care performance data improves quality of care. Ann Intern Med 148(2):111–123
8. Mannion R, Goddard M (2002) Performance measurement and improvement in health care. Appl Health Econ Health Policy 1(1):13–23
9. McGlynn EA (2002) Introduction and overview of conceptual framework for a national quality measurement and reporting system. Med Care 41(suppl 1):1–7
10. Greenfield D, Braithwaite J (2008) Health sector accreditation research: a systematic review. Int J Qual Health Care 20(3):172–183
11. Walshe K (2007) Understanding what works—and why—in quality improvement: the need for theory-driven evaluation. Int J Qual Health Care 19:57–59
12. Pawson R, Greenhalgh P, Harvey G, Walshe K (2005) Realist review—a new method of systematic review designed for complex policy interventions. J Health Serv Res Policy 10 (suppl I):21–34
13. Grol R, Bosch M, Hulscher M, Eccles M, Wensing M (2007) Planning and studying improvement in patient care: the use of theoretical perspectives. Milbank Q 85(1):93–138
14. Leveson ANG (2004) Systems-theoretic approach to safety in software-intensive systems. IEEE Trans Dependable Secure Comput 1(1):66–86
15. Checkland P (1981) Systems thinking, systems practice. Wiley, New York
16. Bailey KD (2006) Living systems theory and social entropy theory Syst Res. Behav Sci 23 (3):291–300 (Special issue: James Grier Miller's living systems theory)
17. Ackoff RL (1971) Towards a system of systems concepts. Manage Sci 17(11):661–671
18. Rasmussen J (1997) Risk management in a dynamic society: a modeling problem. Safety Sci 27(2):183–213

19. Braithwaite J, Healy J, Dean K (2005) The governance of health safety and quality. Commonwealth of Australia, Canberra
20. Rasmussen J, Svedung I (2000) Proactive risk management in a dynamic society Karlstad. Swedish Rescue Services Agency, Sweden
21. Hitchins DK (1992) Putting systems to work. Wiley, Chichester
22. McGarvey B, Hannon B (2004) Dynamic modeling for business management. Springer, New York
23. Braithwaite J, Westbrook J, Pawsey M, Greenfield D, Naylor J, Iedema R, Runciman B et al (2006) A prospective, multi-method, multi-disciplinary, multi-level, collaborative, social-organisational design for researching health sector accreditation. BMC Health Serv Res 6:113–123
24. Joly BM, Polyak G, Davis MV, Brewster J, Tremain B, Raevsky C, Beitsch LM (2007) Linking accreditation and public health outcomes: a logic model approach. J Public Health Manage Pract 13(4):349–356
25. Porter ME, Teisberg EO (2006) Redefining health care: creating value-based competition on results. Harvard Business School Press, Boston

Chapter 20
Paper Batteries

Critika Agrawal, Bhaskar Sharma, Deepak Bhojwani and Shalini Rajawat

Abstract This paper gives a thorough insight on this relatively revolutionizing and satisfying solution of energy storage through paper batteries and provides an in-depth analysis of the same. A paper battery is a flexible, ultra-thin energy storage and production device formed by combining carbon nanotubes with a conventional sheet of cellulose-based paper [1]. A paper battery can function both as a high-energy battery and supercapacitor, combining two discrete components that are separate in traditional electronics. This combination allows the battery to provide both long-term steady power production and bursts of energy. Being biodegradable, lightweight, and non-toxic, flexible paper batteries have potential adaptability to power the next generation of electronics, medical devices, and hybrid vehicles, allowing for radical new designs and medical technologies. The paper is aimed at understanding and analyzing the properties and characteristics of paper batteries, to study its advantages, potential applications, limitations, and disadvantages. This paper also aims at highlighting the construction and various methods of production of paper battery and looks for alternative means of mass production.

Keywords Carbon · Nanotubes · Cellulose · Paper battery

C. Agrawal (✉) · B. Sharma · D. Bhojwani · S. Rajawat
Department of Computer Science, Vivekananda Institute of Technology, Jaipur, India
e-mail: critikaagrawal@yahoo.in

B. Sharma
e-mail: bhaskarshr@live.com

D. Bhojwani
e-mail: deepakonwork@hotmail.com

S. Rajawat
e-mail: shalinirajawat19@gmail.com

V. Vijay et al. (eds.), *Systems Thinking Approach for Social Problems*,
Lecture Notes in Electrical Engineering 327,
DOI 10.1007/978-81-322-2141-8_20

20.1 Introduction

A paper battery is a flexible, ultra-thin energy storage and production device formed by combining carbon nanotubes with a conventional sheet of cellulose-based paper. A paper battery acts as both a high-energy battery and supercapacitor, combining two components that are separate in traditional electronics. This combination allows the battery to provide both long-term, steady power production and bursts of energy. Non-toxic, flexible paper batteries have the potential to power the next generation of electronics, medical devices, and hybrid vehicles, allowing for radical new designs and medical technologies. Paper batteries may be folded, cut, or otherwise shaped for different applications without any loss of integrity or efficiency. Cutting one in half halves its energy production. Stacking them multiplies power output [2]. Early prototypes of the device are able to produce 2.5 V of electricity from a sample of the size of a postage stamp. Specialized paper batteries could act as power sources for any number of devices implanted in humans and animals, including RFID tags, cosmetics, drug delivery systems, and pacemakers. A capacitor introduced into an organism could be implanted fully dry and then be gradually exposed to bodily fluids over time to generate voltage [3]. Paper batteries are also biodegradable, a need only partially addressed by current e-cycling and other electronics disposal methods increasingly advocated for by the green computing movement.

20.2 Uses of Paper Batteries

While a conventional battery contains a number of separate components, the paper battery integrates all of the battery components in a single structure, making it more energy efficient.

- A paper battery is a battery engineered to use a paper-thin sheet of cellulose infused with aligned carbon nanotubes (CNT). Nanotubes act as electrodes, allowing the storage devices to conduct electricity.
- It functions as both a lithium-ion battery and a supercapacitor and can provide a long, steady power output comparable to a conventional battery, as well as a supercapacitor's quick burst of high energy.
- Integrates all of the battery components in a single structure, making it more energy efficient.
- Paper battery has extreme flexibility; the sheets can be rolled, twisted, folded, or cut into numerous shapes with no loss of integrity or efficiency, or stacked, like printer paper (or a Voltaic pile), to boost total output.
- Paper battery has extreme flexibility; the sheets can be rolled, twisted, folded, or cut into numerous shapes with no loss of integrity or efficiency, or stacked, like printer paper (or a Voltaic pile), to boost total output.

- The paper-like quality of the battery combined with the structure of the nanotubes embedded within gives them their light weight and low cost, making them attractive for portable electronics, aircraft, automobiles, and toys.
- Ability to use electrolytes in blood makes them potentially useful for medical devices attractive for portable electronics, aircraft, automobiles, and toys as pacemakers, and they do not contain any toxic material and are biodegradable [4].

20.3 Construction and Working of Paper Battery

20.3.1 Construction

1. Zinc and manganese dioxide-based cathode and anode are fabricated from proprietary links.
2. Standard silkscreen printing presses are used to print the batteries onto paper and other substrates.
3. Power paper batteries are integrated into production and assembly processes of thin electronic devices.
4. The paper is infused with aligned carbon nanotubes, which gives the device its black color [5].
5. The tiny carbon filaments or nanotubes substitute for the electrode used in conventional battery.
6. Use an ionic liquid solution as an electrolyte with the two components which conduct electricity.
7. They use the cellulose or paper as a separator—the third essential component (Fig. 20.1).

20.3.2 Working

1. The nanotubes acting as electrodes allow the storage device to conduct electricity.
2. Chemical reaction in battery occurs between electrolyte and carbon nanotubes.

Fig. 20.1 Construction

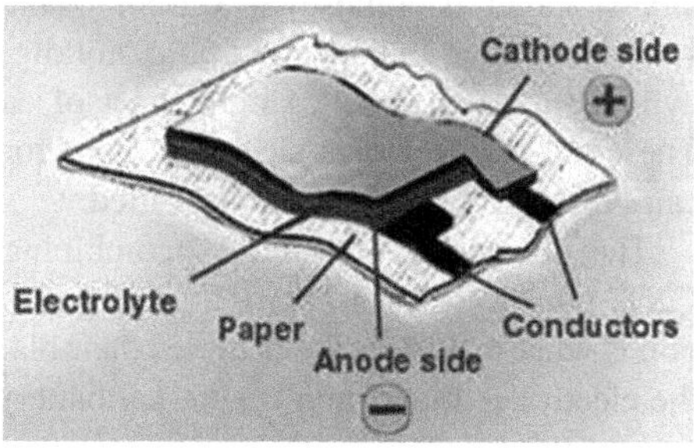

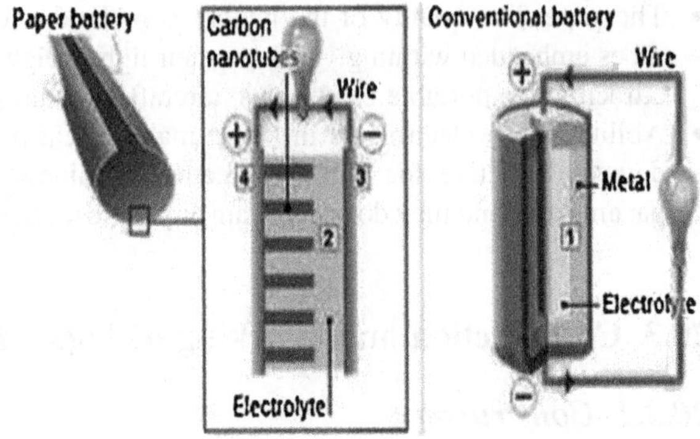

Fig. 20.2 Working

3. Battery produces electrons through a chemical reaction between electrolyte and metal in the traditional battery.
4. Electrons must flow from the negative to the positive terminal for the chemical reaction to continue. Ionic liquid, essentially a liquid salt, is used as the battery electrolyte.
5. The organic radical materials inside the battery are in an "electrolyte-permeated gel state," which is about halfway between solid and a liquid. This helps ions to smooth move to reduce resistance allowing to charge batteries faster.
6. We can stack one sheet on top of another to boost the power output. It is a single, integrated device. The components are molecularly attached to each other: The carbon nanotube print is embedded in the paper, and the electrolyte is soaked into the paper (Fig. 20.2).

20.4 Paper Batteries Basics

A paper battery is flexible, ultra-thin energy storage and production device formed by combining carbon nanotubes with a conventional sheet of cellulose-based paper. A paper battery acts as both a high-energy battery and supercapacitor, combining two discrete components that are separate in traditional electronics. Paper Battery = Paper (Cellulose) + Carbon Nanotubes Cellulose is a complex organic substance found in paper and pulp, not digestible by humans. A CNT is a very tiny cylinder formed from a single sheet of carbon atoms rolled into a tiny cylinder. These are stronger than steel and more conducting than the semiconductors. They can be single walled and multiwalled.

The devices are formed by combining cellulose with an infusion of aligned CNT, which are each approximately one-millionth of a centimeter thick. The carbon is what that gives the batteries their black color [6]. These tiny filaments act like the electrodes found in a traditional battery, conducting electricity when the paper comes into contact with an ionic liquid solution. Ionic liquids contain no water, which means that there is nothing to freeze or evaporate in extreme environmental conditions. As a result, paper batteries can function between −75 and 150 °C.

20.5 Properties of Paper Batteries

The properties of paper Batteries are mainly attributed to the properties of its constituents.

Properties of Cellulose

- Cellulose has low shear strength.
- It is biodegradable.
- It is biocompatible.
- It is non-toxic.

Properties of Carbon Nanotubes

- They have high tensile strength.
- Low mass density and high packing density.
- Low resistance.
- The open circuit voltage of batteries is directly proportional to CNT concentration.
- Thickness: typically about 0.5–0.7 mm.
- Normal continuous current density: 0.1 ma/cm^2/active area [7].
- Shelf life: 3 years.
- No heavy metals (like Hg, Pb, etc.).
- Temperature operating range: −75 to 150 °C.
- No overheating in case of battery abuse or mechanical damage.

Additional properties acquired by Proper Batteries

- Output open circuit voltage (O.C.V): 1.5–2.5 V.
- The O.C.V of paper batteries is directly proportional to CNT concentration.
- Stacking the paper and CNT layers multiplies the output voltage.

The Advantages of Paper Batteries

- Paper battery is cheap, thin, and flexible.
- It can generate a voltage of 1.5 V.
- It is disposable with household wastes.
- Paper batteries are rechargeable.
- Thin-film cells can be stored for decades.

Advantages over existing batteries

- Biodegradable and non-toxic: As its major constituents are of organic origin, it is biodegradable and non-toxic.

- Easily reusable and recyclable: As it is a cellulose-based product, it is easily recyclable and reusable [8].
- Rechargeable: Paper battery can be recharged up to 300 times using almost all electrolytes including biosalts such as sweat and blood.
- It is very lightweight and flexible.
- Easily moldable into desired shape and size.

20.6 Applications

In electronics

- In laptops and mobile phones. The weight of these devices can be reduced by using paper batteries instead of alkaline batteries without compromise in any area.
- In calculators and wrist watches.
- In wireless communication devices like Bluetooth.
- In enhanced printed circuit board.

In medical sciences

- In pacemakers for the heart.
- In cosmetics and drug delivery systems.
- In biosensors, for example, sugar meters.

In automobiles

- In hybrid car system.
- For powering electronic devices in satellite programs.
- For lightweight missiles (Fig. 20.3).

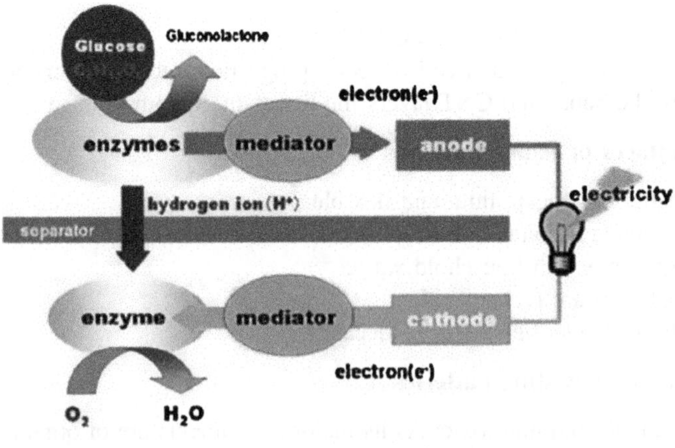

Fig. 20.3 Applications

20.7 Limitations of Paper Batteries

- Paper batteries have low strength that they can be torn easily.
- The techniques and the setups used in the production of carbon nanotubes are very expensive and very less efficient.
- When inhaled, their interaction with the microphages present in the lungs is similar to that with asbestos fibers. Hence, they may be seriously hazardous to human health.

20.8 Result and Conclusion

One of the major problems arising in the world is of energy degradation. And this problem disturbs the developed nations and perturbs developing nations like India to a greater extent. Being at a generation where there cannot be a day without power, paper batteries are path-breaking solutions. As paper batteries are light-weight and non-toxic and flexible, they have the potential to power the upcoming generations of electronics and medical fields by allowing new designs of technologies. Since India is a developing nation, it is a very useful technique which will increase the energy resources and will provide new ways of discoveries in the country as well as world.

References

1. www.technicaljournalsonline.com/ijaers
2. www.seminarsonly.com/Labels/Paper-Battery-Advantages.php
3. www.caelusconsulting.com
4. www.energexbatteries.com
5. www.blogs.siliconindia.com
6. www.seminartopics.in/Mechanical/Paper-Battery.php
7. www.share.pdfonline.com
8. www.https://ilmasto-opas.fi/en/.../hillinta/-/.../kierratys-ja-

Chapter 21
Adaptive Control of Nonlinear Systems Using Multiple Models with Second-Level Adaptation

Vinay Kumar Pandey, Indrani Kar and Chitralekha Mahanta

Abstract Adaptive control of a class of single-input single-output (SISO) nonlinear systems with large parametric uncertainties has been investigated in this paper. Control of nonlinear systems using adaptive schemes suffers from the drawback of poor transient responses in parametrically uncertain environment. The use of multiple models presents a solution to this problem. In this paper, state transformation and feedback linearization have been used to algebraically transform nonlinear system dynamics to linear ones. The unknown parameter vector for the plant is assumed to be bounded within a set of compact parameter space. Indirect adaptive control using multiple identification models has been used to improve transient response and convergence time. The observer-based identifier model is used for all these models. Lyapunov stability analysis is used to obtain tuning laws for estimator parameters. Further, second-level adaptation using combination of all the adaptive estimator models is used. Simulations have demonstrated that multiple models with second-level adaptation yield better transient performance with faster convergence.

Keywords Adaptive control · Feedback linearization · Multiple model · Second-level adaptation

V.K. Pandey (✉) · I. Kar · C. Mahanta
Department of Electronics and Electrical Engineering, Indian Institute of Technology Guwahati, Guwahati 781039, India
e-mail: p.vinay@iitg.ernet.in

I. Kar
e-mail: indranik@iitg.ernet.in

C. Mahanta
e-mail: chitra@iitg.ernet.in

V. Vijay et al. (eds.), *Systems Thinking Approach for Social Problems*, Lecture Notes in Electrical Engineering 327,
DOI 10.1007/978-81-322-2141-8_21

21.1 Introduction

Adaptive control using feedback linearization is a popular method to deal with certain classes of nonlinear systems. Adaptive control is frequently used to control plants with unknown parameters. Direct and indirect methods of adaptive control are widely used [1, 8, 10]. In direct method, adaptive laws for tuning of controller parameters are designed based on output or tracking error between plant and reference model. In indirect method, an explicit identifier is used to estimate parameters of the plant and those estimated parameters are used to find tuning laws for controller parameters. Direct adaptive control is rather simpler to apply, but it requires over parametrization, whereas indirect adaptive control is free from this drawback [1, 8]. Adaptive control provides asymptotic stability in most applications, but its transient response needs some attention in case of large parametric uncertainties [16]. In this context, notion of multiple identification models has been introduced by Han and Narendra [4] and Narendra and Han [14].

The motivation behind using multiple identification models is to cope with uncertainties present in the system. In modern linear time-invariant (LTI) control, it is assumed that parametric uncertainties affecting the system are small. However, large variations in operating conditions, failure of system components, and changes in subsystem dynamics violate this assumption. This prevents the modern control techniques to achieve desired closed-loop behavior as well as stability conditions.

This work is aimed at designing a controller to deal with a class of nonlinear systems with large parametric uncertainties [12]. The design of the controller is based on adaptive control methodologies which can handle large parametric uncertainties by tuning the controller gains with respect to the estimated variations in the model. Multiple model adaptive control (MMAC) is used to design controllers offline which provides a better platform to combine the adaptive and modern robust control techniques [9].

The paper is organized as follows. In Sect. 21.2, the control problem is discussed. Feedback linearization technique which is used in designing the proposed controller is described in Sect. 21.3. Design of estimators for changing environments is explained in Sect. 21.4. The proposed second-level adaptation using combination of all the adaptive estimator models is described in Sect. 21.5. In Sect. 21.6, the proposed adaptive controller with multiple models is described. Section 21.7 presents simulation results. Conclusion is drawn in Sect. 21.8.

21.2 Problem Formulation

Let us consider a class of affine SISO nonlinear system as:

$$\begin{aligned} \dot{x} &= f(x,\theta) + g(x,\theta)u \\ y &= h(x) \end{aligned} \tag{21.1}$$

where $x \in \mathbb{R}^n$ is the state vector, $f \in \mathbb{R}^n$, $g \in \mathbb{R}^n$, and $h \in \mathbb{R}$ are smooth vector fields. Further, θ is the unknown parameter vector with known bounds, $u \in \mathbb{R}$ is the control input, and $y \in \mathbb{R}$ represents the output. It is assumed that the full state vector is available. Considering the number of unknown parameters to be p and the system dynamics being linear in its parameters, the vector fields f and g in (21.1) can be written as

$$
\begin{aligned}
f(x,\theta) &= \sum_{i=1}^{p} \vartheta_i f_i(x) \\
g(x,\theta) &= \sum_{i=1}^{p} \vartheta_i g_i(x)
\end{aligned}
\tag{21.2}
$$

The aim is to design an indirect adaptive controller with different estimator models representing different and continuously changing operating environments.

21.3 Feedback Linearization

Feedback linearization is an effective technique for nonlinear control design. This algebraically transforms the nonlinear system dynamics into linear form such that linear control techniques can be applied [5, 7]. Generally, there are two methods to achieve the desired objective for feedback linearization

- Input/output linearization
- Input/state linearization

When the relative degree r of the system is equal to its order n, both methods are the same. For the system under consideration in (21.1), a transformation $\mathcal{T} = \Psi(x)$ is defined such that the transformed system has linearized states. This transformation is given by the diffeomorphism [6, 15]

$$
\mathcal{T} = \begin{pmatrix} \tau_1 \\ \tau_2 \\ \cdots \\ \tau_r \end{pmatrix} = \begin{pmatrix} L_f^0 h(x) \\ L_f^1 h(x) \\ \cdots \\ L_f^{r-1} h(x) \end{pmatrix}
\tag{21.3}
$$

where $\tau_1, \tau_2, \ldots, \tau_r$ are the linearized states and $L_f h, L_f^2 h, \ldots, L_f^{r-1} h$ are the lie derivatives, and r is the relative degree of the system and $r \leq n$.

Using this diffeomorphism, the system (21.1) can be represented in terms of linearized states $\tau_1, \tau_2, \ldots, \tau_r$ as

$$
\begin{aligned}
\dot{\tau}_1 &= \tau_2 \\
\dot{\tau}_2 &= \tau_3 \\
&\cdots \\
\dot{\tau}_r &= L_f^r h(x) + L_g L_f^{r-1} h(x) u = p(x,\theta) + q(x,\theta) u \\
\dot{\xi} &= \varphi(\tau, \xi) \\
y &= x_1
\end{aligned} \tag{21.4}
$$

where $\dot{\xi} = \varphi(\tau, \xi)$ represents internal dynamics of the system, which exist when relative degree is strictly less than the actual degree of the system. For the original nonlinear system to be asymptotically stable, its zero dynamics $\dot{\xi} = \varphi(0, \xi)$ must be asymptotically stable.

Now a virtual input v can be chosen and defined as $v = p(x,\theta) + q(x,\theta)u$, which yields

$$
u = \frac{1}{q(x,\theta)}(-p(x,\theta) + v) \tag{21.5}
$$

For a tracking problem, v is the control signal to be designed using the required trajectory information as

$$
v = y_d^r + c_r(y_d^{(r-1)} - y^{(r-1)}) + \cdots + c_0(y_d - y) \tag{21.6}
$$

where constants $c_0, c_1, \ldots, c_r$ can be chosen such that

$$
s^r + c_r s^{r-1} + \cdots + c_1 \tag{21.7}
$$

yields a Hurwitz polynomial.

21.4 Estimators for Different Environments

21.4.1 Assumptions for Environment

The changes in operating environment of a real-time process which causes the parameters of the system to vary with time are given as [11],

- Faults in system
- Sensor/actuator failure
- External disturbance
- Changes in system parameters

For using different identification models efficiently to estimate the unknown parameters of the plant, following assumptions are made about the region on which the plant parameters reside [13]:

- The unknown plant parameters are assumed to belong to a set of compact parameter space $\mathcal{S}$.
- The plant parameter vector P and estimator model parameter vector $\widehat{P}_i$ belong to $\mathcal{S}$.
- For N models, the set $\mathcal{S}$ can be divided s.t. $\cup_{i=1}^{N} \mathcal{S}_i = \mathcal{S}$.

At first, a stable estimation model of the plant is selected whose states and output converge to the plant as time $t \to \infty$. Here, the estimators are built for different operating environments in which the system has to operate. All the estimators have identical structures with the same initial states as the plant but with different initial values of the parameter vector. Therefore, the system given by (21.1) can be written in the regressor form as

$$\dot{x} = \omega^T(x, u)\theta \tag{21.8}$$

where ω is the regressor matrix [2]. The observer-based estimation model for the system (21.8) is given as

$$\dot{\hat{x}} = A(\hat{x} - x) + \omega^T(x, u)\hat{\theta} \tag{21.9}$$

where $\hat{x}$ and $\hat{\theta}$ are the estimated values of x and θ, respectively. Here, the matrix $A \in \mathbb{R}^{n \times n}$ can be any stable matrix chosen by the designer based on system requirements. Defining estimation error $e = \hat{x} - x$ and parameter error as $\tilde{\theta} = \hat{\theta} - \theta$, the identifier error dynamics of the system (21.1) is given as,

$$\dot{e} = Ae + \omega^T(x, u)\tilde{\theta} \tag{21.10}$$

Subsequently, to prove that if the nonlinear system (21.1) is bounded-input bounded-output (BIBO) stable, error $\lim_{t \to \infty} e(t) = 0$ and a suitable Lyapunov function [10] $V(e, \tilde{\theta}) = e^T Pe + \tilde{\theta}^T \tilde{\theta}$ are chosen, where P is the positive definite matrix solution of the Lyapunov function $A^T P + PA = -Q$ and $Q = Q^T > 0$. Taking time derivative of V and using adaptive law for θ as

$$\dot{\hat{\theta}} = -\omega(x, u)Pe \tag{21.11}$$

or

$$\dot{\hat{\theta}} = -\omega(x, u)P\hat{x} \tag{21.12}$$

yields,

$$\dot{V}(e,\tilde{\theta}) = -e^T Q e \leq 0 \tag{21.13}$$

Hence, the boundedness of both the identification error as well as the parameter error is guaranteed.

21.5 Second-level Adaptation

For incorporating the idea of N number of models, the same estimator model as in (21.9) can be used with the same initial states as the plant but with N parameter vectors having different starting points. For N models, (21.9) can be written as

$$\dot{\hat{x}}_i = A(\hat{x}_i - x) + \omega^T(x,u)\hat{\theta}_i, \quad i = 1,2,\ldots,N \tag{21.14}$$

Similarly, adaptive law (21.12) can be written as

$$\dot{\hat{\theta}}_i = -\omega(x,u)P\tilde{x}_i \tag{21.15}$$

21.5.1 Selection

The parameter space $\mathcal{S}$ is a closed and bounded set implying that every element of the parameter vector θ has an upper and a lower bound. The parameter vector θ is a p-dimensional vector given as $\theta = [\vartheta_1, \vartheta_2, \ldots, \vartheta_p]$. Based on the assumption above, $\vartheta_1 = [\vartheta_1^{\min}, \vartheta_1^{\max}], \ldots, [\vartheta_p^{\min}, \vartheta_p^{\max}]$. If $\mu \in \mathbb{Z}$ is the number of points between each maximum and minimum values of ϑ (including $\vartheta_1^{\min}$ and $\vartheta_1^{\max}$), then the total number of models is given by $N = \mu^p$ [3]. The space of the models based on the arrangement given above will be the Cartesian product of these sets denoted as,

$$\mathcal{S} = [\vartheta_1^{\min}, \ldots, \vartheta_1^{\max}] \times [\vartheta_1^{\min}, \ldots, \vartheta_1^{\max}] \times \cdots \times [\vartheta_p^{\min}, \ldots, \vartheta_p^{\max}]. \tag{21.16}$$

21.5.2 Combination

The adaptation of parameters of the models as well as the movement of the virtual model toward the actual value of the plant can be seen in Fig. 21.1. Combination of models here implies allocation of some adaptive weights to each of the model based on their identification error. The weighted summation of all the models provides the estimated plant parameter vector to be used in control input equation as per the certainty equivalence principle. It could be noted here that convergence of virtual

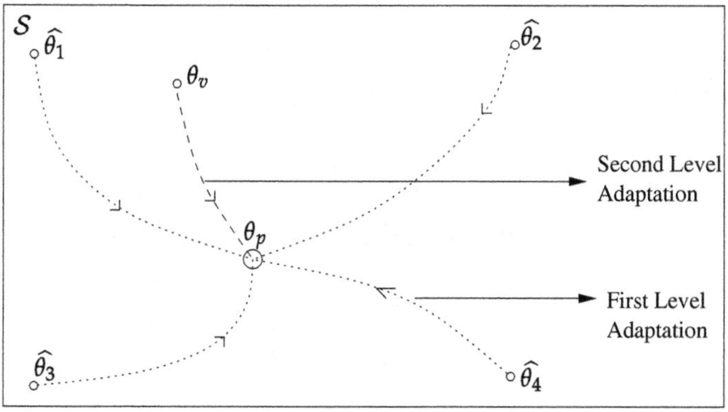

Fig. 21.1 Second-level adaptation

model is faster than the actual models. In Fig. 21.1, $\mathcal{S}$ is the bounded set for parameter vectors, $\hat{\theta}_1, \hat{\theta}_2, \hat{\theta}_3, \hat{\theta}_4$ are the parameters corresponding to four models, and θ_v is the virtual model created by combination of all the models.

$$\theta_v(t) = \sum_{i=1}^{N} w_i(t)\vartheta_i(t) \tag{21.17}$$

Following conditions must be fulfilled by $w_i(t)$,

- The contribution from a model can never be negative, i.e., $w_i \geq 0$.
- Sum of contributions from all the models must be unity, i.e., $\sum_{i=1}^{N} w_i(t) = 1$.

Subtracting θ from both sides of (21.17) and using above properties, it can be shown that $\sum_{i=1}^{N} w_i(t)\vartheta_i(t) = 0$. Now considering the identification error (21.4) and using the property of linearity as well as the fact that the initial state errors are zero, it can be shown that $\sum_{i=1}^{N} w_i(t)e_i(t) = 0$. It can be written as

$$[e_1(t), e_2(t), \ldots, e_N(t)]W = E(t)W = 0 \tag{21.18}$$

where $W(t) = [w_1(t), w_2(t), \ldots, w_N(t)]^{\mathrm{T}}$ and $E(t) = [e_1(t), e_2(t), \ldots, e_N(t)]$. In another way, W can be written as $W = [\underline{W}(t), w_N(t)]^{\mathrm{T}}$, where $\underline{W}(t) = [w_1(t), w_2(t), \ldots, w_{N-1}(t)]$. Using the set of differential equations derived from (21.18), $\underline{W}(t)$ can be computed. Similarly, based on the conditions that must be fulfilled by $w_i(t)$, it can be found that $w_N(t) = 1 - \sum_{i=1}^{N-1} w_i(t)$. Based on above arguments, (21.18) can be written as

$$P(t)\underline{W} = k(t) \tag{21.19}$$

To estimate the values of $\underline{W}$, an estimation model is built as

$$P(t)\underline{\hat{W}} = \hat{k}(t) \tag{21.20}$$

Using (21.19) and (21.20), error dynamics can be written as

$$P(t)\underline{\tilde{W}} = \tilde{k}(t) \tag{21.21}$$

where $\underline{\widehat{W}}$ is the estimate of $\underline{W}$ and can be obtained using the adaptive law

$$\begin{aligned} \dot{\underline{\hat{W}}} &= -P(t)\tilde{k}(t) \\ &= -P^T(t)P(t)\underline{\hat{W}}(t) + P^T(t)k(t) \end{aligned} \tag{21.22}$$

Using the combination of real estimates of plant parameters θ_i and $W(t)$ at every instant, the virtual estimate of plant parameter values $\hat{\theta}$ can be obtained and used to find the control input $u(t)$ at each instant.

21.6 Adaptive Controller Design with Multiple Models

An efficient controller must achieve desired performance in multiple environments. Multiple models are used to represent different environments in which the plant has to operate. Multiple models and controllers are expected to improve the performance in presence of large parametric uncertainties. Figure 21.2 shows the basic

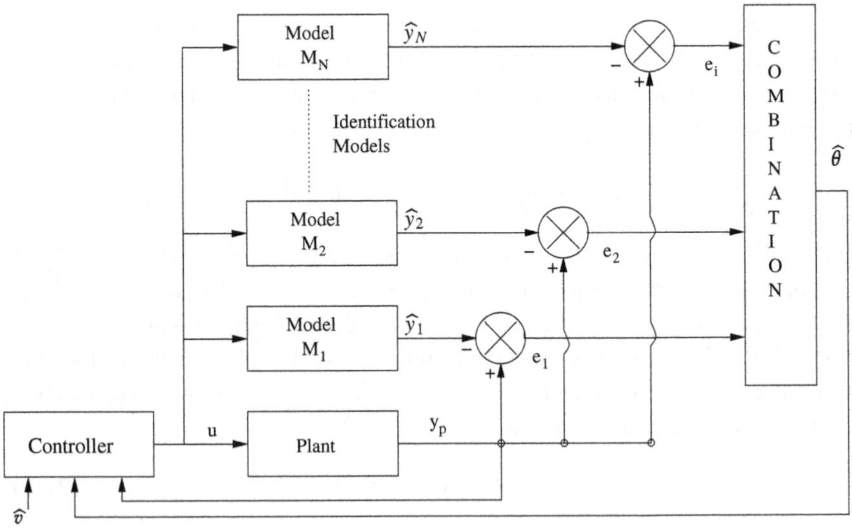

Fig. 21.2 Basic control block diagram using multiple models

block diagram using multiple models. The parameter estimates can be obtained simultaneously by online tuning of model parameters and combination of these parameters by second-level adaptation technique described in Sect. 21.5.

In (21.6), the values of $\dot{y}\ldots y^{r-1}$ will be calculated using Lie derivatives $L_f h, L_f^2 h, \ldots L_f^{r-1} h$, which are functions of unknown parameters. Certainty equivalence principle is applied to acquaint estimated values of parameters $\dot{\hat{y}}\ldots\hat{y}^{r-1}$ in (21.6). Now, let us redefine v as $\hat{v}$ as

$$\hat{v} = y_d^r + c_r(y_d^{(r-1)} - \hat{y}^{(r-1)}) + \cdots + c_2(y_d^2 - \hat{y}^2) + c_0(y_d - y) \tag{21.23}$$

The input control (21.5) can now be obtained using all the known signals as

$$u = \frac{1}{q(x,\hat{\theta})}(-p(x,\hat{\theta}) + \hat{v}) \tag{21.24}$$

21.7 Simulation Results

Let us consider the following nonlinear system [2],

$$\begin{aligned} \dot{x}_1 &= \cos x_2 + x_3 + \theta \tan^{-1} x_1 \\ \dot{x}_2 &= x_1 - x_2 \\ \dot{x}_3 &= u \\ y &= x_1 \end{aligned} \tag{21.25}$$

where θ is the unknown parameter. To obtain the relative degree of the system (21.25), the output y is differentiated until the input u appears. So, the following is obtained,

$$\begin{aligned} y &= x_1 \\ \dot{y} &= \cos x_2 + x_3 + \theta \tan^{-1} x_1 \\ \ddot{y} &= -\sin x_2 (x_1 - x_2) + \frac{\theta}{1 + x_1^2}(\cos x_2 + x_3 + \theta \tan^{-1} x_1) + u \end{aligned} \tag{21.26}$$

As u appears in the second derivative of y, the relative degree r of the system is 2. Using the state transformation as described in (21.3) yields,

$$\begin{aligned} \tau_1 &= h(x) = x_1 \\ \tau_2 &= L_f h(x) = \cos x_2 + x_3 + \theta \tan^{-1} x_1 \end{aligned} \tag{21.27}$$

The system (21.25) can be put into normal form using the diffeomorphism given in (21.27) as

$$\begin{aligned} \dot{\tau}_1 &= \tau_2 \\ \dot{\tau}_2 &= -\sin x_2(x_1 - x_2) + \frac{\theta}{1+x_1^2}(\cos x_2 + x_3 + \theta \tan^{-1} x_1) + u \end{aligned} \tag{21.28}$$

The control input can be found using (21.24) and choosing a virtual control input $\hat{v}$ as

$$u = \hat{v} - \left\{ \frac{\hat{\theta}}{1+x_1^2}(\cos x_2 + x_3 + \hat{\theta} \tan^{-1} x_1) - \sin x_2(x_1 - x_2) \right\} \tag{21.29}$$

where the virtual control or the tracking control for this simulations can be designed using (21.23) as

$$\hat{v} = \ddot{y}_r + c_1(\dot{y}_r - \dot{\hat{y}}) + c_0(y_r - y) \tag{21.30}$$

For designing the estimator model, the system dynamics can be written in regressor form using (21.8) as

$$\begin{aligned} \dot{\hat{x}} &= \omega^T \hat{\theta} \\ &= \begin{pmatrix} \cos x_2 & x_3 & \tan^{-1} x_1 \\ x_1 & -x_2 & 0 \\ u & 0 & 0 \end{pmatrix} \begin{pmatrix} 1 \\ 1 \\ \hat{\theta} \end{pmatrix} \end{aligned} \tag{21.31}$$

The desired trajectory is chosen as $y_d = 12\sin(\pi/2)t$. A value of $\theta = 5$ is chosen for the simulation [2]. For comparison purpose, simulations have also been performed for single adaptive model and six multiple adaptive models with switching. Six adaptive estimator models with starting values of parameters $\hat{\theta} = [-8, -4, -2, 2, 4, 8]$ are used. The parameter variation for the system is assumed in the range of $[-8, 8]$, and this is the reason for selecting estimator models in this range.

Trajectory tracking and tracking error for the system simulated with single adaptive model, six adaptive models with switching and six adaptive models with second-level adaptation and no switching are shown in Figs. 21.3 and 21.4, respectively. It can be seen from the figures that the transient response, in terms of convergence time, is best in case of the second-level adaptation case. This is because the weighted combination of adaptive models leads to reduction in difference between actual and desired output more rapidly than single adaptive model or multiple adaptive models with switching.

Parametric tracking for the above-mentioned three techniques are plotted in Fig. 21.5. Figure also reveals high initial transients and slow progress of estimator

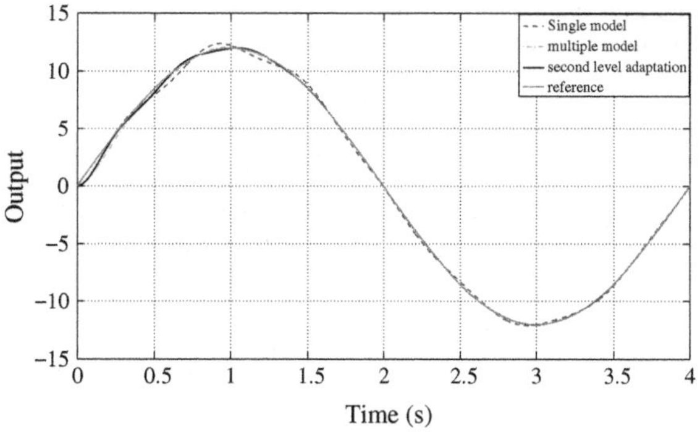

Fig. 21.3 Trajectory tracking

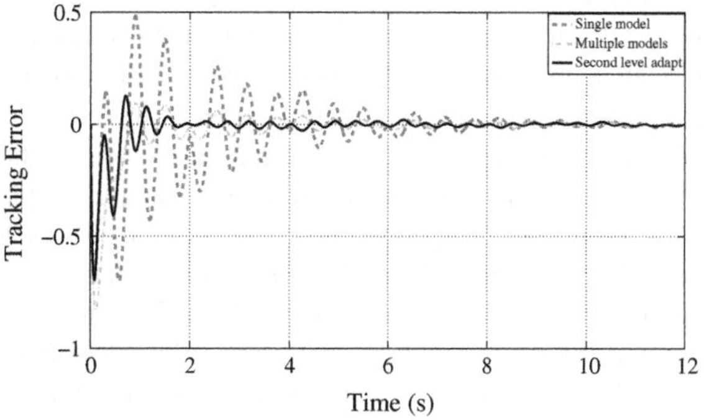

Fig. 21.4 Tracking error

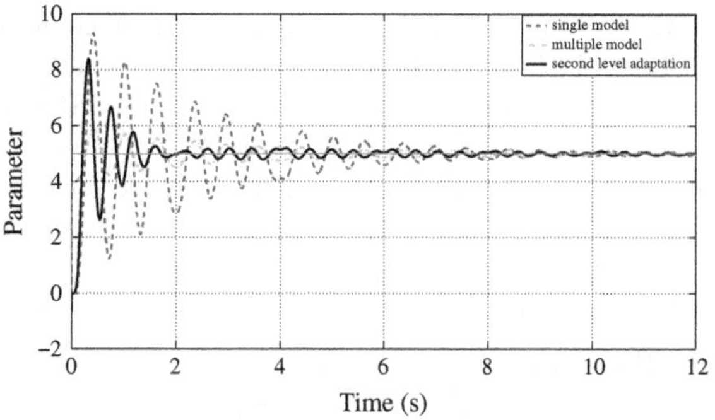

Fig. 21.5 Parameter tracking

model parameter toward the actual value in the case of single adaptive model. In this case also, second-level adaptation shows the best enhanced results. After having some initial transients as is showed, the parameter converges to the actual value after 2 s. It is also observed that change in the frequency of input signal has a direct relation with the transient performance of the estimators. An increase in the frequency of the input signal increases the amplitude of oscillation as well as settling time of the parameters.

Figure 21.6 describes the parameter tracking of all six adaptive models. It shows the convergence of the parameters of all the models to their actual values starting

Fig. 21.6 Parameter tracking for all six models

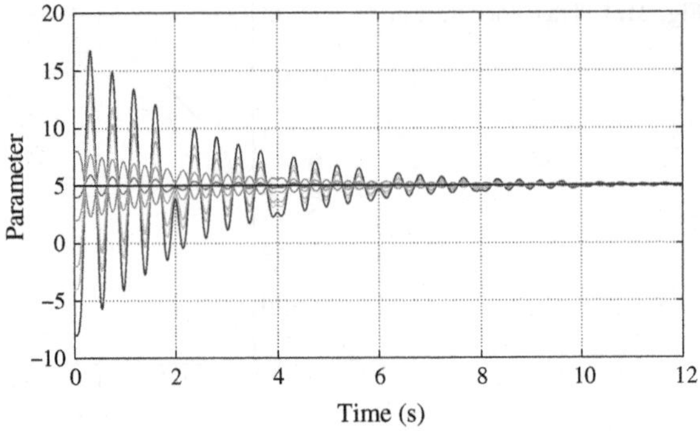

from different initial points. This picturizes the difference between multiple models with switching and multiple models with second-level adaptation. In multiple models with switching, the model corresponding to the minimum identification error at a particular instant is selected, whereas in second-level adaptation, the adaptive weights associated with all the models are tuned on the basis of identification errors of all the models. Parameter tracking using multiple models with switching, as depicted in Fig. 21.5, shows how switching leads to selection of the model closest to the real plant and the adaptation of parameter values leading to the actual plant parameters. A close observation of Figs. 21.5 and 21.6 shows that when unknown parameter space is small, a less number of models in multiple models with switching may give good results. But when unknown parameter space is large, number of models required is high. In this case, second-level adaptation with a small number of models covering the whole parameter space produces better results.

Control inputs for all the techniques described above can be found in Fig. 21.7.

Table 21.1 shows the comparison of the input and output performances for all the methods. The root mean square error (RMSE) accounts for the tracking performance. Similarly, peak overshoot (M_p), settling time (t_s), and steady-state error (e_{ss}) represent the transient and steady-state performance of the output, respectively. Control energy (CE), given by $||u||_2$, accounts for the input performance. One more

Fig. 21.7 Control input

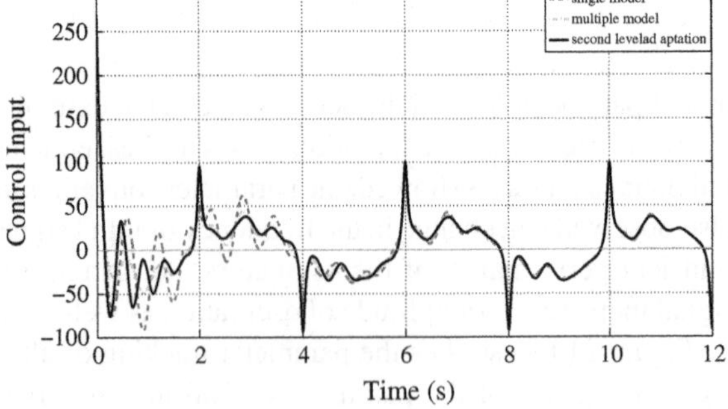

Table 21.1 Comparison between schemes

Performance specifications	Adaptation methodology		
	Single model	Multiple models with switching	Second-level adaptation
RMSE	0.1586	0.1148	0.0843
Peak overshoot (%)	48	9.5	12.5
Settling time (s)	9.5	6	2
Steady-state error	3.273×10^{-3}	2.85×10^{-3}	1.57×10^{-3}
Control energy	3.854×10^{3}	3.42×10^{3}	3.68×10^{3}
Total variation	358.4798	359.1022	360.8522

important input performance specification is the total variation (TV), which accounts for the smoothness of the control signal and input usage and is given as,

$$\mathrm{TV} = \sum_{i=1}^{k} |u_{i+1} - u_i| \tag{21.32}$$

where $u_1, u_2, \ldots, u_k$ is the discretized sequence of the input signal. It is observed from Table 21.1 that settling time and steady-state error are least in case of the proposed MMAC using second-level adaptation. This proves superior transient and steady-state performances of the proposed method over those of single and multiple models with switching. Lower value of RMSE in the proposed scheme depicts better output tracking performance. Further, CE in the proposed method is lesser than single model, but higher than multiple model with switching as expected because presence of second-level adaptation requires high CE. Hence, the proposed method has been found to be more efficient for systems demanding good transient and steady-state performance.

21.8 Conclusion

This paper presents indirect adaptive control of a SISO nonlinear plant using multiple models adaptive control with second-level adaptation. Feedback linearization technique has been used to algebraically transform the nonlinear system dynamics into a linear one. Sufficient number of multiple adaptive models is evenly distributed to cover the complete range of parameter space. Requirement of lesser number of models is another advantage of using second-level adaptation. Use of multiple models with second-level adaptation is highly recommended when better transient performance and faster convergence have high priority in the system under consideration. However, a large number of estimator models may add to the complexity of the closed-loop system. Eminence of using multiple models and second-level adaptation scheme depends on number of models and their distribution in parameter space. Simulation results show that if the position of models selected is not optimal, then the control effort needed is high.

References

1. Astrom KJ, Wittenmark B (1994) Adaptive Control, 2nd edn. Addison-Wesley Longman Publishing Co. Inc, Boston
2. Cezayirli A, Ciliz M (2007) Transient performance enhancement of direct adaptive control of nonlinear systems using multiple models and switching. Control Theory Appl IET 1 (6):1711–1725. doi:10.1049/iet-cta:20060494
3. Cezayirli A, Kemal Ciliz M (2008) Indirect adaptive control of non-linear systems using multiple identification models and switching. Int J Control 81(9):1434–1450
4. Han Z, Narendra K (2012) New concepts in adaptive control using multiple models. IEEE Trans Autom Control 57(1):78–89. doi:10.1109/TAC.2011.2152470
5. Isidori A (1995) Nonlinear control systems, 3rd edn. Springer, Secaucus
6. Kanellakopoulos I, Kokotovic P, Morse A (1991) Systematic design of adaptive controllers for feedback linearizable systems. IEEE Trans Autom Control 36(11):1241–1253. doi:10.1109/9.100933
7. Khalil H (1996) Nonlinear system. Prentice Hall, Englewood Cliffs
8. Krstic M, Kanellakopoulos I, Kokotovic PV (1995) Nonlinear and adaptive control design. Adaptive and learning systems for signal processing, communications and control series, 1st edn. Wiley-Interscience, New York
9. Kuipers M, Ioannou P (2010) Multiple model adaptive control with mixing. IEEE Trans Autom Control 55(8):1822–1836. doi:10.1109/TAC.2010.2042345
10. Narendra K, Annaswamy A (2005) Stable adaptive systems. Dover books on electrical engineering series. Dover Publications, New York
11. Narendra K, Balakrishnan J (1997) Adaptive control using multiple models. IEEE Trans Autom Control 42(2):171–187. doi:10.1109/9.554398
12. Narendra K, George K (2002) Adaptive control of simple nonlinear systems using multiple models. In: Proceedings of the American control conference, vol 3, pp 1779–1784. doi:10.1109/ACC.2002.1023824
13. Narendra KS, Balakrishnan J, Ciliz MK (1995) Adaptation and learning using multiple models, switching, and tuning. Control Syst IEEE 15(3):37–51. doi:10.1109/37.387616
14. Narendra KS, Han Z (2011) The changing face of adaptive control: the use of multiple models. Annu Rev Control 35(1):1–12. doi:10.1016/j.arcontrol.2011.03.010
15. Sastry S, Isidori A (1989) Adaptive control of linearizable systems. IEEE Trans Auto Control 34(11):1123–1131. doi:10.1109/9.40741
16. Ye X (2008) Nonlinear adaptive control using multiple identification models. Syst Control Lett 57(7):578–584. doi:10.1016/j.sysconle.2007.12.007

Chapter 22
Chattering-Free Adaptive Second-Order Terminal Sliding-Mode Controller for Uncertain System

Sanjoy Mondal and Chitralekha Mahanta

Abstract In this paper, an adaptive second-order terminal sliding-mode (SOTSM) controller is proposed for controlling uncertain systems. The design procedure is carried out in two parts. A linear sliding surface is designed first, and then, using the linear sliding surface, the terminal sliding manifold is obtained. Instead of the normal control input, its time derivative is used by the proposed control law. The actual control is obtained by integrating the derivative control input. The discontinuous sign function is contained in the derivative of the control input, and hence, chattering is eliminated in the actual control. An adaptive tuning method is designed to deal with the unknown system uncertainties, and their upper bounds are not required to be known apriori. System stability is proved by using the Lyapunov criterion. Simulation results demonstrate the effectiveness of the proposed controller for both the single-input single-output (SISO) and multi-input multi-output (MIMO) uncertain systems.

Keywords Terminal sliding mode · Non-singularity · Chattering · Adaptive tuning · Finite time convergence

22.1 Introduction

Sliding-mode control has established itself as an effective control technique which has proven to be robust against system uncertainties and external disturbances [1–3]. The sliding-mode technique is designed to drive the system state variables to equilibrium by using a suitable control law. In conventional sliding-mode control,

S. Mondal (✉) · C. Mahanta
Department of Electronics and Electrical Engineering, Indian Institute of Technology Guwahati, Guwahati 781039, India
e-mail: m.sanjoy@iitg.ernet.in

C. Mahanta
e-mail: chitra@iitg.ernet.in

V. Vijay et al. (eds.), *Systems Thinking Approach for Social Problems*,
Lecture Notes in Electrical Engineering 327,
DOI 10.1007/978-81-322-2141-8_22

the convergence of the states is asymptotic usually because of the linear type of switching surfaces leading to the convergence of the system states to the equilibrium. But this convergence can only be achieved in infinite time, although the parameters of linear sliding mode (LSM) can be adjusted to make the convergence arbitrarily fast.

In high-precision control schemes, faster convergence is a priority which can be produced in the case of asymptotic stability only at the expense of a high control input. This may result in the saturation of the actuators which is highly undesirable in practical control applications. The terminal sliding-mode (TSM) control has been developed to achieve finite time convergence and is being widely used to control both linear and nonlinear uncertain systems. The TSM was first proposed by Venkataraman et al. [4] and Zhihong et al. [5]. The TSM contains a nonlinear term to produce fast convergence without spending large control effort. However, due to the presence of the negative fractional power in the discontinuous control law of the TSM [4, 5], singularity may arise around the equilibrium point generating an unbounded control input. In [6], an indirect approach is adopted to avoid the singularity problem by switching the system between the terminal and the LSM. Later on, non-singular terminal sliding-mode (NTSM) control was proposed by Feng et al. [7] and Yu et al. [8] to avoid the singularity problem by the selection of a suitable fractional power in the discontinuous control law.

If the system state is far away from the equilibrium, the convergence of the TSM is not fast like in the linear switching manifold. In order to achieve fast convergence when the system state is far away from the equilibrium, the fast terminal sliding mode (FTSM) was proposed [9]. The FTSM produces fast convergence irrespective of the system states being far away from the equilibrium point or near to it. In [10], an adaptive fuzzy first-order terminal sliding-mode controller was proposed for linear uncertain systems with mismatched time-varying uncertainties. Feng et al. [11] employed a hybrid terminal sliding-mode observer based on non-singular terminal sliding mode and the high-order sliding mode for the rotor position and speed estimation in case of permanent magnet synchronous motor.

The TSM control features the same drawback of chattering as in the case of conventional sliding-mode control. The high-frequency chattering [12] occurs due to the discontinuous signum function present in the control input. In [13], a second-order sliding-mode controller was developed for multi-variable linear systems using the non-singular terminal sliding manifold. The major disadvantage of this method is that the application is restricted to linear uncertain systems only and the upper bound of the system uncertainty must be known in advance.

In this paper, a chattering-free adaptive second-order terminal sliding-mode (SOTSM) controller is proposed for both single-input single-output (SISO) and multi-input multi-output (MIMO) uncertain systems. In the proposed controller, a non-singular terminal sliding manifold is used to design the control law. The time derivative of the control signal is used as the control input instead of the actual control. The derivative control law is a discontinuous signal because of the presence of the sign function. However, its integral which is the actual control is continuous and, hence, the chattering is eliminated. An adaptive tuning law is used here to estimate the unknown uncertainties. This adaptive tuning method does not require prior knowledge about the upper bound of the system uncertainty for designing the TSM controller as was the case with the TSM controllers developed so far [6–8, 13–15].

The outline of this paper is as follows. Sections 22.2 and 22.3 state the control problem and the proposed chattering-free adaptive SOTSM control method. In Sect. 22.4, simulation examples for both the SISO and MIMO uncertain systems are presented to demonstrate the efficacy and advantages of the proposed algorithm. Conclusions are drawn in Sect. 22.5.

22.2 Problem Definition

Let us consider the following state equation,

$$\dot{x} = f(x) + \Delta f(x) + d(t) + bu \tag{22.1}$$

where $x = \begin{bmatrix} x_1 & x_2 & x_3 & \ldots & x_n \end{bmatrix}^{\mathrm{T}} \in R^n$ is the state vector, $u \in R^m$ is the control input, $\Delta f(x)$ is an uncertain term representing the unmodeled dynamics or structural variation of the system (22.1), and $d(t)$ is an external disturbance. The uncertainties of the system (22.1) are assumed to be bounded and matched such that $\Delta f(x)$ and $d(t) \in \text{span } b$. The control objective is to track a given reference signal x_d in finite time from any initial state.

Let the desired state vector be $x_d = \begin{bmatrix} x_{d1} & x_{d2} & x_{d3} & \ldots & x_{dn} \end{bmatrix}^T \in R^n$. The tracking error is defined as

$$\begin{aligned} e &= x - x_d \\ &= \begin{bmatrix} (x_1 - x_{d1}) & (x_2 - x_{d2}) & (x_3 - x_{d3}) & \ldots (x_n - x_{dn}) \end{bmatrix}^{\mathrm{T}} \\ &= \begin{bmatrix} e_1 & e_2 & e_3 & \ldots e_n \end{bmatrix}^{\mathrm{T}}. \end{aligned} \tag{22.2}$$

The goal is to design a chattering-free adaptive SOTSM controller for a given target x_d such that the resulting tracking error satisfies

$$\lim_{t \to \infty} ||e|| = \lim_{t \to \infty} ||x - x_d|| \to 0 \tag{22.3}$$

where $|| \cdot ||$ denotes the Euclidean norm of a vector.

22.3 Design of Chattering-Free Adaptive Second-Order Terminal Sliding-Mode Controller

The controller is designed in two steps. At first, a linear sliding surface is defined, and then, a terminal sliding manifold is obtained using the previously defined sliding surface so that the derivative of the control input occurs at the first derivative of the terminal sliding manifold. The actual control input is obtained by integrating the derivative of the control signal which contains the discontinuous function and thus eliminates the chattering [16–22]. The uncertainty is estimated by using an adaptive tuning law.

The linear sliding surface is defined as

$$s = c^{\mathrm{T}} e \tag{22.4}$$

where $c = \begin{bmatrix} c_1 & c_2 & \ldots c_n \end{bmatrix}^{\mathrm{T}}$. In the sliding mode, the error dynamics will be

$$c_n e_1^{n-1} + c_{n-1} e_1^{n-2} + \cdots + c_1 e_1 = 0 \tag{22.5}$$

The constants $c_1, c_2, \ldots, c_n$ are chosen such that the polynomial

$$\phi(\lambda) = c_n \lambda^{n-1} + c_{n-1} \lambda^{n-2} + \cdots + c_1 \lambda \tag{22.6}$$

is Hurwitz. The choice of c determines the convergence rate to the sliding surface. Let us consider (22.4), where $e = x - x_d$. The first time derivative of (22.4) yields

$$\begin{aligned} \dot{s} &= c^{\mathrm{T}} \dot{e} \\ &= c^{\mathrm{T}} (\dot{x} - \dot{x}_d) \end{aligned} \tag{22.7}$$

Using (22.1) and (22.7) yields

$$\dot{s} = c^{\mathrm{T}} (f(x) + \Delta f(x) + d(t) + bu - \dot{x}_d) \tag{22.8}$$

Taking the derivative of (22.8), we obtain

$$\begin{aligned} \ddot{s} &= c^{\mathrm{T}} \left(\frac{\mathrm{d}}{\mathrm{d}t} f(x) + \frac{\mathrm{d}}{\mathrm{d}t} \Delta f(x) + \dot{d}(t) + b\dot{u} - \ddot{x}_d \right) \\ &= c^{\mathrm{T}} (\dot{f}(x) + \Delta \dot{f}(x) + \dot{d}(t) + b\dot{u} - \ddot{x}_d) \end{aligned} \tag{22.9}$$

A NTSM manifold is first designed as

$$\sigma = s + \beta \dot{s}^{\frac{p}{q}} \tag{22.10}$$

Here, $\beta > 0$ and p/q (p and q are positive odd integers) are chosen in such a way that the condition $1 < \frac{p}{q} < 2$ holds [8, 23]. The linear sliding surface s is combined with the non-singular terminal sliding manifold σ to realize the TSM control. As σ reaches zero in finite time, both s and $\dot{s}$ are bound to reach zero. Then, the tracking error e asymptotically converges to zero.

Taking the time derivative of (22.10), we have

$$\begin{aligned}\dot{\sigma} &= \dot{s} + \beta\left(\frac{p}{q}\right)\dot{s}^{\frac{p}{q}-1}\ddot{s} \\ &= \beta\left(\frac{p}{q}\right)\dot{s}^{\frac{p}{q}-1}\left(\ddot{s} + \beta^{-1}\left(\frac{p}{q}\right)^{-1}\dot{s}^{2-\left(\frac{p}{q}\right)}\right)\end{aligned} \tag{22.11}$$

Assumption 1 The uncertain term $\Delta f(x)$ and the disturbance $d(t)$ in (22.1) are assumed to be bounded, i.e., there exists a positive bounded function $B(x)$ satisfying the following inequalities:

$$||c^{\mathrm{T}}(\Delta f(x) + d(t))|| < B \tag{22.12}$$

Assumption 2 The first time derivative of the uncertain term, $\Delta\dot{f}(x)$, and the first time derivative of the disturbance, $\dot{d}(t)$, are assumed to be bounded, i.e., there exists a positive bounded function $\bar{B}(x)$ such that the following inequalities hold [24]:

$$||c^{\mathrm{T}}(\Delta\dot{f}(x) + \dot{d}(t))|| < \bar{B} \tag{22.13}$$

It will be proven in Theorem 1 that since the derivative of the control input contains the discontinuous term, the actual control signal which will be obtained after the integration operation will not contain any high-frequency switching component. Thus, the proposed TSM controller will be free from the chattering phenomenon [23].

Theorem 1 *Considering the uncertain system* (22.1), *the tracking error dynamics* (22.7) *can asymptotically converge to zero if the non-singular terminal sliding manifold is chosen as* (22.10) *and the control law is obtained as follows*

$$u = u_{\mathrm{eq}} + u_n$$

where

$$\dot{u}_{\mathrm{eq}} = -(c^{\mathrm{T}}b)^{-1}c^{\mathrm{T}}\dot{f}(x)$$

thus

$$u_{\mathrm{eq}} = -(c^{\mathrm{T}}b)^{-1}c^{\mathrm{T}} \int \dot{f}(x)\mathrm{d}\tau \quad (22.14)$$

u_n can be found as

$$\begin{aligned}\dot{u}_n &= -(c^{\mathrm{T}}b)^{-1}(\beta^{-1}(\frac{p}{q})^{-1}\dot{s}^{2-(\frac{p}{q})} \\ &\quad + \bar{B}\mathrm{sgn}(\sigma) + \varepsilon\sigma - c^{\mathrm{T}}\ddot{x}_d) \\ u_n &= -(c^{\mathrm{T}}b)^{-1} \int \left(\beta^{-1}\left(\frac{p}{q}\right)^{-1} \dot{s}^{2-(\frac{p}{q})} + \bar{B}\mathrm{sgn}(\sigma) \right. \\ &\quad \left. + \varepsilon\sigma - c^{\mathrm{T}}\ddot{x}_d\right)\mathrm{d}\tau \end{aligned} \quad (22.15)$$

where $(c^{\mathrm{T}}b)^{-1}$ is non-singular, $\mathrm{sgn}(\sigma)$ is the sign function, and $||c^{\mathrm{T}}(\Delta\dot{f}(x) + \dot{d}(t))|| < \bar{B}$ and $\varepsilon > 0$ are the designed parameters.

In practice, the uncertain term $||c^{\mathrm{T}}(\Delta\dot{f}(x) + \dot{d}(t))||$ is often difficult to know. Hence, an adaptive tuning law is designed to determine $\bar{B}$. So the control law is represented as

$$\begin{aligned} u &= u_{\mathrm{eq}} + u_n \\ u_{\mathrm{eq}} &= -(c^{\mathrm{T}}b)^{-1}c^{\mathrm{T}} \int \dot{f}(x)\mathrm{d}\tau \\ u_n &= -(c^{\mathrm{T}}b)^{-1} \int \left(\beta^{-1}\left(\frac{p}{q}\right)^{-1} \dot{s}^{2-(\frac{p}{q})} + \hat{T}\mathrm{sgn}(\sigma) \right. \\ &\quad \left. + \varepsilon\sigma - c^{\mathrm{T}}\ddot{x}_d\right)\mathrm{d}\tau \end{aligned} \quad (22.16)$$

where $\hat{T}$ estimates the value of $c^{\mathrm{T}}(||\Delta\dot{f}(x) + \dot{d}(t)||)$. Defining the adaptation error as $\tilde{T} = \hat{T} - T$, the parameter $\hat{T}$ is estimated by using the adaptation law

$$\dot{\hat{T}} = \frac{\beta(\frac{p}{q})}{\kappa} ||\dot{s}^{(\frac{p}{q})-1}\sigma|| \quad (22.17)$$

where κ is a positive tuning parameter.

A Lyapunov function is selected as $V(t) = \frac{1}{2}\sigma^2 + \frac{1}{2}\kappa\tilde{T}^2$. Using (22.9), (22.11) and the control law (22.16), the time derivative of the Lyapunov function $V(t)$ becomes

$$
\begin{aligned}
\dot{V}(t) &= \sigma\dot{\sigma} + \kappa\tilde{T}\dot{\tilde{T}} \\
&= \beta\left(\frac{p}{q}\right)\dot{s}^{\left(\frac{p}{q}\right)-1}\sigma\left(\ddot{s} + \beta^{-1}\left(\frac{p}{q}\right)^{-1}\dot{s}^{2-\left(\frac{p}{q}\right)}\right) + \kappa\tilde{T}\dot{\tilde{T}} \\
&= \beta\left(\frac{p}{q}\right)\dot{s}^{\left(\frac{p}{q}\right)-1}\sigma(\bar{B} - \hat{T}\mathrm{sgn}(\sigma) - \varepsilon\sigma) + \kappa(\hat{T} - T)\dot{\hat{T}} \\
&\leq \beta\left(\frac{p}{q}\right)\dot{s}^{\left(\frac{p}{q}\right)-1}(\bar{B}||\sigma|| - T||\sigma||)
\end{aligned} \tag{22.18}
$$

The above inequality holds if $\dot{\hat{T}} = \frac{\beta\left(\frac{p}{q}\right)}{\kappa}||\dot{s}^{\left(\frac{p}{q}\right)-1}\sigma||$ and $T \geq \bar{B}$. Moreover, $||\dot{s}^{\left(\frac{p}{q}\right)-1}|| > 0$ for any $||\dot{s}|| \neq 0$ and $||\dot{s}^{\left(\frac{p}{q}\right)-1}|| = 0$ only when $||\dot{s}|| = 0$. This ensures the convergence of σ and the states to equilibrium.

Remark 1 Theoretically, $||\sigma||$ cannot become exactly zero in finite time, and thus, the adaptive parameter $\dot{\hat{T}}$ may increase boundlessly. A simple way of overcoming this disadvantage is to modify the adaptive tuning law (22.17) by using the dead-zone technique [12, 23] as

$$
\dot{\hat{T}} = \begin{cases} \frac{\beta\left(\frac{p}{q}\right)}{\kappa}||\dot{s}^{\left(\frac{p}{q}\right)-1}\sigma||, & ||\dot{s}^{\left(\frac{p}{q}\right)-1}\sigma|| \geq v \\ 0, & ||\dot{s}^{\left(\frac{p}{q}\right)-1}\sigma|| < v \end{cases} \tag{22.19}
$$

where v is a small positive constant.

The above theorem proves that the states reach the non-singular terminal sliding manifold $\sigma = 0$ in finite time [7]. Suppose that t_r is the time when σ reaches zero from $\sigma(0) \neq 0$, i.e., $\sigma = 0$ for all $t \geq t_r$. Once σ reaches zero, it will stay at zero using the control law (22.14). Thus, the sliding surface s will converge to zero in finite time. The total time from $\sigma(0) \neq 0$ to s_{tf} can be calculated by using the equation $s + \beta\dot{s}^{\left(\frac{p}{q}\right)} = 0$ (22.10) from which the time taken from s_{tr} to s_{tf} [8] is obtained as

$$
t_f = t_r + \frac{\left(\frac{p}{q}\right)}{\left(\frac{p}{q}\right) - 1}\beta^{-\left(\frac{p}{q}\right)}||s_{tr}||^{\left(\frac{p}{q}\right)-1} \tag{22.20}
$$

Therefore, $s = \dot{s} = 0$ and the system states converge to the sliding surface asymptotically.

Assumption 3 The missing derivatives of the functions can be estimated by using the Levant's exact differentiator [25].

Assumption 4 The controller parameters c^{T}, β and $\frac{p}{q}$ should be properly chosen to ensure fast convergence.

22.4 Simulation Results

The proposed chattering-free adaptive SOTSM controller is applied to uncertain SISO and MIMO systems both. The simulations are carried out in the MATLAB–Simulink platform by using ODE 4 solver with a fixed step size of 0.001 s.

22.4.1 Example 1 Linear Uncertain SISO System

The proposed chattering-free adaptive SOTSM controller is applied to an uncertain linear system of second order, and the performance is compared against the NTSM controller proposed by Feng et al. [7]. The second-order uncertain linear dynamical system considered here is given by

$$\begin{aligned}\dot{x}_1 &= x_2 \\ \dot{x}_2 &= 0.1\sin 20t + u\end{aligned} \tag{22.21}$$

The initial states are assumed as $x_1(0) = 1$ and $x_2(0) = 0$.

For the above linear uncertain SISO system, the state response and the control input obtained by using the NTSM control proposed by Feng et al. [7] are shown in Figs. 22.1 and 22.2. It is observed from these figures that although no singularity occurs, high-frequency chattering is present in the control input.

The design parameter of the proposed controller is chosen as $c^{\mathrm{T}} = [2\ 1], \beta = 1$, $\kappa = 1/6$ and $p/q = 5/3$ [7]. ε is chosen as 9. The chattering-free adaptive SOTSM control law is obtained as

$$u = \int \dot{u}\mathrm{d}\tau = \int \dot{u}_{\mathrm{eq}}\mathrm{d}\tau + \int \dot{u}_n\mathrm{d}\tau \tag{22.22}$$

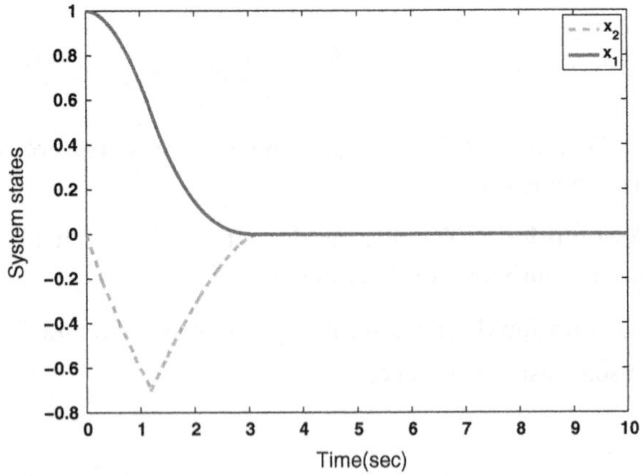

Fig. 22.1 State response with the NTSM controller proposed in [7]

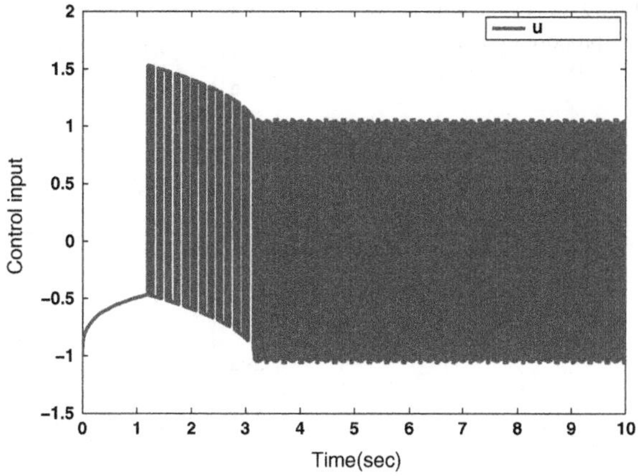

Fig. 22.2 Control input with the NTSM controller proposed in [7]

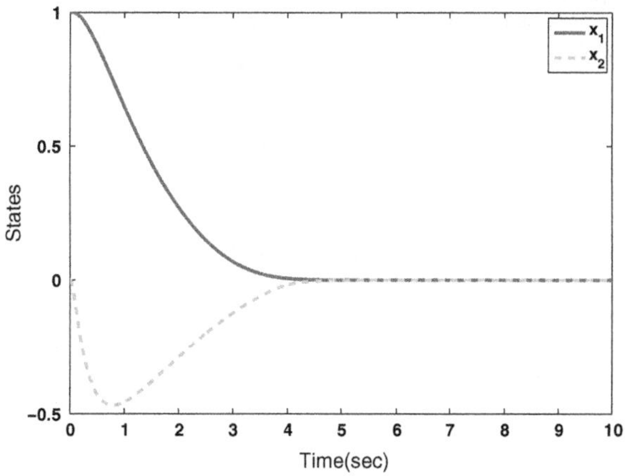

Fig. 22.3 State response with the proposed controller (22.23)

where $\dot{u}$ can be expressed as

$$\dot{u} = -\left(\begin{bmatrix} 2 & 1 \end{bmatrix}\begin{bmatrix} 0 \\ 1 \end{bmatrix}\right)^{-1}\left(\begin{bmatrix} 2 & 1 \end{bmatrix}\right)\begin{bmatrix} 0 & 1 \\ 0 & 0 \end{bmatrix}\begin{bmatrix} \dot{x}_1 \\ \dot{x}_2 \end{bmatrix} + (3/5)\dot{s}^{(1/3)} + \hat{T}\mathrm{sign}(\sigma) + 9\sigma) \tag{22.23}$$

where tuning law is chosen as $\dot{\hat{T}} = 10||\dot{s}^{(\frac{5}{3})-1}\sigma||$ and $\hat{T}(0) = 0$. Here, the sliding surface is chosen as $s = [2\ 1]x$ where $x = [x_1\ x_2]^T$ is the state vector. The non-singular terminal sliding manifold is chosen as $\sigma = s + \dot{s}^{5/3}$.

Figures 22.3, 22.4 and 22.5 show the system states, control input and estimated adaptive gain obtained by using the proposed chattering-free adaptive SOTSM

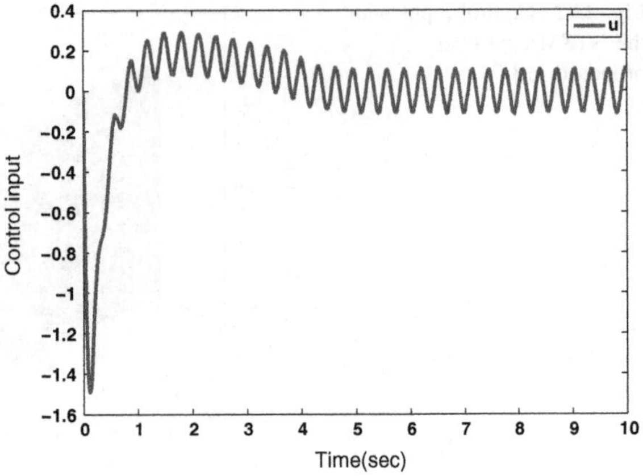

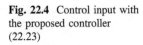

Fig. 22.4 Control input with the proposed controller (22.23)

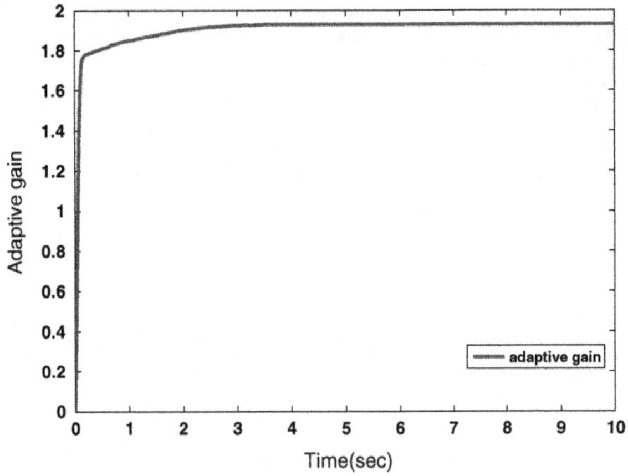

Fig. 22.5 Adaptive gain with the proposed controller (22.23)

control law. It is obvious from these figures that the proposed controller ensures finite time stabilization and provides a control input which is chattering free.

22.4.2 Example 2 Tracking Control of a Nonlinear Uncertain MIMO System

Let us consider a nonlinear uncertain MIMO system given by [26]

$$\begin{aligned} \dot{x} &= f(x) + \Delta f(x) + d(t) + bu \\ y &= x \end{aligned} \tag{22.24}$$

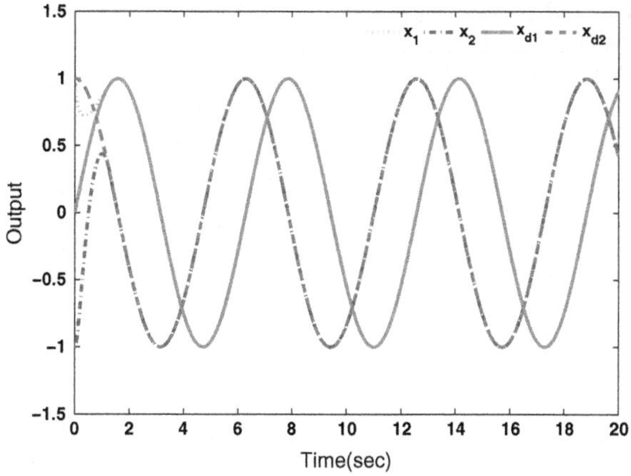

Fig. 22.6 Tracking response with the proposed chattering-free adaptive SOTSM controller

where

$$f(x) = \begin{bmatrix} x_2 \\ (1-\exp(-x_2))/(1+\exp(-x_2)) \end{bmatrix}, \quad \Delta f(x) = \begin{bmatrix} 0 \\ 0.1 \times \sin(4\pi x_1)\sin(2\pi x_2) \end{bmatrix} \tag{22.25}$$

and

$$d(t) = \begin{bmatrix} 0.1 \times \sin(0.86t) \\ 0.2 \times \cos(0.86t) \end{bmatrix}, \quad u = \begin{bmatrix} u_1 \\ u_2 \end{bmatrix}, \quad b = \begin{bmatrix} 1 & 0 \\ 0 & 1 \end{bmatrix} \tag{22.26}$$

The initial states are considered as $x_1(0) = 1, x_2(0) = -1$, and the following parameters are chosen:

$$c^{\mathrm{T}} = \begin{bmatrix} 1 & 0 \\ 0 & 1 \end{bmatrix}, \quad \beta = \begin{bmatrix} 0.4 & 0 \\ 0 & 0.4 \end{bmatrix}, \quad \varepsilon = \begin{bmatrix} 10 & 0 \\ 0 & 10 \end{bmatrix}, \quad \kappa = \frac{1}{3}\begin{bmatrix} 0.4 & 0 \\ 0 & 0.4 \end{bmatrix} \tag{22.27}$$

The value of p/q is assumed as 5/3, and the initial condition of the adaptive tuning parameter is 0. The control objective is to track the reference trajectory $x_d = [\sin t \;\; \cos t]^{\mathrm{T}}$. Thus, x_{1d} and x_{2d} are considered as $x_{1d} = \sin t$ and $x_{2d} = \cos t$.

Figures 22.6, 22.7, 22.8, 22.9 and 22.10 show the simulation results obtained by using the proposed chattering-free adaptive SOTSM controller. The tracking response and the control input are plotted in Figs. 22.6 and 22.7, respectively. It is evident from these figures that the tracking performance is excellent and the control

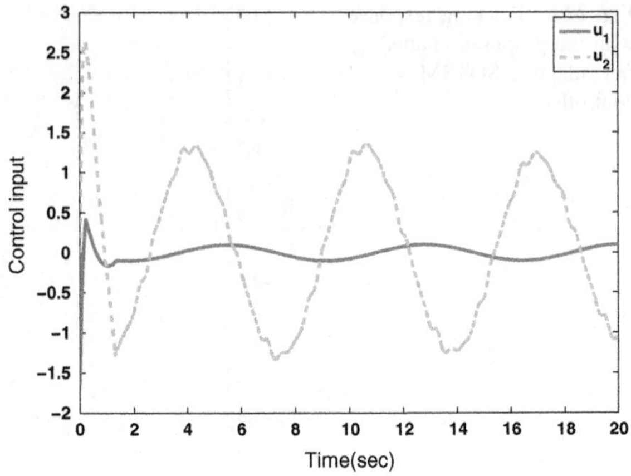

Fig. 22.7 Control input with the proposed chattering-free adaptive SOTSM controller

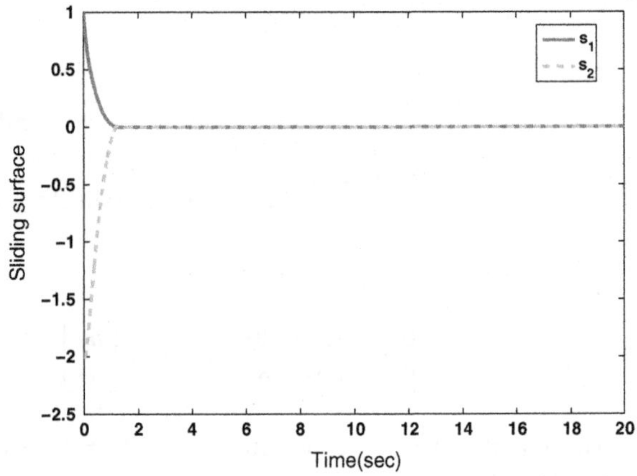

Fig. 22.8 Sliding surface with the proposed chattering-free adaptive SOTSM controller

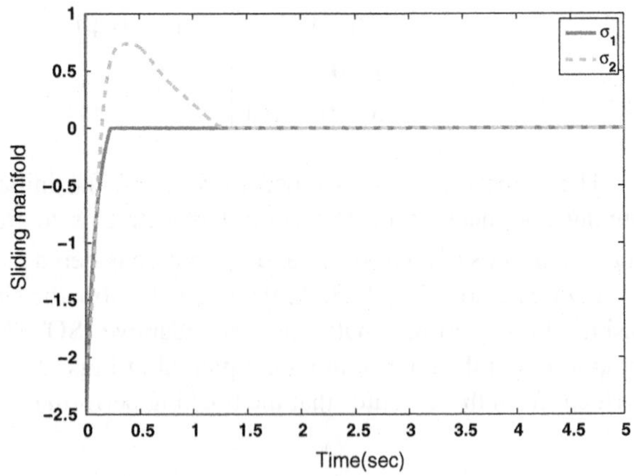

Fig. 22.9 Sliding manifold with the proposed chattering-free adaptive SOTSM controller

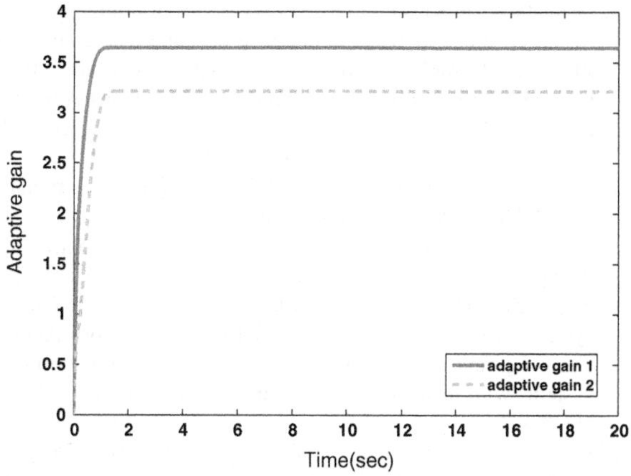

Fig. 22.10 Adaptive gain with the proposed chattering-free adaptive SOTSM controller

input is chattering free. The sliding surface and the sliding manifold obtained in the case of the proposed chattering-free adaptive SOTSM controller are shown in Figs. 22.8 and 22.9. The parameter $\hat{T}$ is shown in Fig. 22.10 to confirm its convergence.

22.5 Conclusion

A chattering-free adaptive SOTSM controller for uncertain systems is proposed in this paper. The controller acts on the first derivative of the control input which contains the switching term involving the sign function. The actual control law is obtained by integrating the discontinuous derivative control signal, and hence, it is continuous. The requirement of prior knowledge about the uncertainty bounds for the design of traditional sliding-mode controller is not a necessary requirement in the proposed controller. Simulation results demonstrate that the proposed control strategy is successful in eliminating the undesired chattering in the control input while ensuring a satisfactory tracking performance. Hence, the proposed controller is suitable for practical applications.

References

1. Emelyanov SV (1967) Variable structure control systems. Nauka 35:120–125
2. Utkin VI, Young KKD (1978) Methods for constructing discontinuity planes in multidimensional variable structure systems. Autom Remote Control 39:1466–1470
3. Young KD, Utkin VI, Özgüner Ü (1999) A control engineers guide to sliding mode control. IEEE Trans Control Syst Technol 7(3):328–342

4. Venkataraman ST, Gulati S (1992) Control of nonlinear systems using terminal sliding modes. In: Proceedings of IEEE conference on American control conference, pp 891–893
5. Zhihong M, Paplinski AP, Wu HR (1994) A robust MIMO terminal sliding mode control scheme for rigid robotic manipulators. IEEE Trans Autom Control 39(12):2464–2469
6. Zhihong M, Yu XH (1997) Terminal sliding mode control of MIMO linear systems. IEEE Trans Circuits Syst I 44(11):1065–1070
7. Feng Y, Yu X, Man Z (2002) Non-singular terminal sliding mode control of rigid manipulators. Automatica 38:2159–2167
8. Yu S, Yu X, Shirinzadeh B, Man Z (2005) Continuous finite-time control for robotic manipulators with terminal sliding mode. Automatica 41:1957–1964
9. Yu X, Zhihong M (2002) Fast terminal sliding-mode control design for nonlinear dynamical systems. IEEE Trans Circuits Syst I Fundam Theory Appl 49(2):261–264
10. Tao CW, Taur JS, Chan M-L (2004) Adaptive fuzzy terminal sliding mode controller for linear systems with mismatched time-varying uncertainties. IEEE Trans Syst Man Cybern Part B 34 (1):255–262
11. Feng Y, Zheng J, Yu X, Truong NV (2009) Hybrid terminal sliding mode observer design method for a permanent-magnet synchronous motor control system. IEEE Trans Industr Electron 56(9):3424–3431
12. Utkin VI (1992) Sliding modes in control and optimization. Springer, Berlin
13. Feng Y, Han X, Wang Y, Yu X (2007) Second-order terminal sliding mode control of uncertain multivariable systems. Int J Control 80(6):856–862
14. Wang Y, Zhang X, Yuan X, Liu G (2011) Position-sensorless hybrid sliding-mode control of electric vehicles with brushless dc motor. IEEE Trans Veh Technol 60(2):421–432
15. Chang F-J, Chang E-C, Liang T-J, Chen JF (2011) Digital-signal-processor-based DC/AC inverter with integral-compensation terminal sliding-mode control. IET Power Electron 4 (1):159–167
16. Moreno JA, Osorio M (2008) A Lyapunov approach to second-order sliding mode controllers and observers. In: 47th IEEE conference on decision and control, pp 2856–2861
17. Mondal S, Mahanta C (2011) Nonlinear sliding surface based second order sliding mode controller for uncertain linear systems. Commun Nonlinear Sci Numer Simul 16:3760–3769
18. Zhang N, Yang M, Jing Y, Zhang S (2009) Congestion control for DiffServ network using second-order sliding mode control. IEEE Trans Industr Electron 56(9):3330–3336
19. Mondal S, Mahanta C (2012) A fast converging robust controller using adaptive second order sliding mode. ISA Trans 51(6):713–721
20. Li H, Liao X, Li C, Li C (2011) Chaos control and synchronization via a novel chatter free sliding mode control strategy. Neurocomputing 74(17):3212–3222
21. Mondal S, Mahanta C (2012) Adaptive integral higher order sliding mode controller for uncertain systems. J Control Theory Appl 10:212–217
22. Mondal S, Mahanta C (2013) Chattering free adaptive multivariable sliding mode controller for systems with matched and mismatched uncertainty. ISA Trans 52(3):335–341
23. Mondal S, Mahanta C (2014) Adaptive second order terminal sliding mode controller for robotic manipulators. J Franklin Inst. 351(4):2356–2377
24. Mondal S, Mahanta C (2012) Adaptive second-order sliding mode controller for a twin rotor multi-input multi-output system. IET Control Theory Appl 6(14):2157–2167
25. Levant A (2003) Higher-order sliding modes, differentiation and output feedback control. Int J Control 76(9/10):924–941
26. Roopaei M, Jahromi MZ (2009) Chattering-free fuzzy sliding mode control in MIMO uncertain systems. Nonlinear Anal Theory Methods Appl 71:4430–4437

Chapter 23
Modelling and Simulation of Mechanical Torque Developed by Wind Turbine Generator Excited with Different Wind Speed Profiles

Parikshit G. Jamdade and Shrinivas G. Jamdade

Abstract Proper representation of the wind speed and the mechanical torque is necessary in designing the wind turbine generator system (WTGS) and its components. This paper presents the effects of different wind speed profiles on mechanical torque output of the WTGS. Mechanical torque model consists of mathematical models of wind speed, mechanical power, mechanical torque and power coefficient of WTGS. For the case study purpose, wind speed data of Chalkewadi, Satara location, are used. Four different wind speed profiles are modelled according to the wind speed data of Chalkewadi, Satara location. WTGS is excited with the modelled wind speed profiles, and response to these wind speed profiles on mechanical torque produced by WTGS is observed. There is a reduction of 56 % in mechanical torque for three cases when compared to wind speed with all four designed components. The presented model, dynamic simulation and simulation results are tested in MATLAB/Simulink and presented.

Keywords Wind turbine generator system (WTGS) · Wind speed model · Power coefficient (C_p) · Mechanical power · Mechanical torque · MATLAB/Simulink interface

23.1 Introduction

The commercialization of renewable energy technologies is needed especially for rural, social and economic development in India. Wind, being a non-polluting and non-toxic energy source, will go a long way in solving our energy requirements.

P.G. Jamdade (✉)
Department of Electrical Engineering, PVG's College of Engineering and Technology, Pune, Maharashtra, India
e-mail: parikshit_jamdade@yahoo.co.in

S.G. Jamdade
Department of Physics, Nowrosjee Wadia College, Pune, Maharashtra, India

V. Vijay et al. (eds.), *Systems Thinking Approach for Social Problems*,
Lecture Notes in Electrical Engineering 327,
DOI 10.1007/978-81-322-2141-8_23

Wind can be utilized in windmills, which in turn drive a generator to produce electricity [1]. In India, few designs of windmills were developed but could not sustain. An important reason could be that wind velocity in India, apart from the coastal region, is relatively low and varies appreciably with the seasons. This low velocity and seasonal winds imply a higher cost of exploitation of wind energy. The solution lies with the proper analysis of wind pattern, and then only, proper and commercial windmill can be designed [2–6].

Wind turbines produce electricity by using the power of the wind to drive an electrical generator. Wind passes over the blades, generating lift and exerting a turning force. The rotating blades turn a shaft inside the nacelle, which goes into a gearbox. The gearbox increases the rotational speed to that which is appropriate for the generator, which uses magnetic fields to convert the rotational energy into electrical energy. The power output goes to a transformer, which converts the electricity from the generator to the appropriate voltage for the power collection system.

An overall wind energy system can be divided into following components:

1. the wind model
2. the turbine model
3. the shaft and gearbox model
4. the generator model and
5. the control system model.

The most important parameter for wind energy is the wind speed. Statistical parameters and methods are useful for estimating wind speed as it is random in nature. For this reason, wind speed probabilities can be estimated by using various modelled wind speed profiles. In previous papers, Weibull distribution, Rayleigh distribution and extreme value distribution models are used for characterization of wind speeds. Knowledge of the statistical properties of wind speed is essential for predicating the energy output of wind turbine generator system (WTGS). An accurate determination of the probability distribution of wind speed values is very important for evaluating wind speed energy potential of a region [4, 7–9]. Most statistical methods assume a specific distribution in the calculation of their result but when we assume that our data are a result of a specific distribution model, we are taking huge risk. If such condition does not hold, then the results obtained may be incorrect [10–12]. To overcome this disadvantage, depending on statistical parameters of the wind, we modelled the wind speed.

The mechanical system of the wind turbines plays a big role in the energy transformation. Most of the simple wind turbine gearbox consists of two main shafts, the low-speed shaft which is basically connected with the turbine blades and the second one which is called the high-speed shaft connected directly to the generator and shown in figure.

Modern large wind turbines are variable speed machines. When the wind speed is below the rated value, generator torque is used to control the rotor speed in order to capture as much power as possible. The most power is captured when the tip speed ratio (TSR) is held constant at its optimum value ($\lambda = 8.1$ at $\theta = 0$). This means that as wind speed increases, rotor speed should increases proportionally. The difference between the aerodynamic torque captured by the blades and the applied generator torque controls the rotor speed. If the generator torque is lower, the rotor accelerates, and if the generator torque is higher, the rotor slows down. In below-rated wind speeds, the generator torque control is active while the blade pitch is typically held at the constant angle that captures the most power, fairly flat to the wind. In above-rated wind speeds, the generator torque is typically held constant while the blade pitch is active.

As torque is applied on generator shaft which is the basic parameter for designing flux controller to reduce transients, gear train/gearbox designing by knowing gearbox ratio by which efficiency of wind turbine generator is going to be decided, deciding suitable applicability by analysing dynamic characteristic, transient analysis and stabilization of generated voltage for unloaded and loaded conditions and also development of mathematical models of machines specially generators. Therefore, studying variation in torque is extremely important so in this paper, we are studying variation in mechanical torque with statistically modelled wind speeds.

The paper is organized as follows. Section 23.2 describes the dynamic system models used in the paper. In this mathematical model of wind speed, mechanical power, mechanical torque and power coefficient of WTGS are presented. Section 23.3 describes the results and discussions obtained in the paper. Section 23.4 is the concluding section.

23.2 System Models

Block diagram of the mechanical torque model of WTGS used in the study is shown in Fig. 23.1. Entire mechanical torque model is subdivided into mathematical models of wind speed, mechanical power, mechanical torque and power coefficient of WTGS.

23.2.1 Wind Speed Model

Data sets of 2008–2012 are obtained containing the mean wind speed of each month in a year of location Chalkewadi, Satara in Maharashtra, India, with an observation height of 20 m above the ground level. The Chalkewadi, Satara location, is located at a latitude of 17° 59′ N and longitude of 73° 84′ E in Maharashtra, India.

We are analysing the wind speed data for variations with the help of statistical parameters such as mean, median, standard deviation and skewness.

Mean—It is an average value of values of a data set.

$$\overline{V} = \frac{1}{n}\sum_{i=1}^{n} V_i \tag{23.1}$$

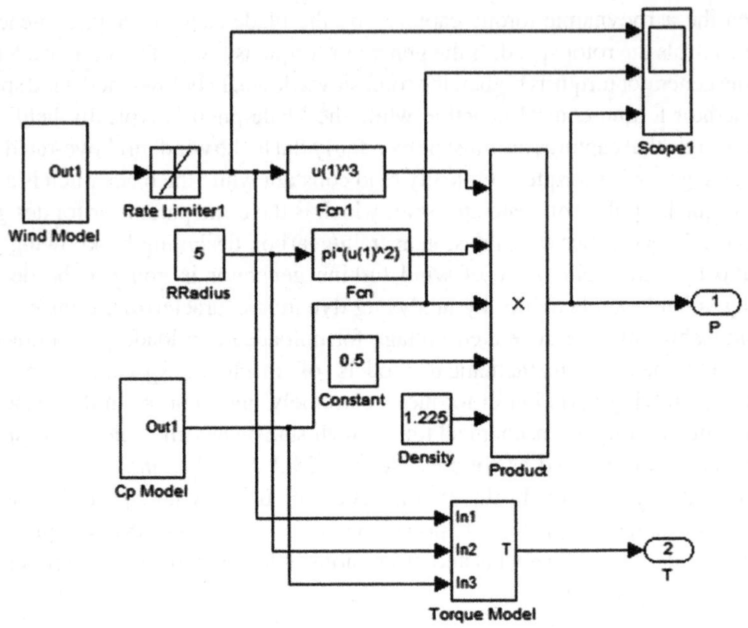

Fig. 23.1 Block diagram of entire mechanical torque model

Mean indicates the wind regime has a potential for selecting wind turbines.

Median—It is a value which divides the entire distribution of data into two equal parts. It is a positional average.

Standard deviation—It is a widely used measure of variability. If the standard deviation is large, the degree of uniformity or homogeneity of the series is less.

$$\sigma(\overline{V}) = \sqrt{\frac{1}{n}\sum_{i=1}^{n}(V_i - \overline{V})^2} \tag{23.2}$$

Skewness—It is a measure of the asymmetry of the distribution of data having real values. The skewness value can be positive or negative. A negative skew indicates that the tail on the left side of the probability density function lies longer than the right side and the bulk of the values (including the median) to the right of the mean. This shows that the mean is smaller than the median.

A positive skew indicates that the tail on the right side is longer than the left side and the bulk of the values lying to the left of the mean. This shows that the mean is longer than the median. A zero value indicates that the values are relatively evenly distributed on both sides of the mean, typically but not necessarily implying a symmetric distribution. Larger the value of skewness, larger the variation in data values. Table 23.1 is showing summarized statistics for Chalkewadi, Satara location, from the years 2008 to 2012, used for case study purpose.

Table 23.1 Summarized statistics for Chalkewadi, Satara

Year	2008–2012
Minimum	3.4
Maximum	9.4
Mean	5.54
Median	4.85
Skewness	1.16
Standard deviation	2.03

In this wind model, step changes of wind speed values can be introduced. They would be represented by means of a step, repeating sequence, rate limiter and uniform and non-uniform random number block, in which wind speed values and the moment when the transition from one value to another one takes place can be fixed. By this, one can also model the wind gust and the repeating sequence in wind speed depending on the site characteristics.

The wind speed model is having four main blocks, base wind speed component, ramp wind speed component, gust wind speed component and base noise wind speed component. Variation in wind speed is obtained by having different combinations of four wind speed components with random number blocks. The model implemented for wind speed in Simulink is shown in Fig. 23.2.

$$V_{\mathrm{w}}(t) = v_{\mathrm{b}}(t) + v_{\mathrm{r}}(t) + v_{\mathrm{g}}(t) + v_{\mathrm{n}}(t) \tag{23.3}$$

where

V_{b} is the constant base wind component
V_{r} is the ramp wind component
V_{g} is the gust wind component
V_{n} is the base noise wind component, all of them in m/s

Fig. 23.2 Wind speed model

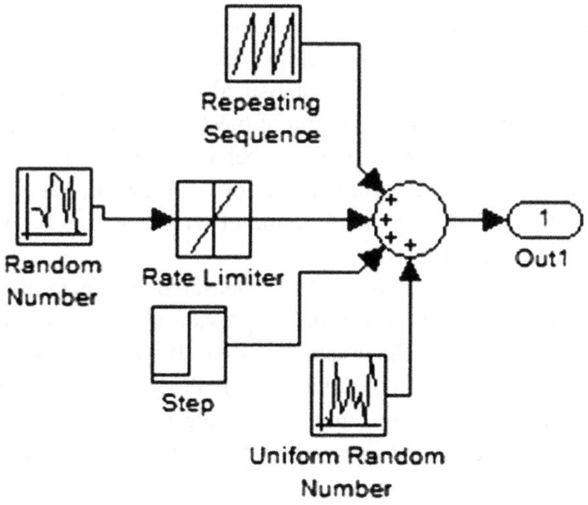

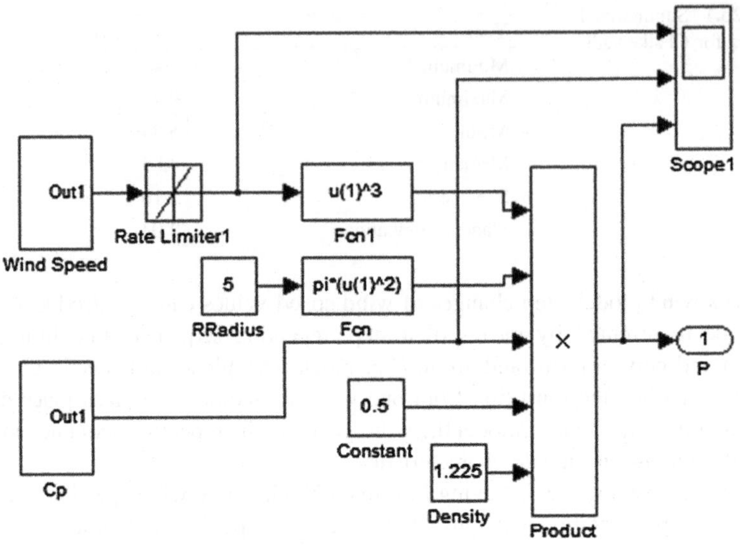

Fig. 23.3 Mechanical power model

The main component is a random number block which represents the base component of the wind speed. The ramp component is used to repeat the value of wind speed which is modelled by repeating sequence block, and gust component is represented for sudden increases in wind speed which is represented by step block. The uniform random number block is expressed to represent the noise component of the wind speed. Rate limiter is used for changing the rate of change of wind speed. By putting rate limiter in the simulation, we are ensuring that wind speed should not be above the cut-out speed (furling speed) of the turbine and should not be under the cut-in speed of the turbine. In both cases where wind speed above cut-out speed and below the cut-in speed, the turbine will not work properly resulting in the underrated operation (Figs. 23.3, 23.4 and 23.5).

With the help of this statistical data, we modelled the wind speed to provide more realistic representation of mechanical power output of practical WTGS and to simulate the effects of wind behaviour properly, including gusting, rapid ramp changes and background noise for Chalkewadi, Satara location. In Figs. 23.6, 23.7, 23.8 and 23.9, maximum value of wind speed is 9.4 with changes in variations in wind speed waveform.

The wind speed profile is measured by anemometer and taken as a constant piecewise function in order to provide more realistic representation of mechanical power output of the practical system. If we keep rate limiters falling rate between 0 and −1, the values of wind speed fed to the turbine are much below the cut-in speed of the turbine so turbine is not going to operate. For those purposes, rate limiters falling rate is kept as a constant and the value is −1.

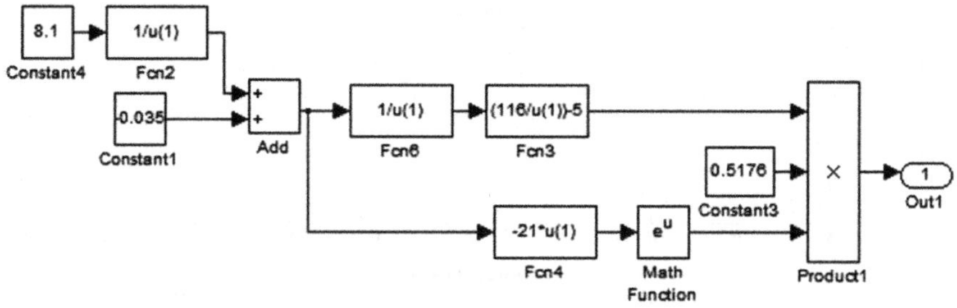

Fig. 23.4 Power coefficient model

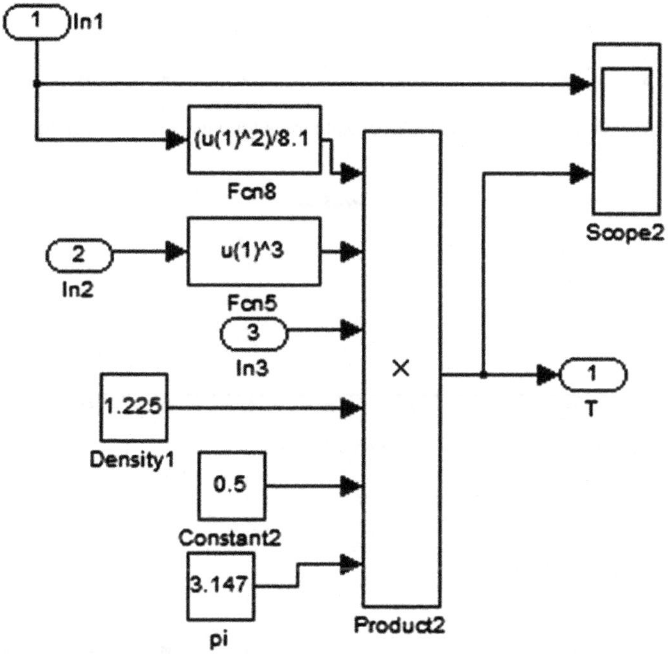

Fig. 23.5 Mechanical torque model

23.2.2 Mechanical Power Model

WTGS converts the kinetic energy of the wind into the electrical energy. The kinetic energy of the wind is given as

$$\mathrm{KE_w} = \frac{1}{2} mv^2 \tag{23.4}$$

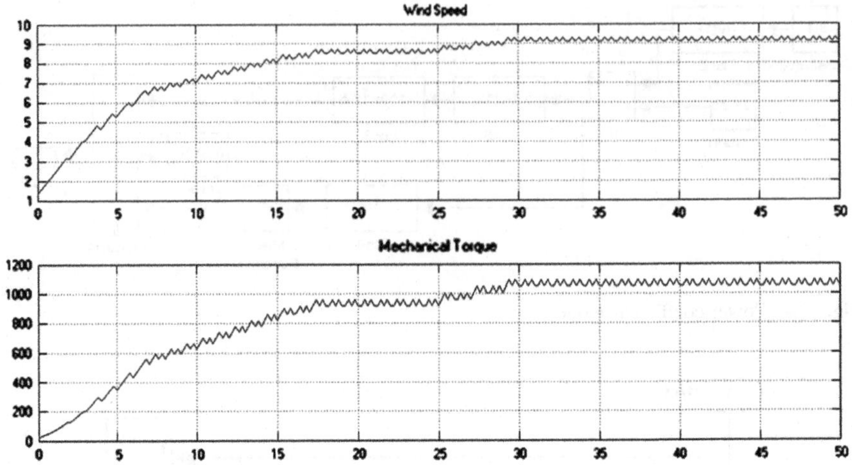

Fig. 23.6 Variation in wind speed and mechanical torque for wind speed with all four components

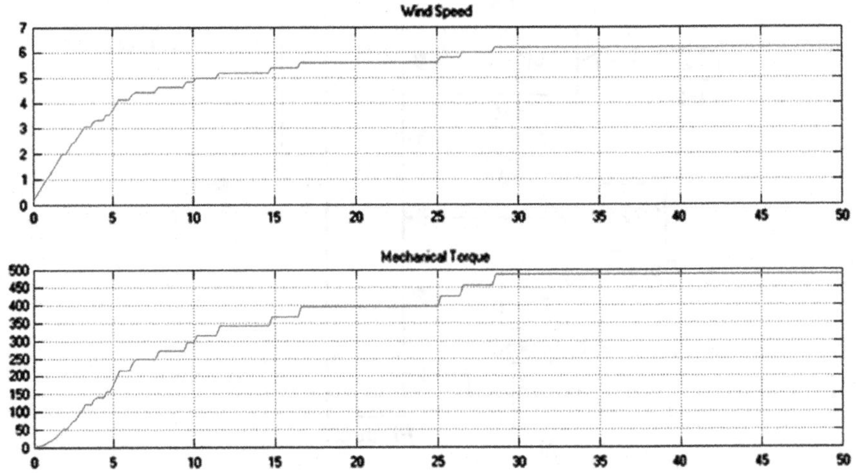

Fig. 23.7 Variation in wind speed and mechanical torque for wind speed without repeating sequence component

where m is the air mass which is given as

$$m = \rho V_o = \rho A v$$

where v is the wind speed, ρ is the air density which is equal to 1.225 kg/m^3, V_o is the volume of air, and A is the area covered by turbine blades.

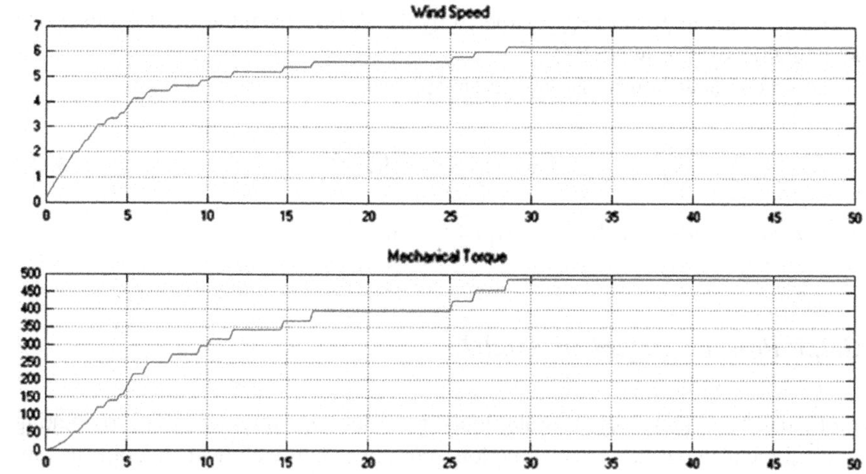

Fig. 23.8 Variation in wind speed and mechanical torque for wind speed without repeating sequence and uniform random number component

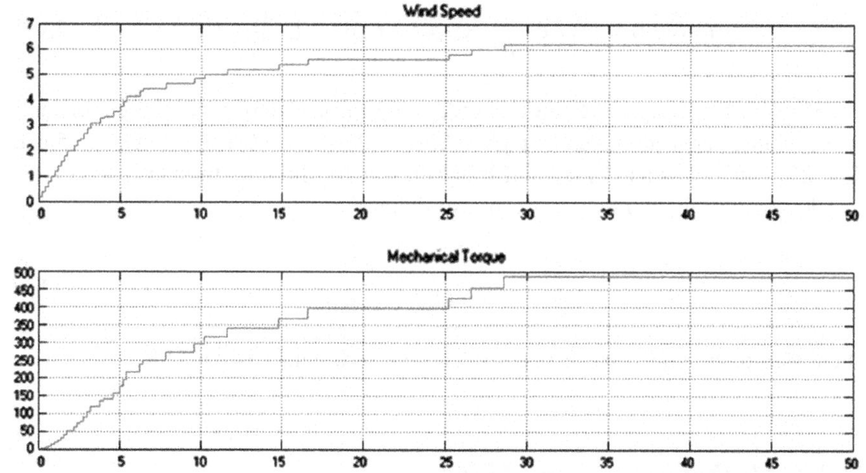

Fig. 23.9 Variation in wind speed and mechanical torque for wind speed without repeating sequence, uniform random number and step component

By performing mathematical calculations, the available wind power (P_w) is expressed as

$$P_w = \frac{d}{dt}(KE_w) = \frac{1}{2}\rho A V^3 \tag{23.5}$$

The mechanical power (P_m) extracted by the WTGS from the wind is inferior to the available wind power (P_w).

The mechanical power (P_m) is given by

$$P_m = \frac{1}{2}\rho\pi R^2 V^3 C_p(\lambda) \tag{23.6}$$

where R is the radius of the rotor equal to 5 m.

23.2.3 Power Coefficient Model

The C_p can be obtained by the data fields in the lookup tables or by approximating the coefficient using analytical function. We are using analytical function to represent the power coefficient (C_p) of the wind turbine which is given by (23.4).

C_p is a function of the TSR of the wind turbine and the blade pitch angle. In this paper, we are taking TSR as a constant equal to 8.1.

$$C_p(\lambda, \theta) = C_1\left(C_2\frac{1}{\beta} - C_3\beta\theta - C_4\theta^x - C_5\right)e^{-C_6(1/\beta)} \tag{23.7}$$

The C_p is the function depends on the wind turbine rotor type, the coefficients C_1–C_6 in this paper are taken as $C_1 = 0.5$, $C_2 = 116$, $C_3 = 0.4$, $C_4 = 0$, $C_5 = 5$ and $C_6 = 21$, x is a constant value, and it is different for different turbines.

Additionally, the parameter β can also defined as

$$\frac{1}{\beta} = \frac{1}{\lambda + 0.08\theta} - \frac{0.035}{1 + \theta^3} \tag{23.8}$$

where θ is pitch angle, and it is considered as zero, and λ is the TSR.

23.2.4 Mechanical Torque Model

The wind turbine torque on the shaft can be calculated in terms of mechanical dynamic equation which is given as:

$$T_m = \frac{P_m}{\omega} = \frac{1}{2}\rho\pi R^2 \frac{V^3}{\omega} C_p(\lambda) \tag{23.9}$$

By introducing $\lambda = \frac{\omega R}{v}$,

$$T_m = \frac{1}{2}\rho\pi R^3 \frac{V^3}{\lambda} C_p(\lambda) \tag{23.10}$$

The resulted mechanical turbine torque is applied as the input torque to the wind generator and makes generator to operate.

23.3 Results and Discussion

We evaluated the performance of the WTGS by performing a simulation using MATLAB/Simulink for four different profiles of wind. The chosen wind speed model for MATLAB/Simulink is shown in Fig. 23.2. With the wind speed profile, variations in mechanical torque are observed and compared.

Figure 23.6 shows the variations in the wind speed, coefficient of performance and mechanical power with respect to time for wind speed with all four components. In this case, mechanical torque curve is continuously zigzag in nature due to zigzag nature of the wind speed. The saturated value of wind speed is 9.31 m/s with a saturated value of mechanical torque of 1,096.2 Nm.

Figures 23.7, 23.8 and 23.9 show the variations in wind speed and mechanical torque with time for wind speed without repeating sequence component, wind speed without repeating sequence and uniform random number component and wind speed without repeating sequence and uniform random number component, respectively. The saturated value of wind speed is 6.2 m/s with a saturated value of mechanical torque of 486 Nm; for all three cases, the only difference in response is the nature of the mechanical torque output waveform. It shows that there is a reduction of 56 % of mechanical torque in all three cases when compared to wind speed with all four components.

For wind speed without repeating sequence component, mechanical torque curve is a partial step waveform with small value of the slope; for wind speed without repeating sequence and uniform random number component, mechanical torque curve is a partial step waveform with a medium value of slope; and for wind speed without repeating sequence, uniform random number and step component, mechanical torque curve is a complete step waveform with sharp value of the slope.

23.4 Conclusion

In this paper, dynamic characteristic of mechanical torque developed by WTGS is studied for four different modelled wind speed profiles. Entire mechanical torque model of WTGS consists of mathematical model of wind speed, mechanical power,

power coefficient and mechanical torque model. For the case study purpose, wind speed data of Chalkewadi, Satara, are used. Different wind speed profiles are modelled according to the wind speed data of Chalkewadi, Satara location. Mechanical torque model of WTGS is excited with the modelled wind speed profiles. The response to these wind speed profiles on mechanical torque produced by WTGS is observed. Proper representation of wind speed and the mechanical torque will help in designing the WTGS and its components.

References

1. Urtasun A, Sanchis P, Martin IS, Lopez J, Marroyo L (2013) Modeling of small wind turbines based on PMSG with diode bridge for sensorless maximum power tracking. Renew Energy 55:138–149
2. Dehghan SM, Mohamadian M, Varjani AY (2009) A new variable speed wind energy conversion system using permanent magnet synchronous generator and Z-source inverter. IEEE Trans Energy Convers 19(2):714–724
3. Kim HW, Kim SS, Ko HS (2010) Modeling and control of PMSG based variable speed wind turbine. Electr Power Syst Res 80(1):46–52
4. Chang TJ, Tu YL (2007) Evaluation of monthly capacity factor of WECS using chronological and probabilistic wind speed data: a case study of Taiwan. Renew Energy 32(12):1999–2010
5. Johnson GL (1985) Wind energy systems. Prentice-Hall Inc., Englewood Cliffs
6. Jamdade PG, Sarode UB (2010) Material science. Tech-Max Publications, India. ISBN 978-81-8492-557-9
7. Albadi MH, El-Saadany EF (2009) Wind turbines capacity factor modeling—a novel approach. IEEE Trans Power Syst 24(3):1637–1638
8. Jamdade SG, Jamdade PG (2012) Extreme value distribution model for analysis of wind speed data for four locations in Ireland. Int J Adv Renew Energy Res 1(5):254–259
9. Jamdade SG, Jamdade PG (2012) Analysis of wind speed data for four locations in Ireland based on Weibull distribution's linear regression model. Int J Renew Energy Res 2 (3):451–455
10. Manyonge AW, Ochieng RM, Onyango FN, Shichikha JM (2012) Mathematical modelling of wind turbine in a wind energy conversion system: power coefficient analysis. Appl Math Sci 6 (91):4527–4536
11. Al-Bahadly I (2011) Wind turbines. InTech Publication. ISBN 978-953-307-221-0
12. MATLAB 7.10.0 (2010) MATLAB 7.10.0 The Math Works Inc., Natick

Chapter 24
A Methodology to Design a Validated Mode Transition Logic

Manjunatha Rao, Atit Mishra, Yogananda Jeppu and Nagaraj Murthy

Abstract Autopilot mode confusion has resulted in many aircraft accidents. This brings out the importance of design validation for the complicated mode transition logic (MTL) for aircraft autopilot systems. A method of implementing mode transition tables in terms of transition and condition tables was developed for the Indian SARAS autopilot. This method is used for designing the mode transition for a generic aircraft. A practical method of reviews, assertions, safe states, and safe transitions is experimented for this design as a process. This practical process has yielded good results and helped the team to come up with the validated MTL. This approach is presented here as an industry experience paper.

Keywords Autopilot · Automatic flight control system · Mode transition logic · Mode confusion · Assertions · Safe states

24.1 Introduction

Very recently, on July 6, 2013, the flight 214 Asiana Airlines crashed on the final approach to San Francisco international airport in the USA [1]. It was a Boeing 777 aircraft with 307 people on board. There were 3 fatalities and 181 others were injured, 12 of them critically. The airplane was cleared for a visual approach, and the pilots were hand flying the airplane since the instrumental landing system's

M. Rao (✉) · A. Mishra · Y. Jeppu · N. Murthy
MOOG India Technology Center Pvt. Ltd., Plot 1, 2 & 3, Electronic City-Phase 1, Bangalore 560100, India
e-mail: rmanjunatha2@moog.com

A. Mishra
e-mail: matit@moog.com

Y. Jeppu
e-mail: jyogananda@moog.com

N. Murthy
e-mail: mnagaraj@moog.com

V. Vijay et al. (eds.), *Systems Thinking Approach for Social Problems*, Lecture Notes in Electrical Engineering 327,
DOI 10.1007/978-81-322-2141-8_24

(ILS) vertical guidance glide slope (GS) mechanism was non-functional. The preliminary investigation by National Transportation Safety Board (NTSB) revealed that during the approach pilots found the airplane to be above the glide path for some duration, then being on the glide path and later being below the glide path. NTSB investigation said the plane's airspeed on the final approach fell to 34 knots below the minimum approach of 137 knots. Although auto-throttle system was engaged, the pilots were unaware if it was able to maintain the speed until 500 ft above the ground level. Pilots manually tried to correct for the deviation below the glide path and also for deviation in the lateral direction. This correction was too late and too little, to the extent that even the go-around (GA) mode for aborting the approach could not be used to the rescue. The investigation also revealed that during the last 2.5 min of the flight, based on the data from the flight data recorder (FDR), there were multiple autopilot modes and multiple auto-throttle modes. It was unclear if the pilots have actually commanded these modes or if the modes got engaged inadvertently based on triggering of flight conditions or if there was a mode confusion experienced by the pilots.

There are growing evidences of similar aircraft accidents and incidents that mainly highlight the automation driving mode changes which are unexpected from pilot's or crew members viewpoint [2–4].

It can be seen that this mode confusion as mentioned in the above accident is ubiquitous and persistent and has a dominant presence in the systems involving the digital avionics computer system and humans. This is very predominant in safety critical systems such as the autopilots. Current trends show a significant rise in complexity where more than one system interacts with the others to determine new modes especially in an automated human machine interface (HMI) environment [5]. This is depicted in Fig. 24.1.

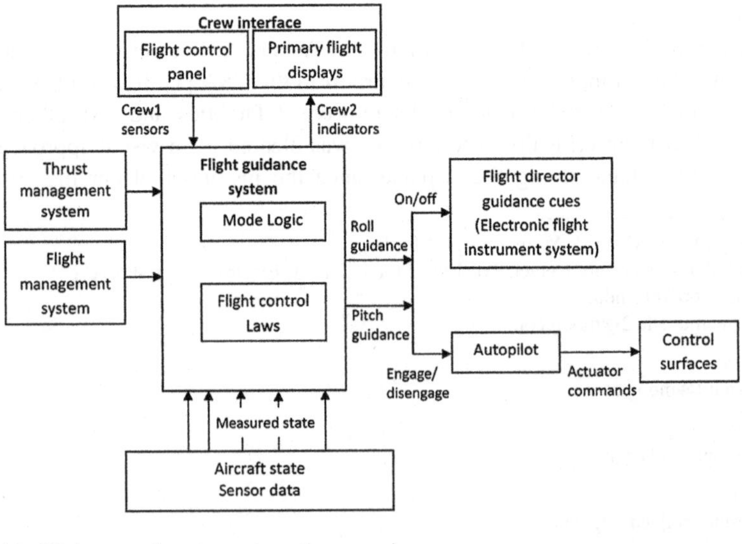

Fig. 24.1 Flight control system automation overview

This puts a special emphasis on system engineers to understand every aspect of operational mode before modeling or designing such a system. The modes in which different systems operate and their interaction with modes from other systems continues to require designing appropriate switching logics and transitions with a prime focus on determining a safe mode and overall flight safety. This should be accomplished without leaving the crew members and pilots in a state of mode confusion. Mode confusion, also referred as lack of mode awareness, has been described in [6]. In simple terms, it means that the system is in a state that can lead from a trivial to catastrophic incidents and puts the crew in a situation where they are unaware of the issue or possibly helpless to take corrective action, like in the mentioned accident.

In the rest of the coming sections, we broadly describe autopilot functionality with emphasis on their modes and respective transitions. We briefly introduce use of formal methods that have been adopted in validating the mode transition logic (MTL). Subsequently, we discuss the systematic methodology for MTL validation from the system design viewpoint through suitably designed assertions. The approach being system centric is capable of bringing ambiguous, conflicting, or missing requirements in the system requirement specification (SRS) document.

24.2 Modes in Autopilot—A General Description

An autopilot system is meant to reduce the workload of the pilot(s). However, it does not eliminate it. It is a mechanical, electrical, or hydraulic system used to guide an airplane with minimal or no assistance from the pilot. As a component within the automatic flight control system (AFCS), it allows the pilot to select different flight modes and controls the aircraft to automatically follow the selected mode. The modes are normally split into vertical and lateral directions. Typically, the vertical and lateral mode sets may contain up to 8 and 4 basic modes, respectively. These sets increase when the other higher modes such as navigational modes including VHF Omni Range (VOR) navigation mode, ILS, or flight management system (FMS) modes are considered in both the vertical and lateral directions for designing the MTL. For instance, vertical mode set sizes swell to contain 12 modes when the ILS GS approach is also considered. The lateral mode sets under such a case increase to contain 8 modes. The autopilot when selected from mode control select panel (MSCP) in the cockpit gets engaged in a coupled sense such that one vertical and one lateral mode is always active. The mode logic design normally allows arming of up to a maximum of two vertical modes and one lateral mode based on the flight conditions.

24.3 Formal Approach to Mode Confusion Analysis

One of the interesting approaches for the mode confusion analysis is to use formal methods as described in [2]. The approach is centered on postulating properties that are false in the model and uses a verification tool to generate insights into why the properties are false. This has been termed as 'exploration' instead of 'verification.' The aim in this approach is to search all the ways by which the mode can be entered and use formal verification tools to identify the ways in which this is not true. Then, suitably make changes to the property to specify that there are no scenarios by which the mode can be entered except for these identified ones and repeat the process until the property is true.

In contrast, the approach proposed in this paper tackles the mode confusion issue from the opposite direction. Essentially, we identified the ways to 'enter' and 'exit' the modes based on the initial knowledge based on similar system designs and pilot manuals. These form an exhaustive list of ways that is documented in the SRS. This is the preliminary list of requirements that describes the way the system shall behave and interact with the pilots. An extensive review of the SRS document is done to ensure that requirements are well captured and unambiguous.

24.4 Modes of Operation

The AFCS operates in various modes as shown in Fig. 24.2. The vertical modes and lateral modes can exist together. The autopilot can exist in only one of the sub-modes in the vertical and lateral modes. Thus, it is perfectly safe for the autopilot to exist in the pitch attitude hold (PAH) in the vertical and heading hold (HH) in the lateral mode.

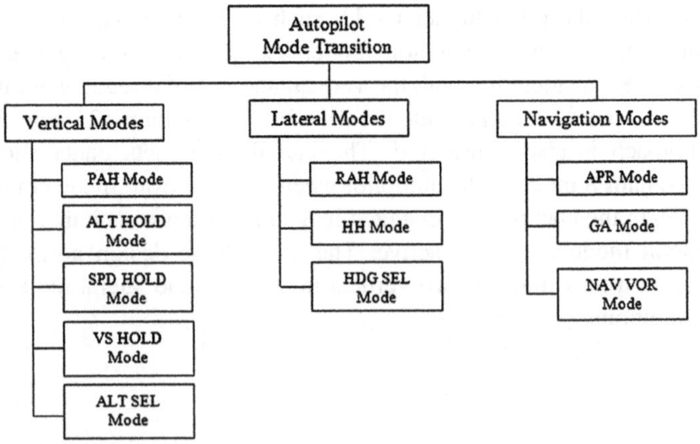

Fig. 24.2 Basic operating mode

The other modes are the altitude hold (ALT), speed hold (SPD), and vertical speed (VS) hold mode. The altitude select mode (ALT SEL) is a complex mode where the aircraft is traveling to a selected higher or lower altitude in one of the PAH, SPD, or VS modes, and then, at a specific altitude, the autopilot enters into an altitude capture mode. In this mode, the autopilot turns the aircraft to level off at the selected altitude. Once the selected altitude is attained, the autopilot enters into the ALT mode. The roll attitude hold (RAH) and HH mode are the basic lateral modes. The heading select (HDG) mode makes the aircraft to go toward the selected heading angle. Once the required heading is achieved, the autopilot enters a HH mode to hold that heading value. The higher modes are the approach mode (APR), the VOR mode, and the GA mode. The GA mode is a special mode where the autopilot is cut off and the pilot is shown cues to fly the aircraft at a particular attitude to abort a landing.

The SRS defines all these modes in detail and defines the entry and exit criteria for each mode. This is converted into a transition table to define the modes and the transition between the modes [7]. The transitions are based on specific conditions which are captured in the condition table. The following section elaborates on the tables used to arrive at the mode transition requirements.

24.5 Transition Table

The transition table indicates the transition from an operational mode to another mode due the occurrence of an event. The event may be triggered by pilot through button press on the MSCP, or it may be due to software trigger as in the case of the ALT SEL mode transferring to the ALT hold mode when the altitude is attained.

A portion of the mode transition table is shown in Fig. 24.3. The MTL is designed in a Microsoft Excel file as it is easy to manage and visualize the numbers. This we find very easy to handle instead of a state chart. There are other subtle advantages of using this method as brought out later in the paper. The vertical, lateral, and AP modes are mutually exclusive set of modes. The other modes that can exist together but are not shown here are FD, ALT SEL, APR LOC, APR GS, NAVVOR, and back course (BC). The first column represents the exclusive modes. The next column has state numbers which represents the sub-modes in the main modes. The state number represents unique numbers assigned to the modes within each set. These numbers are used to mark the transitions. Thus, each row has the states, and each column represents the buttons and software triggers or events that cause a state to transit.

This is shown in the expanded form in Table 24.1. Only the vertical modes are defined for 2 triggers to explain the method. The eight sub-modes are defined as state numbers 1–8. This is always defined as 2 digits to avoid confusion between 10 and 01. The third column defines the transition for the AP button press by the pilot.

When the pilot presses AP with the vertical mode being in 01 (DIS Vertical), it transits to state number 02 (PAH mode). If the current state is PAH, then the pilot

States	Sl. No.	Modes	01 AP	02 FD	03 SPD	04 VS	05 ALT	06 ALTS
Vertical	01	DIS(Vertical)	02	02	00	00	00	00
	02	PAH	010202	010202	03	04	05	00
	03	SPD HOLD	010303	010303	02	04	05	00
	04	VS	010404	010404	03	02	05	00
	05	ALT HOLD	010505	010505	03	04	02	00
	06	GA(Vertical)	02	01	00	00	00	00
	07	ALTS CAP	010707	010707	00	00	05	00
	08	APR GS(CAP & ACT)	010808	010808	00	00	00	00
Lateral	01	DIS(Lateral)	0203	0203	00	00	00	00
	02	RAH	010202	010202	00	00	00	00
	03	HH	010303	010303	00	00	00	00
	04	HDG SEL	010404	010404	00	00	00	00
	05	GA(Lateral)	0203	01	00	00	00	00
	06	APR LOC(CAP & ACT)	010606	010606	00	00	00	00
	07	NAV VOR	010707	010707	00	00	00	00
AP	01	AP ON	02	00	00	00	00	00
	02	AP OFF	01	00	00	00	00	00
	03	AP SYNC	00	00	00	00	00	00

Fig. 24.3 Transition table

Table 24.1 Transition table—expanded

	Modes	AP	FD
01	DIS (vertical)	02	02
02	PAH	010202	010202
03	SPD HOLD	010303	010303
04	VS	010404	010404
05	ALT HOLD	010505	010505
06	GA (vertical)	02	01
07	ALTS CAP	010707	010707
08	APR GS (CAP and ACT)	010808	010808

can transit to three different modes. He can disconnect, i.e., go to 01 based on a condition or remain in the same state based on a specific condition. These conditions are defined in the condition table.

24.6 Condition Table

Condition list consists of set of conditions and is assigned a unique number. The list is named as condition list. The condition table uses these unique numbers from condition list to represent the set of conditions that must be satisfied for the transition shown in transition table to occur. The entries in transition table and condition table have one–one relationship. The condition table is shown in the expanded form in Table 24.2.

Table 24.2 Condition table—expanded

	Modes	AP	FD
01	DIS (vertical)	01	02
02	PAH	070809	090807
03	SPD HOLD	070809	090807
04	VS	070809	090807
05	ALT HOLD	070809	090807
06	GA (vertical)	17	00
07	ALTS CAP	070809	090807
08	APR GS	070809	090807

Thus, as explained above, the autopilot will enter the PAH mode when AP is pressed if condition number 01 is TRUE. Condition number 01 refers to 'No AP or FD inhibit conditions' exists. If the autopilot mode is 02, i.e., PAH and pilot presses AP, the autopilot will transit to disconnect if condition 07 is TRUE. The 010202 in the transition table corresponds to 070809 in the condition table. Condition 07 is defined as "AP is engaged AND FD is OFF". It has been our experience that the SRS could be easily converted to a transition and condition tables in Excel.

24.7 Advantages of Representation

The primary advantage of such a representation is that the table can be loaded into MATLAB software and handled as matrices. The rows correspond to modes and the columns correspond to triggers. It is very easy to pick up a cell corresponding to the specific mode and trigger and do the processing treating the number in the cell mathematically.

It was very easy to write a MATLAB script to convert the table into a text representation in readable English. This helped the team to review the MTL easily and compare it with the SRS document for correctness. The MTL processing in MATLAB is a 20-line script which is easily converted to the implementable C code. Verification is very easy with an all combination of trigger and conditions overnight run and comparing MATLAB and C outputs. The only thing that changes is the data table for transition and conditions.

24.8 Requirements from Mode Transition Logic

The requirements for the MTL are generated using a MATLAB script. The MATLAB script takes transition table; condition table and condition list are the inputs that generate the text requirement as output. The requirements are reviewed to verify their correctness.

```
------------------------------------------
Transitions due to ALT button press
------------------------------------------

240) AP/FD shall remain in DIS(Vertical) when ALT button is pressed
TRIGGER NOT POSSIBLE

241) AP/FD shall transit from PAH mode to ALT HOLD mode when ALT button is pressed
and |VS| is less than 500 ft/min and CWS is not activated

242) AP/FD shall transit from SPD HOLD mode to ALT HOLD mode when ALT button is
pressed  and |VS| is less than 500 ft/min and CWS is not activated

243) AP/FD shall transit from VS mode to ALT HOLD mode when ALT button is pressed
and |VS| is less than 500 ft/min and CWS is not activated

244) AP/FD shall transit from ALT HOLD mode to PAH mode when ALT button is pressed
```

Fig. 24.4 Mode transition requirements

Few of the automatically generated requirements are shown in Fig. 24.4.

It is very easy to read this text and tick away what seem to be correct and what raise a few doubts. Such doubtful cases are revisited and discussed to arrive at a design consensus. The initial SRS document was changed where the transition seemed incorrect. The MTL was updated if there was an error in the transition. A total of 846 such points were generated and reviewed for correctness. In some cases, the trigger cannot happen or is not effective. Such cases are also indicated. This will lead to certain conditions to be placed as safety requirements. This method has successfully implemented the requirements of the SRS and executed the MTL to generate back the requirements actually defined by the MTL. These requirements are generated by executing the MTL in the actual way it would run in the final implementation. This concept is very powerful in the validation of the implementation.

24.9 Assertions

There could be errors in the requirements themselves. This aspect is tested using semiformal means. An assertion technique is used to arrive at the safe state matrix. A total of 23 assertions are identified for the MTL. A MATLAB script implements these assertions.

Few of the assertions are

1. In the ALT HOLD mode, ALT SEL arm is turned OFF
2. In the disconnect mode, both vertical and lateral modes, DIS, AP, and FD, are OFF, and all armed modes are OFF
3. In the GA mode, AP should be OFF and FD should be ON
4. When AP and FD both are OFF, the mode should be disconnected

These assertions define a set of safe states in which the autopilot can exist. Each assertion eliminates a set of states which cannot be possible. If vertical mode is DIS, then lateral mode cannot be RAH or HH. These 23 assertions resulted in 200 safe states from a total combination of 86,016 states, taking into consideration all state. These safe states were reviewed for correctness. The review process resulted in additions of assertions to rectify certain errors.

Some examples of safe states are

1. DIS (Vertical)—DIS (Lateral)—AP OFF—FD OFF—ALTS OFF—LOC OFF —GS OFF—NAVVOR OFF—BC OFF
2. PAH—RAH—AP ON—FD ON—ALTS ARM—LOC OFF—GS OFF—NAVVOR OFF—BC OFF
3. PAH—RAH—AP ON—FD ON—ALTS ARM—LOC OFF—GS OFF—NAVVOR ARM—BC OFF

24.10 Validation of the Requirements

Any event should cause the autopilot to make a transition from one safe state to another safe state. Any combination of safe state, event, and condition that results in a state which is not an element of the safe set is a problem that has to be addressed. The validation procedure is as below

1. Always start with the safe state.
2. Evaluate all the conditions represented in the condition list to TRUE or FALSE
3. Select a trigger in sequence.
4. Go to the respective cells in the condition table corresponding to the safe state and the trigger. Evaluate the condition set. Right most 2 digit number has the higher priority.
5. When the condition in the conditional table is satisfied, the corresponding transition takes place.
6. The previous mode is preserved if the transition condition is not satisfied and a transition cannot occur.
7. Verify that the resultant state is also a safe state.

This process is repeated for all safe states, all triggers, and all condition in the condition table. This simulation is performed in MATLAB and any violation is stored as a data file. This process has given us a good insight into the system behavior and caught many violations that have resulted in the MTL update and in one specific case the SRS update.

24.11 Results and Discussion

In one scenario, SPD button is pressed by the pilot, while the autopilot is operating in a safe state defined by—[ALT HOLD, RAH, AP ON, FD ON, ALTS OFF]. The state transitioned to [ALT HOLD, RAH, AP ON, FD ON, ALTS ARM] which was not falling in safe state matrix. ALTS cannot be armed in the ALT hold mode. The analysis brought out the fact that the airspeed criteria CAS >120 knots (Calibrated Airspeed) did not satisfy the condition to transit to SPD mode. Therefore, the ALT HOLD mode did not change to SPD. There was no condition to check ALTS armed. This was added to the condition table and the MTL prevented the ALTS from becoming armed.

In another scenario, the GA trigger was pressed. There is a condition which prevents GA when the altitude is very low. This was TRUE, and therefore, the vertical and lateral mode did not transit to GA mode. The ALTS however transited to ARM mode when GA was pressed. This resulted in a condition to be added to prevent transition in GA. The multiple simulations take time, but they can be easily put on different PCs and run overnight every day to check for failures.

24.12 Conclusion and Future Work

The MTL is designed for 37 sub-modes, 27 triggers, and 28 condition numbers. This is validated using extensive reviews against the SRS. A practical application of assertions in a semiformal mode has resulted in 23 assertions that should never be violated. The transition from a safe state to a safe state and no assertion violation is a powerful methodology to validate the MTL. This simulation-based validation process ensures that there is a positive proof that the system behaves in a safe manner for all conditions and triggers.

The MTL is coded in C language, and we have to ensure that the tables and code are correct. This is the verification process. At the PC level, all states can be simulated for a test. This is however difficult at a system level. A methodology of optimized system test will be designed as part of this project to implement a MTL for a generic aircraft.

References

1. Asiana 214 flight report ID: DCA13MA120. http://www.ntsb.gov/investigations/2013/asiana214/asiana214.html
2. Joshi A, Miller SP, Heimdahl MPE (2003) Mode confusion analysis of a flight guidance system using formal methods. In: The 22nd digital avionics systems conference (DASC'03), vol 1, pp 2.D.1, 21-12, 12–16 Oct 2003
3. Hughes D, Dornheim MA, Sparaco P, Phillips EH, Scott WB (1995) Automated cockpits special report (part 1). Aviat Week Space Technol 142(5):52–65
4. Rosenkrans W (2008) Autoflight audit. Flight Safety Foundation, pp 30–35
5. Boorman DJ, Mumaw RJ, Pritchett A, Jackson A (2004) A new autoflight/FMS interface: guiding design principles. In: Pritchett A, Jackson A (eds) Proceedings of the international conference on human–computer interaction in aeronautics (CD-ROM), Toulouse

6. Leveson N, Pinnel LD, Sandys SD, Koga S, Reese JD (1997) Analyzing software specifications for mode confusion potential. In: Proceedings of a workshop on human error and system development, pp 132–146
7. Rao PS, Chetty S (2002) Rapid prototyping tools for commercial autopilots. In: AIAA guidance, navigation, and control conference and exhibit. AIAA 2002-4651

Chapter 25
A Model for Systematic Approach Towards Solving Environmental Problems Based on Knowledge from Critical Case Studies

R. Prashant Singh

Abstract Our world is today in a position where the survival of the human race requires a managerial approach towards the protection of the environment and also needs optimisation in the use of available resources as they are now getting limited unlike olden times. But it is time to think different. It is time to make a change by learning from our mistakes, by learning from the cases that we encountered at various places. This paper focuses on creating a model that could provide a plan of action or in some cases a solution to the various critical, unapproachable problems existing in the world, wherein the problem is either due to the topography of the places or maybe due to the resources, and so on. This model provides a new hope, to make the conditions better by providing the optimum solution which may also bring up the utilisation of the unused resources. The applications of this systematically arranged levelled model could lead to the formation of a self-dependant bio-system which utilises itself to feed itself and its components (i.e. us). The model includes the generation of a criticality constant, which could show the relative criticality of the current state (of a particular place) with the standard critical state present in the model. If individual case studies can help to solve critical environmental issues, then imagine what can a systematically structured model, consisting of the knowledge harnessed from these critical cases, and with a prioritised and heuristic approach towards solution do!

Keywords Criticality constant · Resources · Systemic · Heuristic · Bio-systems · Database

---

R. Prashant Singh (✉)
Faculty of Engineering, Dayalbagh Educational Institute, Dayalbagh, Agra 282005, India
e-mail: r.prashant.s@ieee.org

V. Vijay et al. (eds.), *Systems Thinking Approach for Social Problems*, Lecture Notes in Electrical Engineering 327,
DOI 10.1007/978-81-322-2141-8_25

25.1 Introduction

Case studies are descriptive, exploratory and explanatory analysis of a particular individual or a group (in case of psychiatry) or a situation in case of other branches. The understanding of case studies varies based on the branch of which we are talking about. But the common thing in all these case studies that exists is the phenomenon of exploring, understanding and solving the underlying problems. These case studies can be compared to solving a mystery of any situation or an event which might be unique to that particular place or common to other similar events. In our case, we have environmental case studies which have innumerable factors which vary from place to place and make these case studies very essential in case of solving environmental problems and achieving sustainability.

The *applications* of these case studies go beyond the scope of just studies and understanding of nature or the environment. One major application of the same would be the application of case studies in solving the environmental crisis that we are currently under and achieve what we all crave for in this technologically progressing world, environmental sustainability. This involves making decisions and taking actions that are in the favour of the nature, with particular emphasis on consideration of the fact that the nature or the environment after our event remains supportive to life forms. In this paper, further, we will study a few case studies which resulted in either solving a major environmental issue or came up with a better eco-approach to sustain. These case studies (mainly of India) not only help in solving problems but also open a new window to understanding of environmental behaviour in those particular constraints. However, these are not the same for every place. But, by this time, mankind has learnt a lot. We now need a different approach towards solving these environmental issues and also a rapid method, which could directly lead to solutions or conclusions which further could be utilised to achieve a solution.

The difference lies in the approach of attending to a grotesque or a complicated problem. We have been accustomed to the fact that the analysis starts from the scratch, and then, case needs to be built up to understand the current scenario of the problem, and then, these are separately studied, and then, a solution is thought off. This process is a very sluggish. It is time to cut through the shackles of ancient understanding and path towards achieving of goal. Now, we need a better algorithmic or a heuristic approach to accomplish the same and also at a higher rate of displacement towards success, which is possible by having a systems approach towards this problem.

25.2 Concept

Consider this entire environmental biosphere of ours to be a system. Now, isolate the components that are under threat. We find that the entire system is interdependent, so an unresolved issue in any subsystem directly or indirectly affects the other subcomponents adversely. Hence, the one advancement that we need to

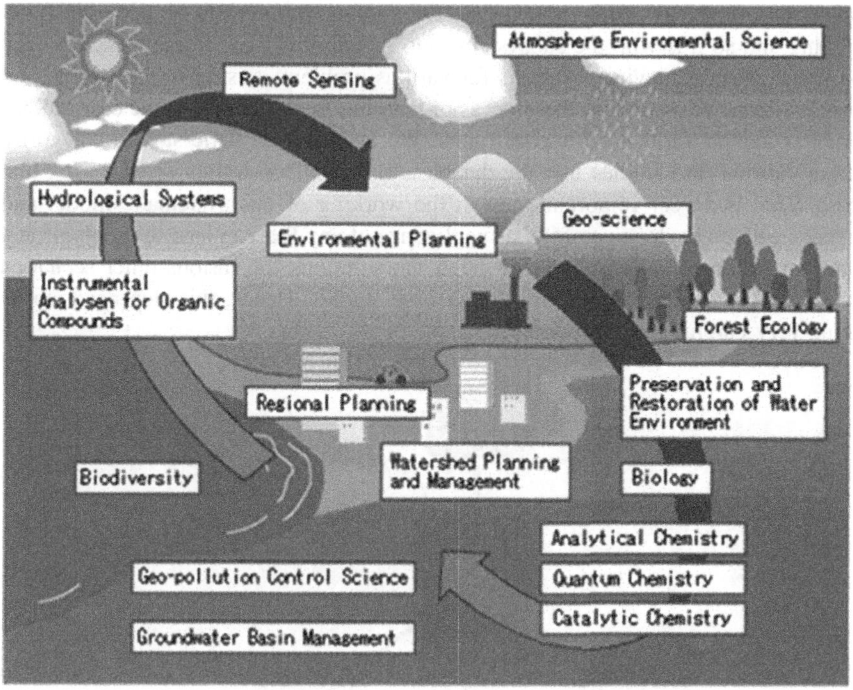

Fig. 25.1 Interdependent subsystems of the environment as a system [1]

achieve is the understanding of the critical components of this system and rating them on a scale, called the criticality scale. Once we have these data, the connected graph approach can be used to analyse the subcomponents which get affected by that particular critical component. The primary aim of managing this system should be prioritizing the components based on their criticality and continue this procedure recursively with their respective subcomponents unless all the units of the entire system have been graded. This grading on the criticality scale should be considering the biomass, the feasibility of the various solutions that could be applied on the subsystems, various environmental factors which might adopt or reject the solution, and so on (Fig. 25.1).

25.3 Procedure

Now, the primary aspect of this model for it to be successful is the creation and management of a huge *database* maintaining these factors and information about the various systems and subsystems of the environmental system. Once this is done, the main purpose of this model would be to keep the most critical systems intact and the law applies to the inner systems and subsystems too. So, if the most critical subsystem of the system is intact, then the system continues to be perfectly functioning and so it is with the subsystems. Now, with this basic framework, we need to know about achieving solutions for various problems arising in the system and that is where we come into the process of learning from analysis of the case studies done so far on various other constraints. We need a systematic step to categorise these various case studies into the database based on constraints faced during that case. This is a very important step in the working of this systematic model and hence needs to be done carefully as negligence of any fact can lead to problems at a later stage where we would be devising the solutions for various other scenarios based on the data harnessed from these case studies. Hence, this step needs to be done very carefully and precisely.

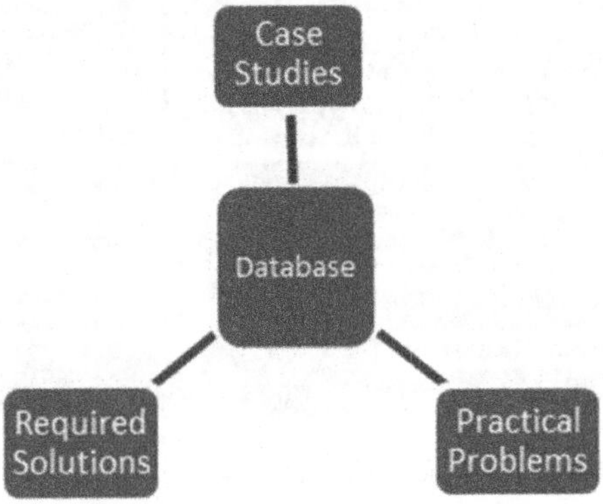

25.4 Database Creation

The ways to organise these data would be using a few fixed set of steps including organisation based on the locations parameters (i.e. if it is a polar place, rainy place), topographical conditions of that place [which can be used to devise solutions which are feasible (at a later stage)], and the resources available at that place, critical identification marks of that place if any. These data if collected from all the case studies and assembled into a large database can be used to optimise the solutions for new problems that we encounter. Further, in this paper, an example of this very model is given where it is clearly seen that an entire case study can be avoided in some cases when we know the properties of that place and can use this database to check in and out the required conditions and obtain a solution for any given problem. However, we also need to specify the set of problems and their solutions in the database so that the optimal solution can be chosen based on the available parameters of that place. Technology, however, can be used to make the application of this model much faster. If we can have database sharing from the various parts of the globe, which consists of all the parameters of that particular place, then obtaining a solution for a scenario which is not local to that place would also not be a problem as if not there, the topographical conditions of that place can be matched to the ones from other places comparing it using the created database. This would ease the complexity of the situation. Now, let us take up a practical example to prove that this model works.

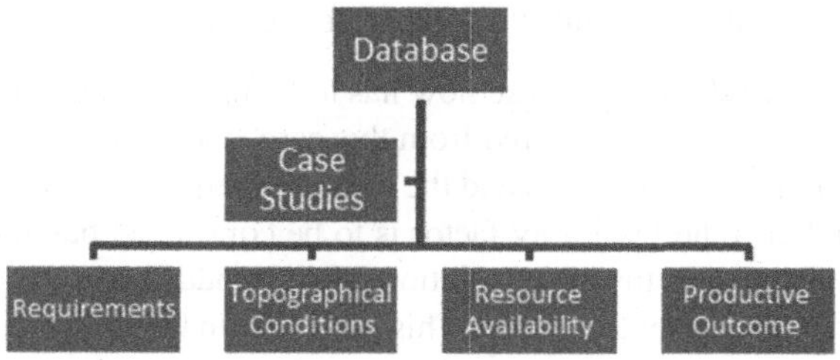

25.5 The Practical Scenario

Consider that the database is prepared thoroughly using the following case studies which were done in various different regions of India. The given example has consideration of only two cases which were enough to solve yet another case without the need for an entire case study on the last problem. Consider the cases:

Case 1: *The Case of Destructive Pine*

The problems arising due to the pine tree in the Himalayan region are pretty lethal as these trees offer very little shade but are a source of fuel wood for villages nearby. However, the pine needles that fall down and form a carpet are combustible and are the reason for the forest fires in that region which cause huge loss. In 1999, the devastating fire burnt more than Rs. 600 crores of forest wealth in Garwhal and Kumaon regions. So, this case was studied and the region was analysed. The effective solution that the organisations converged to was the production of electricity using this biomass, which involved collection and utilisation of these pine needles to produce the producer gas, which was used to generate electricity in the same region where other sources for generation of electricity are not much feasible. For this purpose, a 9 kW gasifier was used which used 1.5 kW of electricity and always an effective of 7.5 kW of power were available for use. This gasifier pyrolyses the pine needles to volatise them into producer gas viz. then passed through various filters to obtain pure producer gas used to produce electricity. [2]

This project has proven to be very advantageous also as the pine needles bed would not retain water and also caused soil erosion inhabiting the recharge of ground water and also the acidic pine needles hampered the growth of other plant species in the region. Now, without the pine needles carpet, the growth occurs as normal and moisture content of soil is maintained for further crop growth. Also, the onset of forest fire due to these combustible forest fires chokes the atmosphere and affects the quality of air available due to the smoke cloud formation in the sky. However, this solution helped providing a cleaner environment. There is a good scope of utilising this also in the central Himalayan region where the pine trees are found in plenty. It was found that pine forest area of 1 m^2 would yield 1.19 kg needles. Now, what does this infer to? What do we understand from this?

Inference for the Database This case now has to be updated into our database. So, the significant things to be considered from this case study is that the topographical conditions are similar to Himalayas and the resources required would be availability of pine trees and also the feasibility factor is to be considered based on the area of pine tree forest. Also, benefits of the solution could be added in as comments so that they are considered whenever needed. This ends the updating database part of the model.

Case 2: *Energy-efficient Steel sector*

India is currently is the world's third largest energy consumer and is ranked third in global greenhouse gas (GHG) emissions at 1.7 billion tCO_2/year out of which 5 % is due to the steel re-rolling and milling (SRRM) industry sector (70 million tCO_2) [3]. As requirement of energy is the primary requirement of this industry, it is very essential to make this energy efficient. The production of 1 ton of steel requires about 44–72 L of furnace oil, 65–143 kg coal (if used) and 52 kWh of electricity [4]. In SRRMs, 25–30 % of overall cost is used for the process. This requirement was a little too high to suffice for. This case was analysed, and requirements were studied. The solution to this problem was the use of better and green technologies. As a solution, the unit switched from furnace oil to biomass-based producer gas as fuel. Furnace was modified to a better design which allowed reheating automatically. Bearings were replaced, automation was achieved in temperature control system, automatic cooling system was installed, and a high speed delivery system was utilised for the process [5].

Outcome After this solution was implemented, the unit production increased significantly from 9.54 TPH to 14.654 tonnes per hour (TPH), resulting in a 53.77 % increased productivity. The use of biomass brought about the savings in fuel consumption also, and the biomass can be produced by the available resources from the surroundings also. Also, as the biomass was being used as the fuel, the output of CO_2 also reduced considerably.

The introduction of this gasifier has caused a lot of advancement in the systems efficiency and output.

Inference for the Database So now, we have to update this case in our database. This involves again the topographical conditions in which this would work, but as this is an industry under study, the most important aspect that we need to enter in our database is the availability of the required resources, power supply information, transport facilities and so on. Also, the benefits could be mentioned for the same and the products that we can obtain as output and their applications in the practical world so that if this database is looking for a solution then this might come in handy.

25.6 Our Problem

Now, we have this database ready for our use. Now consider a practical problem, in which we have lay railway tracks for the public transport passing through the Himalayan region and we need to devise a system for this construction to go on smoothly. Now, what normally would be done is again another case study would be done to understand the requirements and what is needed and all of that has to be repeated. But if we are having a limited time frame, then this project cannot be completed on time if the primitive methods are implemented.

Solution from the Model Now, consider this problem and refer the created database. Here, we know that the Himalayan region has a lot of pine trees which can be used for fuel production. But how is this relevant? Now, we need steel and mostly rolled steel for laying railway tracks. From our database, we find that steel production efficiently can be done using SRRMs and efficiently too. Now, this would require fuel, which we find from Case 1 that can be produced using pine tree needles. This henceforth gives us a solution to produce and lay railway tracks in the region of Himalayas where the difficulty would be very high.

25.7 Conclusion

The systemic approach towards any problem is always better than a random approach which is to be studied from the beginning. This model provides a systematic formation of a database consisting of all the information from various case studies. We could even have a user-friendly user interface on a network to obtain optimal solution for problems given in as input to the software or the program. It is time to think new and fast or we will be left behind, both technologically and structurally. A project this big is not brought into practice in a day or a week. In some cases, we may have possibilities wherein the information from the database is not sufficient. In such cases, we will be developing our database by utilising what is already in our database, leading to the formation of a self-developing or self-growing system. Sustainability is a primary issue these days. This model if used at a large scale can be used to solve huge environmental problems with what are known as *Eco-Ideas* which also do not cause harm to the nature. The building blocks of this model are the studies that have been done, the proper organisation of the available data is what matters, and this peek into the systemic approach can cause greater good if utilised at a larger scale on larger scenarios related to our environment and our survival. This is just the beginning of the evolution of an organised human race, a systematically progressing species towards its success. We have caused enough problems to the nature withholding us. Now, it is time to unleash our true potential, our organisational and understanding skills, the evolved intelligence to set things right, and this model is an example of the same, proving that proper organisational capabilities and understanding can help us reach what we desire the most, a sustainable future!

References

1. Division of Environment as a System and System Understanding. http://www.sss.fukushima-u.ac.jp/en/environment_system_management
2. Jain VK, Srinivas SN (2012) Ministry of new and renewable energy. United Nations Development Programme (UNDP), case study on bio-energy, "From destructive pine to productive fuel" and study of gasifiers from the success stories in empowering India in re-way. Ministry of New and Renewable Energy. Government of India, New Delhi
3. Indian Network for Climate Change Assessment (INCCA) (2011) India and green-house gas emission 2011: ministry of environment and forests. Government of India
4. Global Energy Yearbook (2011)
5. Jain VK, Srinivas SN (2012) Ministry of new and renewable energy. United Nations Development Programme (UNDP), case study on bio-energy, "Making the steel re-rolling mill more efficient" and study CO_2 emissions from the success stories in empowering India in Re-way published by Ministry of New and Renewable Energy, Government of India, New Delhi

Chapter 26
Numerical Study of Variable Lung Ventilation Strategies

Reena Yadav, Mayur Ghatge, Kirankumar Hiremath and Ganesh Bagler

Abstract Mechanical ventilation is used for patients with a variety of lung diseases. Traditionally, ventilators have been designed to monotonously deliver equal-sized breaths. While it may seem intuitive that lungs may benefit from unvarying and stable ventilation pressure strategy, recently it has been reported that variable lung ventilation is advantageous. In this study, we analyze the mean tidal volume in response to different 'variable ventilation pressure' strategies. We found that uniformly distributed variability in pressure gives the best tidal volume as compared to that of normal, scale-free, log-normal, and linear distributions.

26.1 Introduction

Biological systems are inherently adaptive and have evolved to survive stressors and randomness. Beyond being robust to stressors, they are in fact evolutionarily designed to 'gain' from them. This property of systems has been referred to as 'antifragility' [1, 2]. Antifragile systems benefit from stressors or noise. Here, by 'stressors,' one refers to unfavorable abiotic changes in external milieu such as temperature, pH, and pressure. A system's reaction to any external stressor or stimuli could be enumerated through a measurable property that reflects gain or loss in response to the stressor. It has been argued that a system having convex response would benefit from addition of noise [1, 2]. Many of the biological systems have convex responses due to evolutionary selection that they have undergone. Modeling the nature of such system responses would help in better understanding of design principles of biological systems that help them to thrive under stressful circumstances.

Human engineered systems are designed to cope with stable signals. An electrical system is designed assuming unvarying electrical signal, and structural systems are designed for the absence of severe seismic disturbances. Thus, signal variability

R. Yadav · M. Ghatge · K. Hiremath · G. Bagler (✉)
Indian Institute of Technology, Jodhpur, Rajasthan, India
e-mail: bagler@iitj.ac.in

V. Vijay et al. (eds.), *Systems Thinking Approach for Social Problems*,
Lecture Notes in Electrical Engineering 327,
DOI 10.1007/978-81-322-2141-8_26

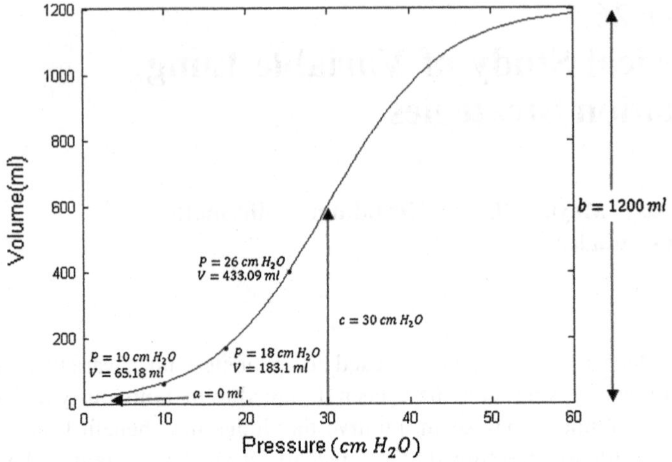

Fig. 26.1 The generic static compliance curve (pressure–volume response curve) obtained from experimental data collected on mechanically ventilated lungs. The *curve* is nonlinear and *roughly sigmoidal* in shape [3]. Various parameters used in Eq. (26.1) have values as shown in the figure

reflects 'noise' and is harmful to the system. Classically physiological systems are thought to be designed to reduce variability and to attain homeostasis. In contrast, signals of a wide variety of physiological systems, such as human heartbeat and brain's electrical activity, fluctuate in a complex manner. In fact, a defining feature of a living organic system is adaptability, the capacity to respond to unpredictable stimuli.

A lung is a physiological system that serves toward respiration, critical for oxidative processes in organisms. Human breathing is driven by pressure generated through spontaneous muscular action. The lungs are ventilated in response to this pressure. The lung volume representing the normal volume of air displaced between normal inspiration and expiration, when no extra effort is applied, is referred to as 'tidal volume.' The pressure–volume response curves for lung are known as 'static compliance curve'. Figure 26.1 shows the standard static compliance curve for normal human lung ventilation [3]. Mathematical modeling of this process results in a sigmoidal equation of the form:

$$v = a + b\frac{1}{1 + \exp^{-((p-c)/d)}}, \tag{26.1}$$

where v is volume and p is pressure.

The equation has been shown to best fit the curve for both the convex region and the concave region [3]. This equation not only comprehensively characterizes the *P–V* curve but also provides various parameters essential for clinical experimentation and studies [3]. The four parameters given by a, b, c, and d in the equation are fitting parameters. The parameter a is the lower asymptote volume, here 0 ml, and *b* corresponds to difference between lower and higher asymptote; here, it is 1,200 ml. The parameter *c* depicts the pressure at the inflection point which is given by 30 cm of H_2O. Finally, *d* is proportional to the pressure range within which most of the volume change takes place, i.e., it is the index of linear compliance with a value of 7 cm of H_2O [3, 4].

When lungs are incapable of ventilating spontaneously, mechanical ventilators are used. Historically, mechanical ventilators are designed to deliver equal-sized breaths. It has been observed that such conventional monotonously regular ventilation has negative consequences for critically ill patients [5–7]. For instance, patients suffering from acute respiratory distress syndrome (ARDS) can have negative impact from this conventional mode of mechanical ventilation due to alveolar collapse and airway damage [5, 6, 8]. Natural healthy ventilation is characterized by its variability. Biologically variable ventilation emulates healthy variation and has been shown to prevent deterioration of gas exchange [9] and increase arterial oxygenation [10] and, in general, is reported to improve respiratory mechanics under various lung pathologies [11–14]. The reasons for the advantageous effects of variable ventilation are not entirely clear and needs to be explored further.

In this study, we explore whether by adding suitably designed noise in ventilation pressure of mechanical ventilators, one can obtain better tidal volume without increasing the mean airway pressure. Apart from ventilation pressure distributed as uniform distribution [4], we studied Gaussian, log-normal, linear, and power-law distributions. In contrast to uniform distribution which gives equal weightage to convex and concave parts of the static compliance curve, these distributions preferentially span the curve. This allowed us to focus on different parts of the curve. For example, Gaussian distribution gives more emphasis on the central part of the curve. The log-normal and linear stress on the latter half of the applied pressure range, whereas power-law emphasizes on the first half of the pressure range while introducing a few high-pressure spikes. Studying various variable ventilation strategies could provide us an insight into the best possible method for operating mechanical ventilators so as to benefit the most from the convexity of the static compliance curve.

Jensen's inequality provides the central argument to explain benefits of convexity and could be used to identify conditions under which addition of noise will be beneficial [4]. Jensen's inequality states that if $f(x)$ is a real-valued convex function in the interval $[a, b]$ and X is a random variable within the range $[a, b]$, then

$$f(E[X]) \leq E[f(X)],$$

and here, $E[X]$ is the expected value (i.e., mean) of the (random) variable.

As a consequence of Jensen's inequality, varying the pressure within the convex region, before the inflection point, of the static compliance curve, i.e., from 10 cm H_2O to 26 cm H_2O (Fig. 26.1), would result in higher mean tidal volume in contrast to constant pressure strategy. The gain in mean tidal volume, as compared to application of constant pressure, depends on the range of the static compliance curve spanned and the frequency.

26.2 Studies

We studied various variable ventilation strategies to study their effect on mean tidal volume in contrast to the conventional mode of mechanical ventilation. 'Constant-mode ventilation' monotonously delivered air into the lungs at regular intervals and at the same pressure. In recent studies [4], it has been shown that uniformly varying pressure values in the range from 10 cm H_2O to 26 cm H_2O results in better tidal volume compared to 'constant-mode ventilation.' Figure 26.2a shows the tidal volume output for the uniformly distributed pressure values and the corresponding volume distribution obtained as shown in Fig. 26.2b. The mean tidal volume improves to 205.72 ml from 183.13 ml obtained using the monotonous strategy, even though both strategies use the same mean pressure of 18 cm H_2O.

The uniform pressure distribution function can be mathematically written as:

$$\mathcal{U} = \begin{cases} \frac{1}{16} & \text{if } 10\,\text{cm}\,H_2O \leq p \leq 26\,\text{cm}\,H_2O \\ 0 & \text{otherwise} \end{cases} \tag{26.2}$$

Variably distributed pressure values are sampled from the interval of (10 cm H_2O, 26 cm H_2O), which is the physiological limit of minimum and maximum pressure values for human lungs [8]. Also, as can be seen from the Static Compliance Curve in Fig. 26.1, this interval lies in the convex region of the curve covering the output volume range of 65.18–433.09 ml as opposed to a single value of 183.13 ml corresponding to 18 cm pressure, used in constant-mode ventilation strategy.

We further explored the effect of variable ventilation in the form of a few canonical distributions on the tidal volume. The mean value of pressure is always

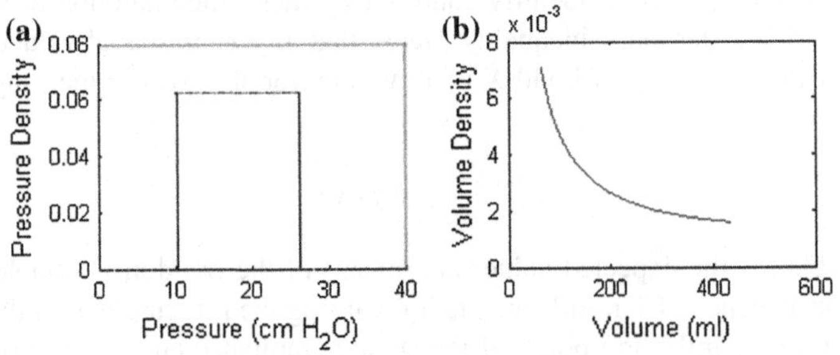

Fig. 26.2 The nature of tidal volume distribution **b** in response to uniformly distributed variable ventilation pressure **a**, as given by Eq. (26.2)

kept constant at 18 cm in all the distributions discussed throughout our studies. The density function equations for various pressure distributions are as follows.

- **Gaussian Pressure Distribution**

$$\mathscr{N} = \frac{1}{\sigma\sqrt{2\pi}} \mathrm{e}^{-(x-\mu)^2/2\sigma^2} \tag{26.3}$$

where μ is the mean of the distribution and σ represents standard deviation.

- **Power-Law Pressure Distribution**

$$\mathscr{P} = \frac{\alpha}{x_{\min}} \left(\frac{x}{x_{\min}}\right)^{-\alpha-1} \tag{26.4}$$

where $x_{\min}$ is the minimum possible value of x and α is the power-law exponent with value greater than 1.

- **Log-Normal Pressure Distribution**

$$\mathrm{Log} - \mathscr{N} = \frac{1}{x\sigma\sqrt{2\pi}} \mathrm{e}^{-(\ln(x)-\mu)^2/2\sigma^2} \tag{26.5}$$

where μ is the mean and σ is standard deviation.

Figure 26.3 depicts the nature of pressure density functions for linear and power-law distributions, and Fig. 26.4 depicts that for Gaussian and log-normal distributions.

The linear pressure distribution in Fig. 26.3a gives volume density function which is only marginally different from that of uniform distribution. The maximum

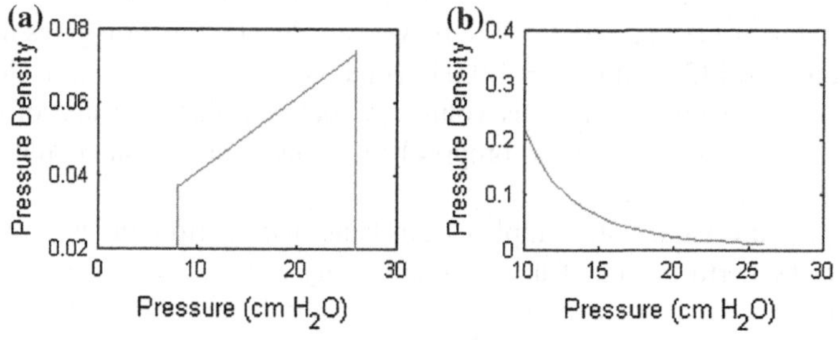

Fig. 26.3 **a** Linear pressure variation with 8 cm $\leq p \leq$ 26 cm. **b** Pressure variation following power-law distribution, given by Eq. (26.4) with α = 2.25 and $x_{\min}$ = 10 cm H_2O

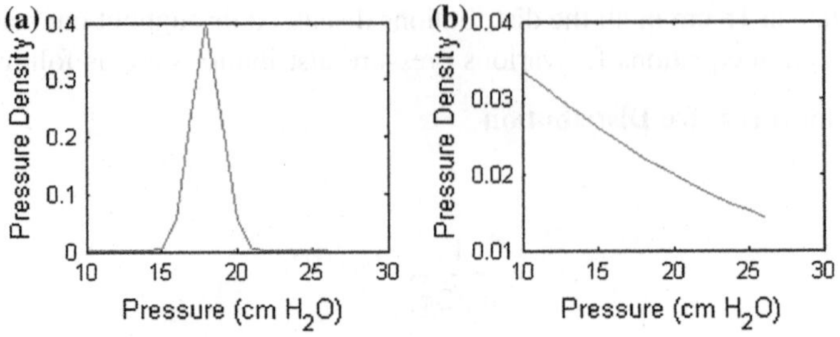

Fig. 26.4 **a** Gaussian distribution given by Eq. (26.3) and **b** log-normal distribution represented by Eq. (26.5) with $\mu = 18$ cm H_2O and $\sigma = 1$

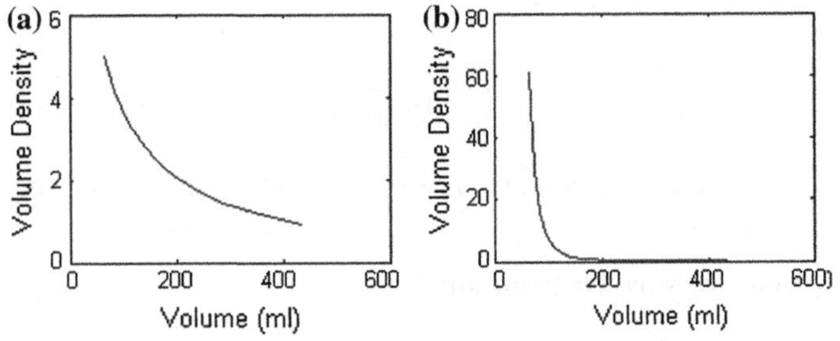

Fig. 26.5 Tidal volume output for **a** linear and **b** power-law pressure distributions

mean tidal volume produced is 205.59 ml with minimum pressure reduced to 8 cm from that of 10 cm used in all other cases to maintain the average pressure at 18 cm. Thus, linear distribution comes close to uniform in terms of the mean output volume. The slope and volume intercept in Fig. 26.3a are 1/486 and 10/486, respectively. The minimum pressure applied was changed to 5 cm while keeping the maximum pressure at 26 cm H_2O to see the effect of change in slope to the volume output. However, it was observed that even after changing the minimum pressure value, the mean tidal volume does not vary much from that observed with the pressure of 8 cm (Fig. 26.5).

The pressure values in Fig. 26.3b follow power-law distribution within the same interval (10 cm H_2O, 26 cm H_2O). The power-law density function gives maximum volume output of 112.3 ml for $\sigma = 1.93$, beyond which the volume starts decreasing as is evident from Fig. 26.7a. This volume output is worse than that of constant-mode ventilation, contrary to the proposal that noisy ventilation should be beneficial in general (Fig. 26.6).

When pressure values are sampled from Gaussian distribution as shown in the Fig. 26.4a, the performance of the volume density function improves from that of constant mode, though not significantly. The mean volume varies with σ and has the maximum value of 190.8 ml at $\sigma = 3.15$ after which the volume starts decreasing sharply as shown in Fig. 26.7b.

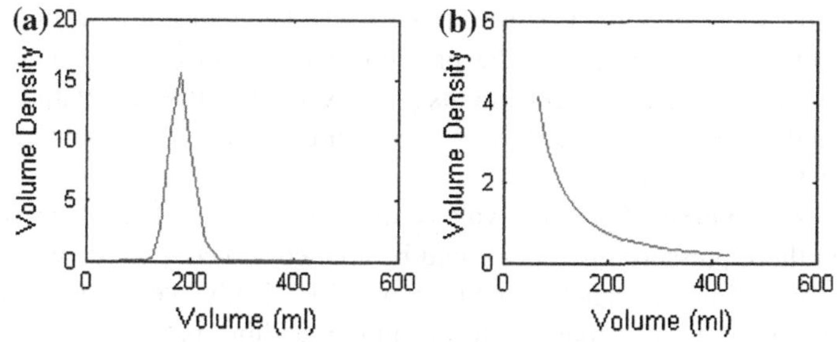

Fig. 26.6 Tidal volume output for **a** Gaussian and **b** log-normal pressure distributions

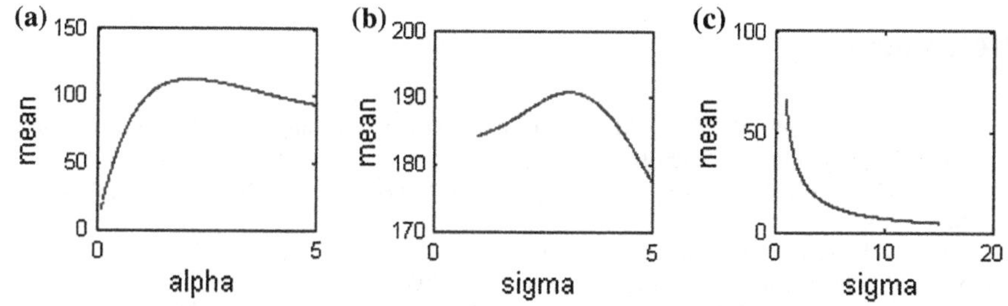

Fig. 26.7 Effect of parameters α [exponent in Eq. (26.3)] and σ [standard deviation in Eqs. (26.3) and (26.5)] on the mean tidal volume output. **a** Power-law distribution. **b** Gaussian distribution. **c** Log-normal distribution

When the probability density function of pressure follows log-normal distribution (Fig. 26.4b), the mean tidal volume output was found to be 70.44 ml, worse than even that of 'constant-mode ventilation'. Thus, log-normal strategy resulted in the worst performance among all the five canonical distributions studied. Figure 26.7c shows that the maximum value for volume output in the log-normal case comes at $\sigma = 325.1$ after which the mean volume decrease sharply.

26.3 Conclusions

Variable ventilation strategy in general is reported to be beneficial in terms of mean tidal volume, without increasing the ventilation pressure. It is desirable to characterize the effect of various variable ventilation strategies, which can help in the identification of best possible strategies to implement in mechanical ventilators. From our studies with five canonical distribution strategies, we found that uniform pressure distribution is the most favorable 'variable pressure strategy.' This distribution exploits the convexity of the generic static compliance curve the most, to emerge as the best variable ventilation strategy among the canonical distributions studied in this paper.

This leads to an important question as to whether uniform distribution is the 'best possible' noisy strategy for the pressure density function or could there be better strategies. Within the constraints of physiologically acceptable range of pressures and without increasing the mean pressure, this question could be posed as an optimization problem.

For any distribution of pressure values, there could be multiple time sequences with which these pressure instances could be applied. It needs to be studied further which of these time sequences could be physiologically meaningful. Also, one could study empirical distributions for various breathing patterns such as normal breathing, panting, and *pranayama*.

References

1. Taleb NN, Douady R (2013) Mathematical definition, mapping, and detection of (anti) fragility. Quant Financ 13(11):1677–1689
2. Taleb NN (2012) Antifragile: things that gain from disorder. Random House Digital, Inc., New York
3. Venegas JG, Harris RS, Simon BA (1998) A comprehensive equation for the pulmonary pressure-volume curve. J Appl Physiol 84(1):389–395
4. Brewster JF, Graham MR, Mutch WAC (2005) Convexity, Jensen's inequality and benefits of noisy mechanical ventilation. J R Soc Interface 2(4):393–396
5. Official Conference Report (1999) International consensus conferences in intensive care medicine: ventilator-associated lung injury in ARDS. Intensive Care Med, 25:1444–1452
6. Albert RK (2012) The role of ventilation-induced surfactant dysfunction and atelectasis in causing acute respiratory distress syndrome. Am J Respir Crit Care Med 185(7):702–708
7. Dos Santos CC, Slutsky AS (2000) Invited review: mechanisms of ventilator-induced lung injury: a perspective. J Appl Physiol 89(4):1645–1655
8. Rouby J-J, Brochard L (2007) Tidal recruitment and overinflation in acute respiratory distress syndrome: yin and yang. Am J Respir Crit Care Med 175(2):104–106
9. Mutch WA, Eschun GM, Kowalski SE, Graham MR, Girling LG, Lefevre GR (2000) Biologically variable ventilation prevents deterioration of gas exchange during prolonged anaesthesia. Brit J Anaesth 84(2):197–203
10. Mutch WAC, Harms S, Lefevre GR, Graham MR, Girling LG, Kowalski SE (2000) Biologically variable ventilation increases arterial oxygenation over that seen with positive end-expiratory pressure alone in a porcine model of acute respiratory distress syndrome. Crit Care Med 28(7):2457–2464
11. McMullen MC, Girling LG, Graham MR, Mutch WAC (2006) Biologically variable ventilation improves oxygenation and respiratory mechanics during one-lung ventilation. Anesthesiology 105(1):91–97
12. Boker A, Haberman CJ, Girling L, Guzman RP, Louridas G, Tanner JR, Cheang M, Maycher BW, Bell DD, Doak GJ (2004) Variable ventilation improves perioperative lung function in patients undergoing abdominal aortic aneurysmectomy. Anesthesiology 100(3):608–616
13. Mutch WAC, Harms S, Graham MR, Kowalski SE, Girling LG, Lefevre GR (2000) Biologically variable or naturally noisy mechanical ventilation recruits atelectatic lung. Am J Respir Crit Care Med 162(1):319–323
14. Mutch WAC, Buchman TG, Girling LG, Walker EK, McManus BM, Graham MR (2007) Biologically variable ventilation improves gas exchange and respiratory mechanics in a model of severe bronchospasm. Crit Care Med 35(7):1749–1755

Chapter 27
Development of a Technique for Measurement of High Heat Flux

Ram Niwas Verma, P.K. Jaya Kumar, Laltu Chandra and Rajiv Shekhar

Abstract High heat flux measurement is necessary for various experiments and industrial applications. For example, to estimate a very high irradiance onto a surface as in solar thermal and in testing of plasma facing component in ITER. In order to estimate high heat flux of the order of MW/m^2, an experiment with plasma jet of non-transferred 9MBM type is employed. This paper describes the design of a non-intercepted calorimeter. This is for estimation of average incident heat flux of the plasma jet along the axial direction. Temperature at different axial locations on the surface of calorimeter and at the outlet is measured. The heat flux associated with the incident plasma is estimated from these measured values of temperature. Rate of heat transfer from plasma jet to the employed target surface at different axial position is measured with the help of intercepted calorimetric method. Based on the estimated heat flux and heat transfer rate, the electrothermal efficiency of 9MBM plasma torch is estimated. Torch heat efficiency (THE), plasma heat efficiency (PHE) and heat transfer effectiveness (HTE) of the plasma jet are also estimated for copper materials. PHE and HTE are required for mathematical modelling of plasma surface interaction.

Keywords Heat flux · Calorimeter · Plasma torch · THE · PHE · HTE

R.N. Verma (✉) · L. Chandra · R. Shekhar
Indian Institute of Technology Jodhpur, Old Residency Road, Jodhpur, India
e-mail: ramniwasv@iitj.ac.in

L. Chandra
e-mail: Chandra@iij.ac.in

R. Shekhar
e-mail: vidtan@iitj.ac.in

P.K. Jaya Kumar
Non Ferrous Material Technology Development Center, Kanchanbagh, Hyderabad, India
e-mail: pkjayg@gmail.com

V. Vijay et al. (eds.), *Systems Thinking Approach for Social Problems*,
Lecture Notes in Electrical Engineering 327,
DOI 10.1007/978-81-322-2141-8_27

27.1 Introduction

Due to increasing demand of energy and continues depletion of fossil fuels, there is a need of another sources of energy. There are many alternative sources of energy such as solar energy, wind energy, tidal energy and geothermal energy. Scientists are also looking towards fusion energy as a neat and clean energy for future. International thermonuclear experimental reactor (ITER) is the device where scientists and engineers have been working to produce fusion energy up to 480 s [1]. Objective of ITER is to develop technologies and process needed for a fusion power plant. Plasma is an artificial environment in ITER to occur fusion reaction. Helium is produced as a by-product during the fusion reaction in plasma. Helium is removed from plasma with the help of divertor [1]. During this process, plasma comes in contact with the divertor plate. Expected heat load on the divertor plate due to disruption and edge localised modes (ELM) is 20 and 3 MW/m^2, respectively. Heat flux transferred to the divertor plate material and finally removed by the coolant. Scientists have been working for synthesis of divertor plate material in Non Ferrous Material Technology Development Centre (NFTDC). Testing of the developed divertor plate materials is performed using a 9MBM plasma torch in NFTDC. 9MB is the series of plasma torch designed and engineered by SULZER METCO. 9MBM (Machine-Mounted) Plasma torch is used for automatic operation [2]. 9MBM is type of plasma torch in NFTDC is a non-transferred-type plasma torch. Efficiency of non-transferred plasma torch is nearly about 40 % [3].

High heat flux characterisation is required for processing and testing of materials under extreme condition. Williams and White [4] estimated high heat flux (MW/m^2) of plasma jet by measuring temperature rise of copper disc, which is exposed in plasma jet. High heat flux is measured using semi-infinite method [5, 6]. In both the cases, heat flux is estimated, which is transferred to the surface of target. A fraction of the incident heat is transferred from plasma jet to the target surface. Cold and thermal both type of plasma are used to synthesis of material [7–12]. Synthesis of material depends on the content of plasma gas and heat transfer rate to the target material [8–13]. Temperature of the target material is one of the most important factors for processing of it. Temperature of target depends on the heat transfer rate from plasma jet to it. Heat transferred depends on the heat flux incident on it. For a given plasma torch, heat flux incident on the target depends on the distance from plasma torch (stand-off distance). For mathematical modelling and analysis of plasma target interaction, experimental value is required [7–13].

27.2 Principle of Measurement of Heat Flux of Plasma Jet

Plasma jet enters into one end of calorimeter (copper pipe) and flows out from the other end of the calorimeter as shown in Fig. 27.1. Cross-sectional area of plasma jet expands along the direction of flow. Hence, after travelling some distance,

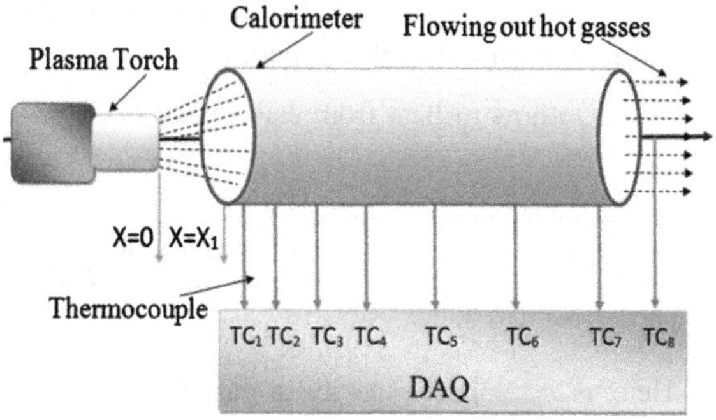

Fig. 27.1 Schematic diagram of heat flux measurement of plasma jet

plasma jet (produced from a mixture of argon and hydrogen gas in the plasma torch) comes in contact with the inner surface of calorimeter. Due to temperature difference between incident plasma jet and the surface of calorimeter, heat is transferred from plasma to the calorimeter. Outer surface of calorimeter is insulated. Hence, the convective and radiation losses to ambient are neglected from the outer surface of calorimeter. Let E_1 be total plasma heat (Joule) flow in calorimeter in t second from plasma torch, E_2 be the heat (Joule) flowing out from the other end of calorimeter and E_3 be the heat (Joule) stored in the calorimeter. In Fig. 27.1, TC_1 to TC_7 are indicating the installed thermocouples to measure temperature of outer surface of calorimeter (pipe).

The thermocouple TC_8 measures the temperature of outflow gasses from the calorimeter.

From conservation of energy:

$$E_1 = E_2 + E_3 \tag{27.1}$$

Average heat flux (HF) incident at the inlet of calorimeter

$$\text{HF} = \frac{E_1}{At} \tag{27.2}$$

Here, A is cross-sectional area of calorimeter, and t is time (plasma torch on). This method to measure high heat flux is referred as "non-intercepted calorimetric method".

27.2.1 Measurement of Heat Flux of Plasma Jet with the Help of Calorimeter

Plasma torch is operated for t second at 40 V and 420 A. In order to estimate the heat flux from plasma jet, a metallic calorimeter of inner diameter d and length L is considered (see Fig. 27.1). Plasma jet enters the calorimeter from inlet and leaves from the outlet of calorimeter. This heat is absorbed by calorimeter, which results in rise of temperature. For this experiment,

E_2 = Outflow of heat from calorimeter during t seconds:

$$E_2 = \dot{m}_g c_{p_g} \int_{T_{2i}}^{T_{2f}} T_2 \mathrm{d}t \tag{27.3}$$

Here, $m_g c_{p_g}$ = heat capacity of gas, T_2 = instantaneous temperature out-flowing plasma.

E_3 = Total heat stored in the pipe during t seconds:

$$E_3 = \frac{m_{\text{cal}}}{L} c_{p\_\text{cal}} \int_0^L T_3 \mathrm{d}x \tag{27.4}$$

Here, T_3 = Rise in temperature of the calorimeter at different positions during the operation time t, m_{cal} = mass of calorimeter and c_{p_cal} = specific heat capacity of material of calorimeter. Once the values of E_2 and E_3 are estimated, value of E_1 can be obtained.

27.3 Heat Transfer to the Target Surface

When the target surface is exposed to plasma, a fraction of the incident energy is transferred to the target surface. This is estimated by the performed experiments with plasma jet. The schematic of this experiment is shown in Fig. 27.2. Copper cylindrical block (target surface) is placed at different axial positions and denoted by stand-off distance, from the outlet of plasma torch. The transferred heat to the target surface is estimated by the rise in temperature with k-type thermocouple mounted at the bottom. This method is referred as "intercepted calorimetric method". In this experiment, mass of the employed target copper block is 0.082 kg, target surface area of copper block is 0.00198 m^2, convective heat transfer coefficient is 13 W/m^2 K (assuming natural convection), specific heat capacity is 0.385 kJ/kg/K, and maximum time of heating is 3 s. Convective heat transfer is calculated using Grashoff number, Prandtl number and Nusselt number for cylindrical copper block in still air. Convective heat loss is 300 times less than heat stored in copper block. Hence, convective heat loss is neglected in this experiment. If it is assumed emissivity of surfaces is 1 to calculate maximum heat loss, then radiation heat loss is $\sim$300 times less than the heat absorbed in target material. Hence, the heat loss is neglected in this experiment.

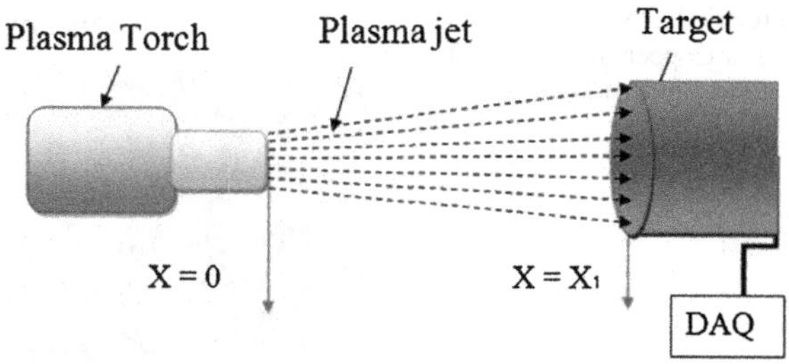

Fig. 27.2 Plasma target interaction

Heat transferred to target surface = Heat absorbed in target material:

$$E_t = m_t c_{pt} \Delta T \tag{27.5}$$

Here, m_t = mass of target material, c_{pt} = specific heat capacity of target material, ΔT = temperature rise of the target material during the experiment (t = 1–3 s).

Power transferred to target surface:

$$P_t = \frac{m_t c_{pt} \Delta T}{t} \tag{27.6}$$

Rate of heat transferred per unit area to target surface:

$$HF_t = \frac{m_t c_{pt} \Delta T}{At} \tag{27.7}$$

27.4 Results and Discussion

27.4.1 Heat Flux Measurement of Plasma Jet with Non-intercept Method

In order to estimate the heat flux of employed plasma jet, a non-interception method is used. In this approach, aluminium (diameter 27 and 35 mm, length 1,000 mm) and copper pipe (diameter 25 and 31 mm, length 1,000 mm) as calorimeter are used. These pipes are place at stand-off distance X_1 = 20, 40, 60 and 80 mm, respectively. Experimental set-up is shown in Fig. 27.3. Plasma jet generated by the plasma torch enters into calorimeter (at the pipe inlet) and leaves from the outlet of calorimeter. In the present paper, the results for copper calorimeter are discussed. The temperature measurements are performed with k-type thermocouples indicated by TC_1, TC_2, TC_3, TC_4, TC_5, TC_6 and TC_7 in Fig. 27.1. These are mounted at 25, 75, 175, 325, 525, 725 and 975 mm distance from the inlet of plasma denoted by $X = X_1$ in Fig. 27.1.

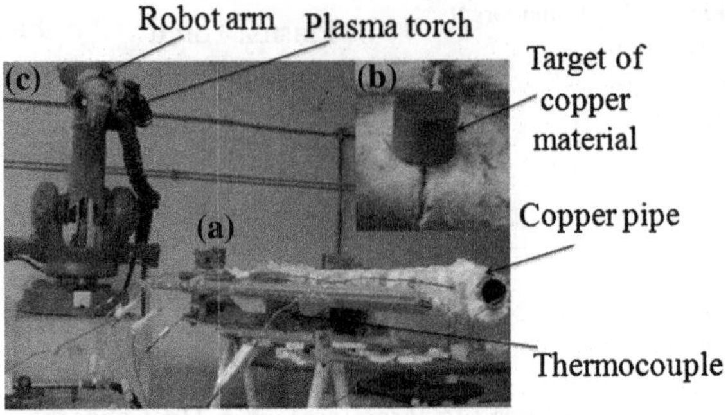

Fig. 27.3 Experimental set-up. **a** Copper pipe, **b** target surface, **c** plasma torch

Temperature rise of the copper calorimeter surface with respect to time at the described positions (indicated by T1, T2, T3, T4, T5, T6 and T7) is shown in Fig. 27.4. The measurements are shown for stand-off distance as 0 mm and duration of operating plasma torch as 20 s. T8 is the temperature of out-flowing plasma (gas) at the outlet.

All the installed thermocouples exhibit simultaneous increase in temperature. It shows that the plasma jet heats up the calorimeter. For stand-off distance of 0 mm, the measured temperature TC_2 is higher than that of TC_1 and TC_3, see Fig. 27.6. It can be inferred that, perhaps, plasma comes in contact with calorimeter (copper pipe) near the location of TC_2 or about 75 mm from the outlet of plasma torch, as indicated in Fig. 27.5.

Temperature rise of pipe at different positions at zero and 20 mm stand-off distances is shown in Fig. 27.6. Temperature rise of the pipe at these positions is shown in Fig. 27.6. In Fig. 27.6, a pipe is put at zero stand-off distance, and here, T1 is less than T2; this is due to lower diameter of jet compared to that of pipe as shown in Fig. 27.5.

Figure 27.6 clearly reveals that the temperature rise, as expected, decreases with increasing axial position of calorimeter (copper pipe) towards the outlet. It can be inferred that the rate of heat transfer from plasma jet to inner surface of pipe

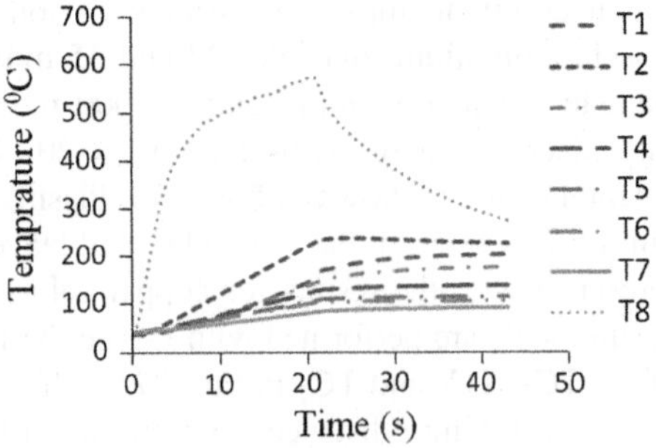

Fig. 27.4 Plasma jet experiment with copper calorimeter and temperature profile of the calorimeter (pipe)

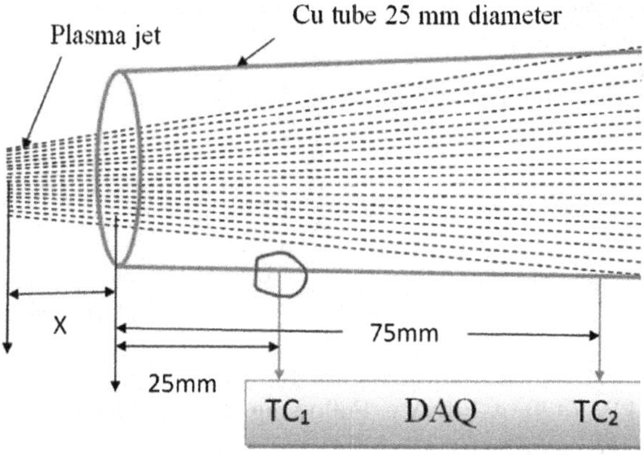

Fig. 27.5 Schematic diagram of temperature measurement of calorimeter

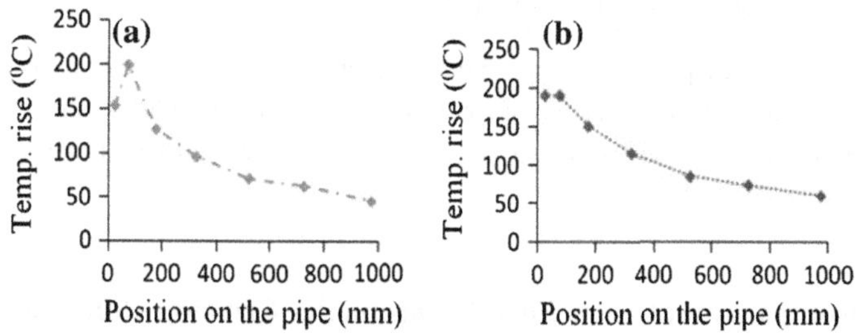

Fig. 27.6 Temperature profile of copper pipe (diameter 25 mm, length 1 m). **a** 0 mm stand-off distance. **b** 20 mm stand-off distance

decreases with increasing distance. A fraction of the received heat by calorimeter is stored in calorimeter, and the rest is transferred with the out-flowing plasma. This is produced by a mixture of argon and hydrogen gas in the plasma torch. Mass flow rate of argon and hydrogen gas mixture is 2.8 and 0.007 gm/s, respectively. Specific heat capacities of argon and hydrogen gas are 0.52 and 14 kJ/kg K, respectively. Estimation from measured surface temperature of calorimeter reveals that 90 % of the incident power is absorbed by the calorimeter. The remaining 10 % leaves as plasma jet emanating from the outlet of calorimeter. The loss is much smaller than 1 % and is, therefore, neglected. The incident power on the calorimeter and, therefore, the corresponding heat flux of plasma jet decrease with increasing stand-off distance. The heat flux decreases with increasing diameter of calorimeter.

27.4.1.1 Average Output Power and Efficiency of Plasma Torch

Ignoring losses, the output power of plasma torch is estimated as the received power by calorimeter at zero stand-off distance. Input power given to the plasma

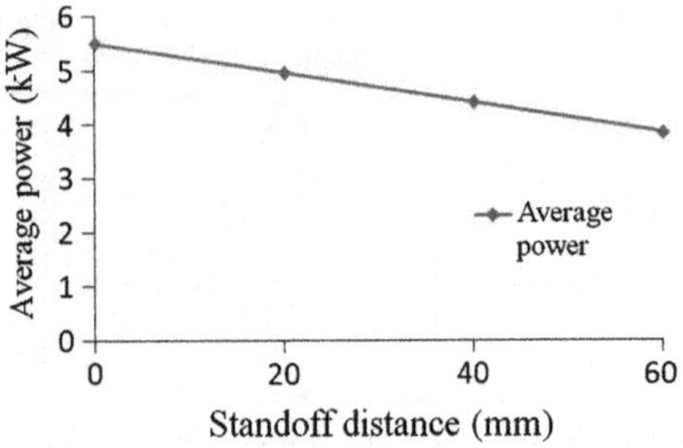

Fig. 27.7 Average output power of plasma torch versus stand-off distance

torch is 16.8 kW (420 A, 40 V). Following are estimated from the measured temperature of the calorimeter:

Average output power of plasma torch = 5.5 kW
Average heat flux at nozzle tip = 109.38 MW/m^2
Efficiency of plasma torch = 32.74 %.

Average power of plasma jet normal to the cross-sectional area, as in Fig. 27.9, at different stand-off distance is shown in Fig. 27.7. It clearly shows that the incident power at the considered cross section decreases with increasing stand-off distance.

27.4.1.2 Comparison of Heat Flux of Plasma Jet Crossing Different Cross-sectional Area Around the Axis of Plasma Jet

In order to estimate heat flux on different area of cross sections, two copper pipes (calorimeter) of internal diameter 6 and 25 mm are used. These pipes are placed co-axially at 20, 40, 60, 80,120, 160, 200 and 240 mm stand-off distance. Comparison of the estimated heat flux of plasma jet is shown in Fig. 27.8.

Heat flux ratio through cross-sectional area of diameter 6 and 25 mm decreases with increasing stand-off distance. This reveals that distribution of heat flux

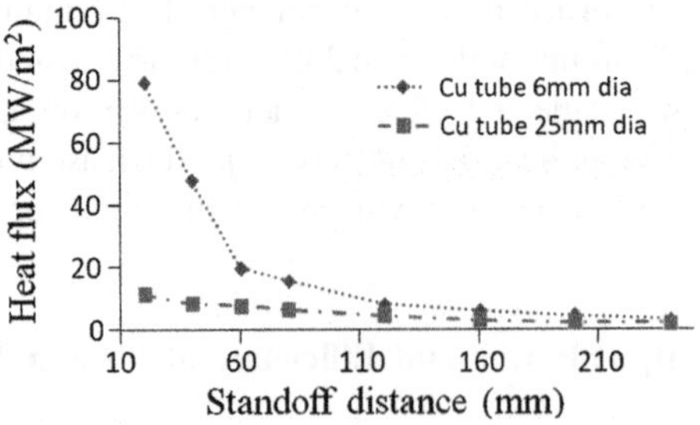

Fig. 27.8 Comparison of heat flux between 6-mm-diameter copper pipe and 25-mm copper calorimeter at different stand-off distances

Fig. 27.9 Plasma jet profile

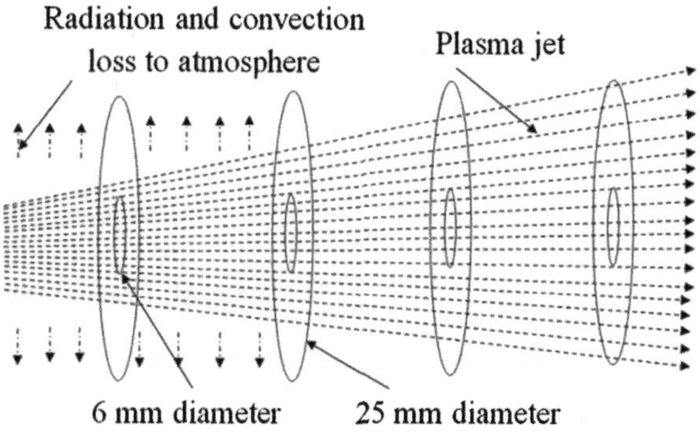

becomes uniform with increasing stand-off distance of the cross section. This is shown in Fig. 27.9.

Total heat loss (radiation and convection) by 20, 40 and 60 mm length of plasma jet is 0.54, 1.09 and 1.65 kW, respectively, as shown in Figs. 27.7 and 27.9. It reveals that heat flux incident on the target surface decreases as stand-off distance increases. Expected error of k-type thermocouple is plus/minus 1 °C. Hence, percentage error in measurement of temperature rise of calorimeter is 2.5–7.5. Hence, inaccuracy of measurement of this experiment is between 2.5 and 7.5 %.

27.4.2 Heat Transferred to Target Surface by Plasma Jet

Copper is used as a target of diameter 25 mm as shown in Fig. 27.3. In order to estimate heat transferred to the target, it is subjected to plasma jet at different positions.

Figure 27.10 shows the amount of heat flux incident on the target surface and the amount transferred to the target. Absorbed heat in the target is the heat transferred by plasma to the target. To estimate heat flux, heat absorbed in target is divided by time and area of target surface. Heat flux incident on the target surface is measured with the help of non-intercepted calorimeter (copper pipe of same cross section) by

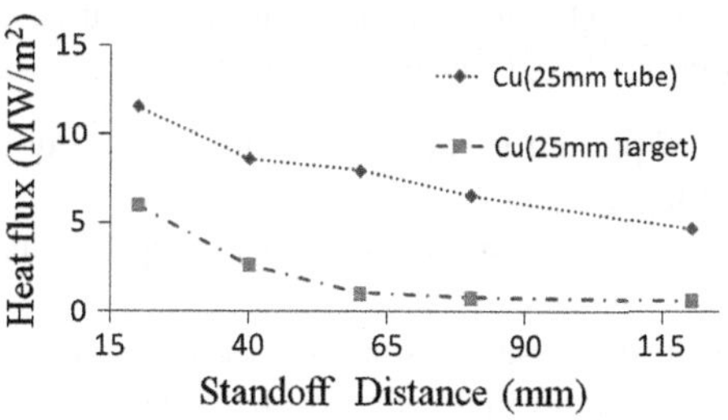

Fig. 27.10 Comparison of incident heat flux on target surface and heat transferred to copper target

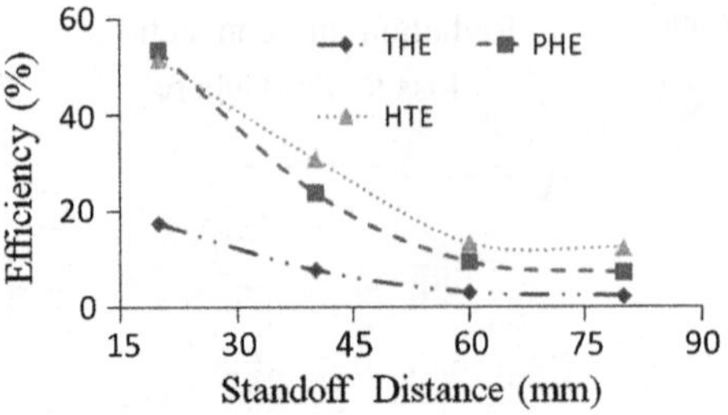

Fig. 27.11 Percentage of heat transferred to copper target

putting at the same position. Figure 27.10 clearly shows heat transferred to the target surface is less than incident on it. This happens due formation of shielding layer on the target and heat transfer characteristic. Shielding layer prevents heat transfer from plasma to target surface [14]. In this paper, ratio of heat transfer to the input power from plasma torch is defined as "torch heat efficiency" (THE). Ratio of heat transfer to output power of plasma torch is defined as "plasma heat efficiency" (PHE). Ratio of heat transfer to the target surface to heat incident on the target surface is defined as "heat transfer effectiveness" (HTE). THE, PHE and HTE versus stand-off for copper target is shown in Fig. 27.11. PHE is more than THE, because electrothermal efficiency of plasma torch itself 33 %. HTE is more than PHE because plasma jet losses heat due to convection and radiation to reach the target surface. THE, PHE and HTE all are decreased as stand-off distance increases.

27.5 Conclusion

In this paper, heat flux estimation from a plasma jet is presented with the help of non-intercepted calorimetric method and calorimetric method. Axial power profile of plasma jet is analysed by non-intercepted calorimetric method. Heat transfer to target surface is estimated with the help of intercepted calorimetric method. The principle is based on the measured temperature on the calorimeter (copper pipe) and on a copper solid target. Non-interception calorimeter (copper pipe) provides the estimate of heat flux, and interception method (copper solid target) is employed for estimation of heat transferred from plasma jet to target surface.

Following are observed based on measurement:

- Average heat flux of plasma jet decreases with increasing stand-off distance of the measurement plane.
- Heat transferred to target surface decreases rapidly as stand-off distance increases.
- THE, PHE and HTE decreases as stand-off distance increases.
- HTE is more than PHE and PHE is more than THE.

Acknowledgments The authors of this paper acknowledge the provided support and guidance by Non Ferrous Technology Development Centre Hyderabad, Indian Institute of Technology Jodhpur and Ministry of New and Renewable Energy for providing technical, infrastructure and financial support.

References

1. www.iter.org
2. www.sulzer.com/en/Products-and-Services/Coating
3. Venkatramani N (2002) Industrial plasma torches and applications. Current Sci-Bangalore 83 (3):254–262
4. Williams AW, White KJ (2001) Plasma-propellant interactions studies: measurements of heat flux produced by hydrocarbon ablation-supported plasmas. IEEE Trans Magn 37(1):203–206
5. Löhle S, Battaglia J, Gardarein J, Jullien P, Ootegem BV (2008) Heat flux measurement in a high enthalpy plasma flow. J Phys Conf Ser 135(1):012064
6. Kishigami T, Heberlein JV, Pfender E (1991) Experimental investigation of heat transfer between plasma jet and substrate. IPSC-10, Bochum
7. Zagorski A, Szuecs F, Balaschenko V, Siegmann S, Margandant N, Ivanov A (2003) Experimental study of substrate thermal conditions at APS and HVOF. In: Proceeding of the 2003 international thermal spray conference, pp 1255–1260
8. Wang H, Chen X, Cheng K, Pan W (2007) Modeling study on the characteristics of laminar and turbulent argon plasma jets impinging normally upon a flat plate in ambient air. Int J Heat Mass Transf 50(3):734–745
9. Kim K, Kim G, Hong YC, Yang SS (2010) A cold micro plasma jet device suitable for bio-medical applications. Microelectron Eng 87(5):1177–1180
10. Tyata RB, Subedi DP, Shrestha A, Baral D (2012) Development of atmospheric pressure plasma jet in air. Kathmandu Univ J Sci Eng Technol 8(1):15–22
11. Xian M, Wen-xia P, Wen-hong Z, Cheng-kang W (2001) Heat flux characterization of DC laminar-plasma jets impinging on a flat plate at atmospheric pressure. Plasma Sci Technol 3 (5):953
12. Li G, Pan W, Meng X, Wu C (2005) Application of similarity theory to the characterization of non-transferred laminar plasma jet generation. Plasma Sources Sci Technol 14(2):219
13. Bonizzoni G, Vassallo E (2002) Plasma physics and technology; industrial applications. Vacuum 64(3):327–336
14. Chuvilo AA, Garkusha IE, Makhlaj VA, Petrov YV (2012) Calorimetric studies of the energy deposition on a material surface by plasma jets generated with QSPA and MPC devices. Nukleonika 57(1):49–53

Chapter 28
A Defaultable Financial Market with Complete Information

I. Venkat Appal Raju

Abstract We consider a Markov-modulated defaultable Brownian market and price the defaultable contingent claims with the intensity-based methodology using the fair price concept under the benchmark approach. We also derive the locally risk-minimizing hedging strategy for defaultable contingent claims under the benchmark approach. We assume that the default intensity and the stock price parameters are modulated by a Markov process. The recovery processes are assumed to have random payments at default time as well as at the maturity of the claims.

28.1 Introduction

In this paper, our goal is to price and hedge defaultable contingent claims, using benchmark approach under the reduced-form methodology of credit risk modeling. Credit risk is the risk associated with any kind of credit-linked event, such as changes in the credit quality, variations of credit spreads, and the default event. Here, we consider only the default event for our modeling of credit derivatives, i.e., the default risk. A default risk is a possibility that a counterparty in a financial contract will not fulfill a contractual commitment to meet its obligation stated in the contract. If default occurs, the creditor will only receive the amount recovered from the debtor, called recovery payment. There are two kinds of recovery payoffs: one recovery payoff Z at the time of default if default occurs prior to or at the maturity date T, and the recovery payoff at time T if default occurs prior to or at the maturity date T. Naturally, one of the recovery payoffs occurs in the real market. But to generalize the model, we consider both together.

The modeling of credit risk may be divided into two broad classes: structural models and reduced-form models. In the former approach, the total value of the firm's assets is directly used to determine the default event, which occurs when the

I.V.A. Raju (✉)
Indian Institute of Technology Jodhpur, Jodhpur 342 011, India
e-mail: i.raju@iitg.ac.in

V. Vijay et al. (eds.), *Systems Thinking Approach for Social Problems*,
Lecture Notes in Electrical Engineering 327,
DOI 10.1007/978-81-322-2141-8_28

firm's value falls through some boundary. The random time of default is announced by an increasing sequence of stopping times. By contrast, in the latter approach, the firm's value process either is not modeled at all, or it plays only an auxiliary role of a state variable. The default time is modeled as a stopping time that is not predictable; the default event thus arrives as a total surprise. They are called reduced form since they are abstracted from the explicit economic reasons for the default, i.e., they do not induce the asset–liability structure of the firm to explain the default as in a structural model approach. Rather, reduced-form models use debt prices as a main input to model the bankruptcy process. Default is modeled by a stochastic process with an exogenous default intensity as hazard rate which we consider to be modulated by a Markov process X. The Markov process X is modeled as a finite-state Markov chain.

The classical Black–Scholes formula for option pricing is based on a geometric Brownian motion model to capture the price dynamics of the underlying security. The parameters involved in the model are assumed to be deterministic constants. It is well known, however, that the Black–Scholes model fails to reflect the stochastic variability in the market parameters. Another widely acknowledged shortfall of this model is its failure to capture what is known as "volatility smile." We consider a regime-switching model, which stems from the need of more realistic models that better reflect random market environment. The market regime could reflect many scenarios starting from the state of the underlying economy, general mood of investors in the market, and to other economic factors. Therefore, along with the default intensity, we also suppose that the stock price parameters are modulated by the Markov process X. The regime-switching model was first introduced by Hamilton [9] to describe a regime-switching time series. Di Masi et al. [5] discuss mean–variance hedging for European option pricing under the regime-switching model. Raju and Selvaraju [15] used the benchmark approach to price European options under a regime-switching model.

Due to the additional randomness in the market, i.e., the default process and the Markov process, the market is incomplete. To date, there is no clear agreement in the literature about the appropriate approach to the pricing of derivatives subject to credit risk. In the risk-neutral method, there will be more than one martingale measures and there have been different approaches introduced in literature to choose an appropriate martingale measure, such as the minimal entropy martingale measure, the minimal martingale measure. Biagini and Cretarola [2] have used minimal martingale measure approach to price the defaultable contingent claims. For credit risk, however, it is not clear what the appropriate pricing measure should be, or even if such a pricing measure can actually be found. Furthermore, it has been demonstrated in Heath and Platen [10] that any realistic parsimonious financial market model is unlikely to have an equivalent risk-neutral martingale measure. Even if an appropriate risk-neutral martingale measure exists, the H-hypothesis may not be stable under the measure transformation. So to remove all these limitations, in the present paper, we consider the defaultable contingent claim beyond the standard risk-neutral approach. Using the growth optimal portfolio (GOP) as numeraire, defaultable claims are priced under real-world probability measure itself (for more detail on benchmark approach see, Platen [14]).

The paper is structured as follows: In Sect. 28.2, we introduce the general setup of the defaultable market that we are going to analyze in this paper. In Sect. 28.3, first we introduce the GOP. Using GOP as numeraire, we derive the price and the locally risk-minimizing hedging strategy for the defaultable claims with the assumption that the stock price process and the default time are mutually independent with each other.

28.2 The Model

Let $(\Omega, \mathscr{F}, \bar{\mathscr{F}} = \{\mathscr{F}_t, t \in [0, T]\}, P)$ be a complete filtered probability space with the usual assumptions of right-continuous filtration. We assume that $\mathscr{F} = \mathscr{F}_T$ where $T > 0$ is a fixed time horizon, with historical (or market or real-world) probability measure P. We consider defaultable state of the firm as described by the process $H = \{H_t\}_{t \geq 0}$ with $H_t = I_{\{\tau \leq t\}}$ where the stopping time $\tau : \Omega \to \mathbf{R}^+$ is a nonnegative random variable indicating the default time of the firm and I is an indicator function, taking the value 1 if the event happens, otherwise 0. For convenience, we assume that $P\{\tau = 0\} = 0$ and $P\{\tau > t\} > 0$ for any $t \in \mathbf{R}^+$. The model is driven by a factor process $X = \{X_t\}_{t \geq 0}$, modeled as a finite-state Markov chain with state space $\mathrm{S} = \{1, 2, \ldots, k\}$ where k is a positive integer. Denote by $Q = \{q_{i,j}(t), i, j \in \mathrm{S}, t \geq 0\}$ the generating matrix of the Markov process. As in [1], we consider the martingale representation of the Markov process X_t given by

$$X_t = X_0 + \int_0^t Q(s, X_s)\mathrm{d}s + M_t^X, \quad t \geq 0. \tag{28.1}$$

In the representation (28.1), $Q : [0, T] \times \mathrm{S} \to \mathbb{R}$ is defined in a natural way as

$$Q(s, i) = \sum_{j \in \mathrm{S}} q_{i,j}(s) j,$$

for $(s, i) \in [0, T] \times S$ and $M^X = \{M_t^X\}_{t \geq 0}$, is an $\mathscr{F}^{S,X}$-martingale. The default time τ has $\mathscr{F}$-adapted intensity $\lambda(X_t)$ where $\lambda(\cdot)$ is given deterministic function. Then, $M_t^H = H_t - \int_0^t \lambda(X_s)\mathrm{d}s$ is a $\mathscr{F}^H$-martingale. In this setup, the process $\{X, H\}$ is jointly Markov.

We analyze market under the consideration that all factors that drive the market are observable. Here, $\mathscr{F}^S$, $\mathscr{F}^H$, and $\mathscr{F}^X$ indicate the filtration generated by the stock price process $S = \{S_t\}_{t \geq 0}$, the default process $H = \{H_t\}_{t \geq 0}$, and Markov process $X = \{X_t\}_{t\geq 0}$, respectively. We consider two primary security accounts S and B. B is a riskless asset given by the differential equation

$$dB_t = B_t r\,dt, \tag{28.2}$$

where the constant r is the risk-free rate. S satisfies the stochastic differential equation (SDE)

$$dS_t = S_t(\alpha(X_t)dt + \sigma(X_t)dW_t), \tag{28.3}$$

with a Brownian motion $W = \{W_t\}_{t\geq 0}$, independent of X. Here, $\alpha(\cdot)$ and $\sigma(\cdot)$ are given deterministic functions of X. For the sake of simplicity, we consider $r = 0$, i.e., $B_t = 1$ for all $t \geq 0$. The consideration implies that all the price processes that are going to be considered are discounted one.

Let F stands for the (right-continuous) cumulative conditional distribution function of τ for given information about the stock price process S and Markov process X up to the present time, i.e., $F_t = P\{\tau \leq t | \mathscr{F}_t^{S,X}\}$ for every $t \in \mathbf{R}^+$. The survival function G of τ is defined by the formula: $G_t = 1 - \mathrm{F}_t = P\{\tau > t | \mathscr{F}_t^{S,X}\}$ for every $t \in \mathbf{R}^+$. We also introduce the $\mathbf{R}^+ \vee \{+\infty\}$-valued hazard process $\Gamma_t = -\log\{G_t\}$. $\mathscr{F}$ represents the global filtration to which all processes are adapted, and throughout the paper, we work under the real-world probability measure P, without any kind of measure transformations. We consider that the investor has the filtration $\mathscr{F}$. Specifically, we consider $\mathscr{F} = \mathscr{F}^S \vee \mathscr{F}^X \vee \mathscr{F}^H$, i.e., the filtration generated by the stock price process, the Markov process, and the default process.

If the martingale nature of any stochastic process remains invariant under the extension of a filtration, then we say that the H-hypothesis holds. In the next section, we consider processes that independent with each other, so the H-hypothesis holds by default. In the subsequent sections, we analyze the market under the consideration that the stock price process and the default time are dependent with each other. In such scenarios, the H-hypothesis may or may not hold and we consider both these cases and derive the price and the hedging strategy.

28.3 Pricing and Hedging

In this section, we consider $W = \{W_t\}_{t\geq 0}$ to be independent of H. The market participants can form self-financing portfolios with primary security accounts as constituents. A portfolio value V_t^π at time $t \geq 0$ is described by the ratio π_t of the wealth held in the risky asset S_t and the ratio $1 - \pi_t$ in riskless asset B_t. For simplicity, assume that the units of the nondividend paying primary security accounts are perfectly divisible and that for all $t \geq 0$, the value π_t for any given strategy $\pi = \{\pi_t\}_{t\geq 0}$ depends only on information available at time t. π is called a self-financing strategy if the corresponding wealth process satisfies the SDE

$$dV_t^{\pi} = V_t^{\pi}\left((1-\pi_t)\frac{dB_t}{B_t} + \pi_t\frac{dS_t}{S_t}\right) \tag{28.4}$$

for $t \in [0, T]$, where $V_0^{\pi} = v_0$ is the initial wealth. Investor chooses the portfolio strategy π_t at time t based on his observation of prices up to the current time t, for $t \in [0, T]$. The set of admissible strategies is given by

$$\mathscr{U} = \left\{\pi : [0, T] \to \mathbf{R} \mid \pi \textit{ is } \mathscr{F} - \text{adapted and } \int_0^T \pi_s^2 ds < \infty\right\}. \tag{28.5}$$

The GOP was originally introduced by Kelly [11] and latter analyzed by Platen [13]. It is the portfolio which maximizes the expected logarithmic growth of the portfolio wealth over all finite time intervals.

Definition 3.1 For a pair of stopping times $\tau_1 \geq 0, \tau_1 \leq \tau_2$, the GOP $\pi^* \in \mathscr{U}$ is a self-financing strategy which is a solution of the problem

$$\sup_{\pi\in\mathscr{U}} E\left(\log\left(\frac{V_{\tau_2}^{\pi}}{V_{\tau_1}^{\pi}}\right) | \mathscr{F}_{\tau_1}\right),$$

where $E(\cdot|\mathscr{F}_{\tau})$ indicates the conditional expectation given complete information about the market till time $t \geq 0$.

In a continuous setting, the existence of a GOP implies no arbitrage in the strong sense of Platen [13]. GOP enjoys the numeraire property, i.e., a price process dominated by the wealth process of the GOP will be a supermartingale, (see Christensen and Platen [4]). In the sequel, we use the term GOP for both the portfolio strategy π^* and its wealth process V^{π^*}

Definition 3.2 A process $\pi' \in \mathscr{U}$ is called a numeraire portfolio if, for every $\pi \in \mathscr{U}$, the relative wealth process $\frac{V^{\pi}}{V^{\pi'}}$ is a supermartingale.

The term "numeraire portfolio" was introduced by Long [12]. He, after defining it as a portfolio π' that makes $\frac{V^{\pi}}{V^{\pi'}}$ a martingale for every portfolio π, went on to show that this requirement is equivalent, under some additional assumptions, to the absence of arbitrage for discrete-time models.

The unique GOP, $V_t^{\pi^*}$, for the model we consider satisfies the SDE

$$dV_t^{\pi^*} = V_t^{\pi^*}\theta(X_t)(\theta(X_t)dt + dW_t), \tag{28.6}$$

for all $t \geq 0$ with $V_0^{\pi^*} = 1$ (see, Raju and Selvaraju [15]). Here, $\theta(X_t)$ indicates the market price of diffusion risk modulated by the Markov process.

The direct observation of the GOP by the formula (28.6) allows us to generalize, in a practical way, the well-known arbitrage pricing theory (APT). Under the benchmark approach introduced by Platen [14], one can use the GOP V^{π^*} as numeraire along the lines of Long [12].

In the sequel, we use the GOP as numeraire or benchmark and call a price process $\hat{A} = \{\hat{A}_t\}_{t\geq 0}$ benchmarked price process if the price process $A = \{A_t\}_{t\geq 0}$ is expressed in units of GOP. The benchmarked values are supermartingales, and in the set of all supermartingale price processes that matches a given future payoff, the martingale is the one with the minimal value.

Definition 3.3 A price process is called fair when it forms a martingale when benchmarked.

In practice, it appears that fair pricing is appropriate for determining the competitive price of a contingent claim and it allows us to price a claim without using the risk-neutral martingale measure.

Definition 3.4 The fair price P_t at time $t \in [0, T]$ of an $\mathscr{F}_T$-measurable contingent claim P_T payable at a maturity time T is defined by the fair pricing formula

$$P_t = E\left(\frac{V_t^{\pi^*}}{V_T^{\pi^*}} P_T | \mathscr{F}_t\right).$$

The market model we have considered is incomplete, as we have only one tradable risky asset and three independent random processes. Even in incomplete markets, fair prices are uniquely determined. Under the existence of a minimal equivalent martingale measure (see, Föllmer and Schwiezer [7]), fair prices have been shown to correspond to locally risk-minimizing process (see, Raju and Selvaraju [15]).

In next section, we use the reduced-form models for pricing the derivatives in the presence of credit risk, which are also known as hazard rate models or intensity-based models.

28.3.1 Reduced-Form Valuations

Bielecki and Ruthkowski [3], Elliott et al. [6], Frey and McNeil [8], and many others have used the reduced-form methodology for pricing the contingent claims. The main tool in this approach is an exogenous specification of the conditional probability of default, given that default has not yet occurred.

We fix a finite horizon date $T > 0$, and we consider $\mathscr{F}$-adapted processes

- the promised contingent claim Y representing the firm's liabilities to be redeemed at time T,

- the recovery claim $\bar{Y}$, which represents the recovery payoff received at time T, if default occurs prior to or at the maturity date T,
- the recovery process $Z = \{Z_t\}_{t \geq 0}$, which specifies the recovery payoff at time of default, if it occurs prior to or at the maturity date T.

If default occurs after time T, the promised claim Y is paid in full at time T. Otherwise, depending on the adopted model, either the amount Z_τ is paid at default time τ, or the amount $\bar{Y}$ is paid at the maturity date T or both. Most practical situations deal with only one type of recovery payoff, i.e., either $\bar{Y} = 0$ or $Z = 0$. But here, we consider the general setting, i.e., simultaneously both kinds of recovery payoff exist, and thus, the defaultable claims formally defined as a quadruple $\mathrm{DCT} = (Y, \bar{Y}, Z, \tau)$.

The financial interpretation of the components of a defaultable claim, that we are going to analyze, is clear from the definition of total cash flow D_t up to time $t \in [0, T]$ given by

$$D_t = YI_{\{\tau > T\}}I_{\{t=T\}} + \bar{Y}I_{\{\tau \leq t\}}I_{\{t=T\}} + Z_\tau I_{\{\tau \leq t\}}. \tag{28.7}$$

Definition 3.5 The benchmarked value $\hat{D} = \{\hat{D}_t\}_{t \geq 0}$ of the total cash flow $D = \{D_t\}_{t \geq 0}$, which is $\mathscr{F}_t$-measurable, is given as

$$\hat{D}_t = \frac{Y}{V_T^{\pi^*}} I_{\{\tau > T\}}I_{\{t=T\}} + \frac{\bar{Y}}{V_T^{\pi^*}} I_{\{\tau \leq t\}}I_{\{t=T\}} + \frac{Z_\tau}{V_\tau^{\pi^*}} I_{\{\tau \leq t\}}, \quad t \in [0, T] \tag{28.8}$$

The benchmarked value of the cash flow between the time interval $[t, T]$ is given by

$$\hat{D}_{[t,T]} = \hat{D}_T - \hat{D}_t = \frac{Y}{V_T^{\pi^*}} I_{\{\tau > T\}} + \frac{\bar{Y}}{V_T^{\pi^*}} I_{\{\tau \leq t\}} + \int_t^T \frac{Z_u}{V_u^{\pi^*}} \mathrm{d}H_u = \hat{D}_T - \int_0^t \frac{Z_u}{V_u^{\pi^*}} \mathrm{d}H_u, \tag{28.9}$$

where $\int_t^T f(u)\mathrm{d}H_u = f(\tau)I_{\{t < \tau \leq T\}} = f(\tau)I_{\{\tau \leq t\}} - f(\tau)I_{\{\tau \leq t\}}$ and $t < T$. From Definitions 3.4 and 3.5, we obtain the fair pricing formula as given below.

Proposition 1 *For a defaultable claim* $\mathrm{DCT} = (Y, \bar{Y}, Z, \tau)$, *the fair price at time* $t \in [0, T]$ *is given by the fair pricing formula*

$$P_t = V_t^{\pi^*} \hat{P}_t, \tag{28.10}$$

where the corresponding fair, benchmarked defaultable claim price process $\hat{P} = \{\hat{P}_s\}_{s \in [0,T]}$ *has the value*

$$\hat{P}_t = E(\hat{D}_{[t,T]}|\mathscr{F}_t) = E(\hat{D}_T|\mathscr{F}_t) - \frac{Z_\tau}{V_\tau^{\pi^*}} I_{\{\tau \le t\}}. \tag{28.11}$$

If we add the benchmarked fair price process $\hat{P} = \{\hat{P}_t\}_{t\in[0,T]}$ with the cash flow up to the given time, i.e., $\hat{P}_t + \frac{Z_\tau}{V_\tau^{\pi^*}} I_{\{\tau\le t\}}$, it is an $\mathscr{F}$-martingale, from (28.11). The seller of claim always tries to hedge the risk he is going to face in future due to the market uncertainty. In next section, we try to hedge the defaultable claim.

28.3.2 Hedging

As we have considered that all the market factors cannot be traded, the market is incomplete. There will not be any self-financing hedging strategy to replicate the contingent claims. Several different approaches have been developed in the literature for hedging in incomplete markets, such as utility indifference method, mean–variance hedging, locally risk-minimizing hedging. Here, we use the locally risk-minimizing hedging introduced in Föllmer and Schweizer [7]. Biagini and Cretarola [2] for the first time used the locally risk-minimizing method to the pricing and hedging of defaultable derivatives under the risk-neutral measure. They considered the particular case of a default put option with random recovery rate and solve explicitly the problem of finding the pseudo-locally risk-minimizing strategy and the portfolio with minimal cost with the minimal martingale measure. We wish to find a portfolio "with minimal cost" that perfectly replicates DCT according to the locally risk-minimizing criterion under the real-world probability measure.

Definition 3.6 For the payment stream $D = \{D_t\}_{t\in[0,T]}$ given by (28.7) the cumulative cost process $C^D(\pi) = \{C_t^D(\pi)\}_{t\in[0,T]}$ of a strategy π *is*

$$C_t^D(\pi) = D_t + V_t^\pi - \int_0^t \frac{V_s^\pi \pi_s}{S_s} dS_s, \tag{28.12}$$

π is called self-financing if $C^D(\pi)$ is constant and mean self-financing if $C^D(\pi)$ *is* $\mathscr{F}$-martingale. π is called 0-achieving if $V_T^\pi = 0$. The risk process $R^D(\pi) = \{R_t^D(\pi)\}_{t\in[0,T]}$ *of* π is

$$R_t^D(\pi) = E((C_T^D(\pi) - C_t^D(\pi))^2|\mathscr{F}_t).$$

From Definition 3.6 and (28.11) and the martingale property of a benchmarked, fair, replicating wealth process, we directly obtain the following result.

Proposition 2 *For the given payment stream D, if* V^{π^D} *is the replicating wealth process, then it must have the value*

$$V_t^{\pi^D} = P_t,$$

at time $t \in [0,T]$, *with* P_t *satisfying the fair pricing formula* (28.11).

The question now is how to reduce the intrinsic risk which arises due to the incompleteness of the market.

The quantity $C_t^D(\pi)$ describes the cumulative cost on $[0,t]$ from paying according to D and trading according to π, (for more detail see Schweizer [16]). He has proved that if S satisfies the structural condition and if the mean–variance trade-off process is continuous, which is true in this model, then the existence of π locally risk-minimizing strategy for D is equivalent to a strategy π which is 0-achieving and mean self-financing and the cost process $C^D(\pi)$ is strongly orthogonal to W. Schweizer derived the link between locally risk-minimizing and the Follmer—Schweizer decomposition.

Definition 3.7 An $\mathscr{F}_T$-measurable random variable D_T admits a Follmer–Schweizer decomposition if it can be written as

$$D_T = D_0 + \int_0^T \phi_s^D \mathrm{d}S_s + L_T^D, \quad P - a.s., \tag{28.13}$$

where D_0 *is* $\mathscr{F}_0$-measurable, $\phi^D = \{\phi_t^D\}_{t\geq 0}$ is such that the corresponding $\pi^D = \{\pi_t^D = \frac{\phi_t^D S_t}{V_t^{\pi^D}}\}_{t\geq 0}$ is in $\mathscr{U}$, and the process $L^D = \{L_t^D\}_{t\geq 0}$ is a right-continuous square-integrable martingale null at 0 and strongly orthogonal to W.

Proposition 3 *A payment stream D admits a locally risk-minimizing strategy if and only if* D_T *admits a Follmer–Schweizer decomposition. In that case, the locally risk-minimizing strategy* π *is given by*

$$\pi_t = \frac{\phi_t^D S_t}{V_t^D}, \quad t \in [0,T],$$

with

$$V_t^D = D_0 + \int_0^t \phi_s^D \mathrm{d}S_s + L_t^D - D_t \tag{28.14}$$

and the mean self-financing cost process is

$$C_t^D(\pi) = D_0 + L_t^D.$$

Let us consider a 0-achieving strategy π for D, with cost process $C^D(\pi)$ satisfying the SDE

$$\mathrm{d}V_t^\pi = V_t^\pi \pi_t \frac{\mathrm{d}S_t}{S_t} + \mathrm{d}C_t^D(\pi) - \mathrm{d}D_t.$$

Its benchmarked value is

$$\mathrm{d}\hat{V}_t^\pi = \hat{V}_t^\pi(\pi_t\sigma_t + \theta(X_t))\mathrm{d}W_t + \frac{\mathrm{d}C_t^D}{V_t^{\pi^*}} - \frac{\mathrm{d}D_t}{V_t^{\pi^*}}. \tag{28.15}$$

So now on our aim will be to derive the Follmer–Schweizer decomposition of the claim to get the risk-minimizing hedging strategy. In next proposition, we derive representation for an $\mathscr{F}$-martingale process.

Proposition 4 *Let K be a $\mathscr{F}_T$-measurable and integrable random variable. Then, the $\mathscr{F}$-martingale $M^K = \{M_t^K = E(K|\mathscr{F}_t)\}_{t\in[0,T]}$ admits the following representation*

$$M_t^K = M_0^K + \int_0^t \xi_s^K \mathrm{d}W_s + \int_0^t \zeta_s^K \mathrm{d}M_s^X + \int_t^0 \gamma_s^K \mathrm{d}M_s^H,$$

where ξ^K, ζ^K and γ^K are $\mathscr{F}$-predictable stochastic processes. Moreover, all the right-hand-side martingale terms are orthogonal with each other.

Proof As in Bielecki and Rutkowski [3], we can consider the $\mathscr{F}_T$-measurable random variable of the form $K = (1 - H_s)Z$ for some $s \leq T$ and some $\mathscr{F}^{S,X}$-measurable random variable Z. We can write

$$K = (1 - H_s)Z = (1 - H_s)\mathrm{e}^{\Gamma_s}\tilde{Z} = L_s\tilde{Z}$$

where $\tilde{Z} = \frac{Z}{\mathrm{e}^{\Gamma_s}}$ is an $\mathscr{F}_T^{S,X}$-measurable, integrable random variable. From martingale representation theorem, the $\mathscr{F}^{S,X}$-martingale can be written as

$$U_t = E(\tilde{Z}|\mathscr{F}_t^{S,X}) = E(\tilde{Z}) + \int_0^t \xi_u \mathrm{d}W_u + \int_0^t \zeta_u \mathrm{d}M_u^X$$

Now from Itô's formula,

$$K = L_0U_0 + \int_0^T L_{t-}\mathrm{d}U_t + \int_0^T U_{t-}\mathrm{d}L_t + [L, U]_s$$
$$= L_0U_0 + \int_0^T L_{t-}\xi_t\mathrm{d}W_t + \int_0^T L_{t-}\zeta_t\mathrm{d}M_t^X + \int_0^T U_{t-}\mathrm{d}L_t.$$

So the asserted formula holds, with the processes $\xi_t^K = \xi_t L_{t-}$, $\zeta_t^K = \zeta_t L_{t-}$ and $\gamma_t^K = U_t I_{[0,s]}$. The orthogonality can be shown easily. □

From (28.11) and Preposition 4, there exists $\mathscr{F}$-predictable processes Φ_1, Φ_2, and Φ_3 such that

$$\hat{P}_t + \frac{Z_\tau}{V_\tau^{\pi^*}} I_{\tau \le t} = \int_0^t \Phi_1(u)\mathrm{d}W_u + \int_0^t \Phi_2(u)\mathrm{d}M_u^X + \int_0^t \Phi_2(u)\mathrm{d}L_u \tag{28.16}$$

i.e.,

$$\hat{P}_t = \int_0^t \Phi_1(u)\mathrm{d}W_u + \int_0^t \Phi_2(u)\mathrm{d}M_u^X + \int_0^t \Phi_2(u)\mathrm{d}L_u + \int_0^t \frac{Z_u}{V_u^{\pi^*}}\mathrm{d}H_u \tag{28.17}$$

and

$$\mathrm{d}\hat{P}_t = \Phi_1(t)\mathrm{d}W_t + \Phi_2(t)\mathrm{d}M_t^X + \Phi_3(t)\mathrm{d}L_t + \frac{Z_t}{V_t^{\pi^*}}\mathrm{d}H_t. \tag{28.18}$$

If we compare (28.15) and (28.18), we will get the required values of risk-minimizing hedging strategy π and the mean self-financing cost process $C^D(\pi)$,

$$\pi_t = \frac{\frac{\Phi_1(t)}{\hat{V}_t} - \theta(X_t)}{\sigma_t}$$
$$\mathrm{d}C_t^D(\pi) = V_t^{\pi^*}(\Phi_2(t)\mathrm{d}M_t^X + \Phi_3(t)\mathrm{d}L_t).$$

Now, we try to derive the predictable terms Φ_i, for $i \in \{1, 2, 3\}$, more explicitly. From (28.8), we have

$$E(\hat{D}_T|\mathscr{F}_t) = E(\hat{Y}(1 - H_T) + \widehat{\widehat{Y}}H_T + \hat{Z}_\tau H_T|\mathscr{F}_t). \tag{28.19}$$

Let $\hat{\hat{Y}} = h(\tau \wedge T)\hat{Y}$ where $h(\tau \wedge T) \in [0, 1)$ is the recovery rate at the maturity and is a fraction of the claim which has to be paid at maturity if there is no default prior or at maturity T. Then,

$$\begin{aligned} E(\hat{D}_T|\mathcal{F}_t) &= E(\hat{Y}((1 - H_T) + h(\tau \wedge T)H_T)|\mathcal{F}_t) + E(\hat{Z}_\tau H_T|\mathcal{F}_t) \\ &= E(\hat{Y}|\mathcal{F}_t)E(1 + (h(\tau \wedge T) - 1)H_T|\mathcal{F}_t) + E(\hat{Z}_\tau H_T|\mathcal{F}_t). \end{aligned}$$

Here, $E(\hat{Y}|\mathcal{F}_t)$ is an $\mathcal{F}^{S,X}$-adapted martingale and admits the decomposition

$$E(\hat{Y}|\mathcal{F}_t) = c + \int_0^t \xi_s \mathrm{d}W_s + \int_0^t \zeta_s \mathrm{d}M_s^X. \tag{28.20}$$

We can write $1 + (h(\tau \wedge T) - 1)H_T = f(\tau)$ for some integrable Borel function $f : \mathbf{R}^+ \to [0, 1]$, by Proposition 4.3.1 of [3], we have

$$E(1 + (h(\tau \wedge T) - 1)H_T|\mathcal{F}_t) = c_h + \int_0^t \tilde{f}(s)\mathrm{d}M_s^H, \tag{28.21}$$

where $c_h = E(f(\tau))$ and the function $\tilde{f} : \mathbf{R}^+ \to \mathbf{R}$ is given by the formula

$$\tilde{f}(t) = f(t) - \mathrm{e}^{\Gamma_t}E(I_{\{\tau > T\}}f(\tau)).$$

It remains to find the representation for $E(\hat{Z}_\tau H_T|\mathcal{F}_t)$. From [3], it can be decomposed as

$$\begin{aligned} E(\hat{Z}_\tau H_T|\mathcal{F}_t) &= H_t E(\hat{Z}_\tau H_T|\mathcal{F}_t^{S,X} \vee \mathcal{F}_T^H) + (1 - H_t)\mathrm{e}^{\Gamma_t}E((1 - H_t)\hat{Z}_\tau H_T|\mathcal{F}_t^{S,X}) \\ &= H_t\hat{Z}_\tau + (1 - H_t)\mathrm{e}^{\Gamma_t}E(\int_t^T \hat{Z}_u \mathrm{e}^{-\Gamma_u}\mathrm{d}\Gamma_u|\mathcal{F}_t^{S,X}) \\ &= H_t\hat{Z}_\tau + I_{\{\tau > T\}}Q_t, \end{aligned} \tag{28.22}$$

where

$$\begin{aligned} Q_t &= \mathrm{e}^{\Gamma_t}E\left(\int_t^T \hat{Z}_u \mathrm{e}^{-\Gamma_u}d\Gamma_u|\mathcal{F}_t^{S,X}\right) \\ &= \mathrm{e}^{\Gamma_t}\left(E\left(\int_0^T \hat{Z}_u \mathrm{e}^{-\Gamma_u}\mathrm{d}\Gamma_u|\mathcal{F}_t^{S,X}\right) - \int_0^t \hat{Z}_u \mathrm{e}^{-\Gamma_u}\mathrm{d}\Gamma_u\right). \end{aligned}$$

Then,

$$\begin{aligned} dQ_t &= Q_t d\Gamma_t + e^{\Gamma_t}[dm_t - \hat{Z}_t e^{-\Gamma_t} d\Gamma_t] \\ &= [Q_t - \hat{Z}_t] d\Gamma_t + e^{\Gamma_t} dm_t, \end{aligned}$$

with $\mathscr{F}_t^{S,X}$-martingale $m_t = E(\int_0^T \hat{Z}_u e^{-\Gamma_u} d\Gamma_u | \mathscr{F}_t^{S,X})$. Therefore, we have

$$Q_t = m_0 + \int_0^t e^{\Gamma_s} dm_s + \int_0^t (Q_s - \hat{Z}_s) d\Gamma_s. \tag{28.23}$$

Furthermore, since Q is continuous

$$\begin{aligned} I_{\{\tau > T\}} Q_t &= m_0 + \int_0^{t\wedge\tau} dQ_u - I_{\{\tau \le t\}} Q_\tau \\ &= m_0 + \int_0^{t\wedge\tau} e^{\Gamma_s} dm_s + \int_0^{t\wedge\tau} (Q_s - \hat{Z}_s) d\Gamma_s - \int_0^t Q_s dH_s \\ &= m_0 + \int_0^{t\wedge\tau} e^{\Gamma_s} dm_s - \int_0^t Q_s dM_s^H - \int_0^{t\wedge\tau} \hat{Z}_s d\Gamma_s. \end{aligned}$$

Consequently, we can rewrite (28.22) as follows:

$$\begin{aligned} E[\hat{Z}_\tau H_T | \mathscr{F}_t] &= m_0 + \int_0^{t\wedge\tau} e^{\Gamma_s} dm_s - \int_0^t Q_s dM_s^H - \int_0^{t\wedge\tau} \hat{Z}_s d\Gamma_s + H_t \hat{Z}_\tau \\ &= m_0 + \int_0^{t\wedge\tau} e^{\Gamma_s} dm_s + \int_0^t (\hat{Z}_s - Q_s) dM_s^H. \end{aligned} \tag{28.24}$$

As m is an $\mathscr{F}^{S,X}$-martingale, it will have the decomposition

$$m_t = m_0 + \int_0^t \xi_s' dW_s + \int_0^t \zeta_s' dM_s^X, \tag{28.25}$$

for some $\mathscr{F}^{S,X}$-predictable processes ξ' and ζ'. Hence,

$$E[\hat{Z}_\tau H_T|\mathscr{F}_t] = m_0 + \int_0^{t\wedge\tau} e^{\Gamma_s}\xi'_s dW_s + \int_0^{t\wedge\tau} e^{\Gamma_s}\zeta'_s dM_s^X + \int_0^t (\hat{Z}_s - Q_s) dM_s^H. \quad (28.26)$$

Finally, the decomposition of $E(\hat{D}|\mathscr{F}_t)$ is

$$\begin{aligned} E(\hat{D}_T|\mathscr{F}_t) &= cc_h + m_0 + \int_0^t (c_h\xi_s + I_{\{\tau\geq s\}} e^{\Gamma_s}\xi'_s) dW_s \\ &+ \int_0^t (c_h\zeta_s + I_{\{\tau\geq s\}} e^{\Gamma_s}\zeta'_s) dM_s^X \\ &+ \int_0^t (c\tilde{f}(s) + (\hat{Z}_s - Q_s)) dM_s^H. \end{aligned} \quad (28.27)$$

Proposition 5 *The* 0*-achieving locally risk-minimizing portfolio for DCT is given by*

$$V_t^\pi = \int_0^t \Phi_u^1 dS_u + C_t^D(\pi) - D_t, \quad (28.28)$$

where the locally risk-minimizing strategy is

$$\Phi^1 = \frac{V_t^\pi \pi_t}{S_t},$$

with

$$\pi_t = \frac{\frac{(c_h\xi_s + I_{\{\tau\geq s\}} e^{\Gamma_s}\xi'_s)}{\hat{V}_t^\pi} - \theta(X_t)}{\sigma_t},$$

and the mean self-financing cost is

$$\begin{aligned} C_t^D(\pi) = C_0^D(\pi) &+ \int_0^t V_s^{\pi^*} (c_h\zeta_s + I_{\{\tau\leq s\}} e^{\Gamma_s}\zeta'_s) dM_s^X \\ &+ \int_0^t V_s^{\pi^*} (c\tilde{f}(s) + (\hat{Z}_s - Q_s)) dM_s^H. \end{aligned}$$

Proof If we compare the Eqs. (28.15) and (28.27), we get the required result. □

28.4 Conclusion

We have used the numeraire change approach, the benchmark approach, for pricing the defaultable claim. A pricing formula has been derived for a model, where parameters are modulated by a Markov process. Total cash flow has been defined for the defaultable claim and benchmark fair price formula has been given. Using the martingale representation, we have derived the formula for the hedging strategy which may be computed numerically by setting up a suitable simulation-based experiment as in the work of Tebaldi [17].

References

1. Bain A, Crisan D (2009) Fundamentals of stochastic filtering, Springer, Berlin
2. Biagini F, Cretarola A (2007) Quadratic hedging methods for defaultable claims. Appl Math Optim 56:425–443
3. Bielecki TR, Rutkowski M (2004) Credit risk: modelling, valuation and hedging, vol 2. Springer, Berlin
4. Christensen MM, Platen E (2005) A general benchmark model for stochastic jump sizes. Stoch Anal Appl 23(5):1017–1044
5. Di Masi GB, Kabanov YM, Runggaldier WJ (1994) Mean-variance hedging of options on stocks with Markov volatilities. Theory Probab Appl 39(1):172–182
6. Elliott RJ, Jeanblanc M, Yor M (2000) On models of default risk. Math Finan 10:179–195
7. Föllmer H, Schweizer M (1991) Hedging of contingent claims under incomplete information. In: Elliot RJ, Davis MHA (eds) Applied stochastic analysis. Gordon and Breach, New York, pp 389–414
8. Frey R, McNeil AJ (2000) Modelling dependent defaults. University of Zurich and ETHZ (preprints)
9. Hamilton J (1989) A new approach to the economic analysis of nonstationary time series and the business cycle. Econometrica 57:357–384
10. Heath D, Platen E (2002) Consistent pricing and hedging for a modified constant elasticity of variance model. Quant Finan 2(6):459–467
11. Kelly JR (1956) A new interpretation of information rate. Bell Syst Technol J 35:917–926
12. Long JB (1990) The numeraire portfolio. J Finan Econ 26(1):29–69
13. Platen E (2002) Arbitrage in continuous complete markets. Adv Appl Probab 34(3):540–558
14. Platen E, Heath D (2006) A benchmark approach to quantitative finance. Springer Finance. Springer, Berlin
15. Raju IVA, Selvaraju N (2012) Growth optimal portfolio for unobservable Markov modulated markets. Int J Math Oper Res 4(1):31–40
16. Schweizer M (2008) Local risk-minimization for multidimensional assets and payment streams. Banach Center Publ 83:213–229
17. Tebaldi C (2005) Hedging using simulation: a least squares approach. J Econ Dyn Control 29 (8):1287–1312

Chapter 29
Application of Voltage Source Converter for Power Quality Improvement

Bhim Singh, Sabha Raj Arya, Chinmay Jain, Sagar Goel, Ambrish Chandra and Kamal Al-Haddad

Abstract This article presents an application of a four-leg voltage source converter (VSC) in a distribution network for the mitigation of power quality problems such as burden reactive power, unbalanced load, waveform distortion, and flow of high neutral currents. Fast and accurate compensation depends upon the used control algorithm for estimation of fundamental load currents. Classical theory-based enhanced phase-locked loop (EPLL) control algorithm is applied for estimation of fundamental components of distorted load current. In this algorithm, extracted fundamental load current components are applied for extraction of reference supply currents. The VSC is controlled using a digital signal processor (DSP) for implementation of control algorithm for the compensation of reactive linear and nonlinear loads. Test results on a developed VSC-based system are presented to prove the functions for improving the power quality.

Keywords Load balancing · Neutral current · Power quality · Power factor · VSC

B. Singh · S.R. Arya · C. Jain · S. Goel
Department of Electrical Engineering, Indian Institute of Technology Delhi, New Delhi 110016, India
e-mail: bsinghrajput56@gmail.com

S.R. Arya
e-mail: sabharaj1@gmail.com

C. Jain
e-mail: Chinmay31jain@gmail.com

S. Goel
e-mail: sagargl@gmail.com

A. Chandra (✉) · K. Al-Haddad
Department of Electrical Engineering, École de Technologie Supérieure, 1100 Notre-Dame, Montreal, QC H3C 1K3, Canada
e-mail: ambrish.chandra@etsmtl.ca

K. Al-Haddad
e-mail: kamal.al-haddad@etsmtl.ca

V. Vijay et al. (eds.), *Systems Thinking Approach for Social Problems*, Lecture Notes in Electrical Engineering 327,
DOI 10.1007/978-81-322-2141-8_29

29.1 Introduction

The voltage source converter (VSC) is generally used for mitigation of power quality problems [1]. Due to development of self commutating power semiconductor devices, the decrease of the cost and advancement in processor technology, its application in the area of power conversion is continuously increasing. Application of VSC includes the area of power quality [2], HVDC system and FACTs devices [3], and electric drives system [4]. In general, VSC is used for mitigation of power quality problem as an active filter or distribution static compensator (DSTATCOM) in distribution system [5]. Power electronic-based loads such as ac/dc converters, ac controllers, and switch mode power supplies are considered as harmonics generating source due to switching and nonlinear behavior of used semiconductor devices [6]. Zeng et al. [7] have reported review on application of VSC in grid-integrated renewable power generation system such as wind, solar, and their control algorithms. In these VSC applications, it works as a power quality conditioner. Medina et al. [8] have discussed basic technique for harmonics analysis in power conversion technology and developed frame of reference for frequency, time, and frequency-time bases. Lin et al. [9] have discussed neutral point-clamped VSC for shunt connected compensator. It has multi-functions such as power factor regulator and active filter and if both are connected then it works as integrated power quality compensator. Some of other topologies of VSC for power quality improvement are six switches VSC with single capacitor and midpoint capacitor, four-leg VSC in the application of low- and high-voltage supply system, and isolated topology such as three legs with star/star-delta system [10]. Jain et al. [11] have discussed implementation issue of active filter for harmonics compensation and design of proportional integral (PI) controller gains. Recently, one new application of VSC is reported on electric vehicle charging station [12].

In all these applications, the control of VSC is another factor to get desired performance. So selection of control algorithm is also one of the criteria to get fast response of VSC. Hsieh and Hung [13] have discussed a review on phase-locked loop (PLL) techniques in the area of control system. It can be in form of analog as integrated circuits (ICs) or digital processor. Singh and Sharma [14] have discussed implementation of voltage frequency controller using EPLL control algorithm in wind energy system. Saez et al. [15] have discussed implementation of FPGA-based system in the application of grid synchronization. It is based on various delay cancelation, generalized integrator, and PLL. Main advantages are improved sample rate and parallel processing to get information related to magnitude, phase, and frequency of generated harmonics. Enhanced phase-locked loop (EPLL) is also an improved version of PLL, and it is adaptive with supply voltage variation and performance not affecting due to double frequency ripples [16]. It is also reported in many other applications such as three-wire active filter and distributed generation system. Ghartemani and Ziarani [17] have discussed mathematical analysis, structure, and tuning guide lines of EPLL internal constants.

In this article, a four-leg VSC is used for power quality improvement employing EPLL control algorithm. It is designed as a compensator to compensate power quality problems such as load reactive power, load balancing, harmonics elimination, and neutral current under balanced and unbalanced loads.

29.2 VSC System Configuration

A four-leg VSC connected to grid supply system is presented in Fig. 29.1. The supply system is feeding to consumer loads. It consists of self commutating semiconductor devices and a capacitor on the DC bus. Other supporting elements of this system are interfacing inductors (L_f), neutral inductor (L_n), and ripple filters. Ripple filter is a series arm of capacitor (C_f) and a resistor (R_f). It is connected at common coupling position for filtering action of noise due to switching. The compensating currents (i_{Ca}, i_{Cb}, i_{Cc}) are supplied by VSC to cancel out harmonics/reactive power components of load currents and to maintain balanced supply currents under unbalanced loads.

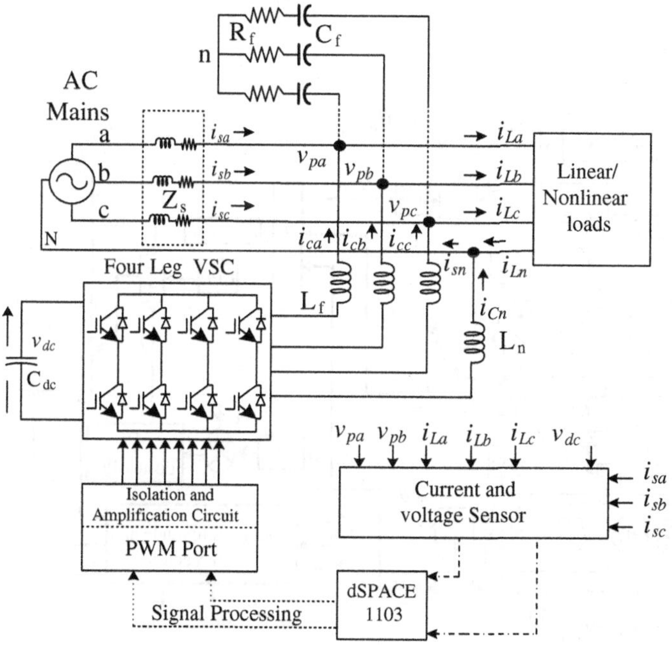

Fig. 29.1 Four-leg VSC-based system

29.3 Control Algorithm of VSC

The formulation of control algorithm is demonstrated in Fig. 29.2. It is based on advanced PLL considered as EPLL. Basic mathematical formulation for extraction of reference supply currents is described as:

29.3.1 Voltage Unit Templates

After sensing three-phase PCC voltages, these are converted into phase voltage (v_{pa}, v_{pb}, v_{pc}).

The amplitude of PCC voltages (V_t) is estimated as [14],

$$V_t = \left(\frac{2(v_{pa}^2 + v_{pb}^2 + v_{pc}^2)}{3} \right)^{0.5} \tag{29.1}$$

It is processed through low-pass filter (LPF).

Inphase (sine) unit templates of PCC voltages are estimated as,

$$\sin\theta_{spa} = \frac{v_{pa}}{V_t}, \quad \sin\theta_{spb} = \frac{v_{pb}}{V_t}, \quad \sin\theta_{spc} = \frac{v_{pc}}{V_t} \tag{29.2}$$

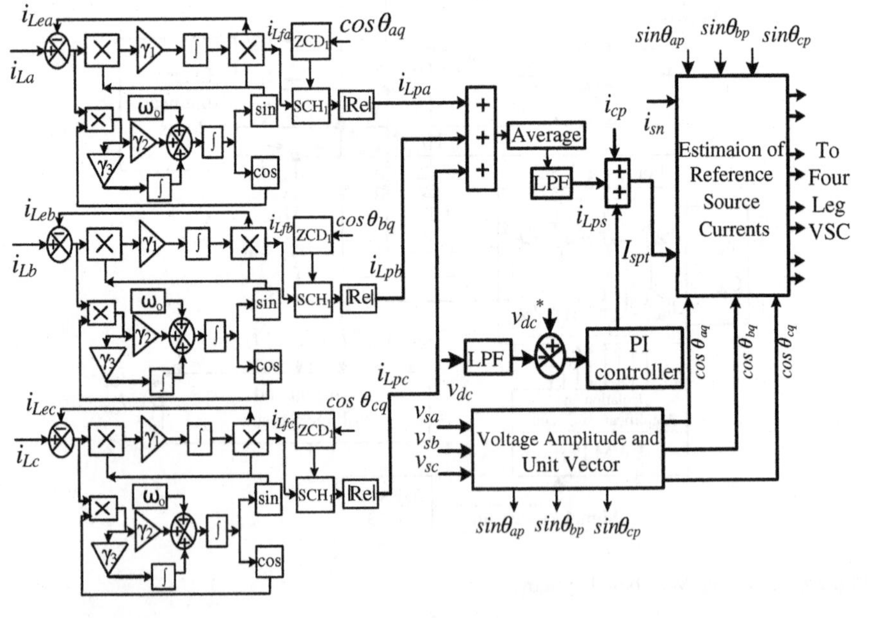

Fig. 29.2 Block diagram of EPLL-based control algorithm

From above equation, the quadrature unit templates (cos) are arranged as,

$$\begin{aligned}\cos\theta_{sqa} &= \frac{\left(-\sin\theta_{spb}+\sin\theta_{spc}\right)}{\sqrt{3}}\\ \cos\theta_{sqb} &= \frac{\left(3\sin\theta_{spa}+\sin\theta_{spb}-\sin\theta_{spc}\right)}{2\sqrt{3}}\\ \cos\theta_{sqc} &= \frac{\left(-3\sin\theta_{spa}+\sin\theta_{spb}-\sin\theta_{spc}\right)}{2\sqrt{3}}\end{aligned} \tag{29.3}$$

29.3.2 Fundamental Active Power Components and Average of Load Currents

EPLL is applied for estimation of fundamental frequency active power components of sensed load currents in each phase of three-phase connected loads. It is fed by distorted current as an input signal. In phase 'a,' distorted signal and estimated fundamental components are represented as 'i_{La}' and 'i_{Lfa},' respectively. Difference between these two components is written in form of error components as 'i_{ea}.' Performance of used EPLL depends upon the value of internal constants. In this implementation, selected value of internal constants of EPLL γ_1, γ_2, and γ_3 are 18, 9, and 1.5, respectively. Guidelines for selection of these constants are described in [14, 16, 17]. Similar procedure is used in other phases for extraction of fundamental components of load currents. For estimation of the peak magnitude of the fundamental active power component of load currents, same phase of the PCC phase 'a' voltage (v_{sa}), a zero crossing detector (ZCD_1), quadrature unit template ($\cos\theta_{aq}$), and sample and hold block (SHC_1) are used as shown in Fig. 29.2. Output of sample and hold block (O_{SHC1}) is considered as amplitude of phase 'a' fundamental active power component of load current (i_{Lpa}). Similarly, amplitude of fundamental active power component of other two phases (i_{Lpb} and i_{Lpc}) is also extracted. Average amplitudes of these extracted three-phase components are written as,

$$i_{Lps} = \frac{1}{3}\left(i_{Lpa}+i_{Lpb}+i_{Lpc}\right) \tag{29.4}$$

29.3.3 Amplitude of Active Power Component of Reference Supply Currents

A PI voltage controller is placed for control of DC bus voltage of VSC and its output is considered as a loss components of VSC. It is denoted as i_{cp}. An addition of required active power component of current for DC bus of the VSC and average magnitude of active power components of load currents is taken as amplitude of active power component of the reference supply current (i_{spt}). It is written as follows:

$$I_{spt} = i_{cp} + i_{Lps} \tag{29.5}$$

29.3.4 Reference Supply Currents and Gating Signals

Reference supply currents are calculated using amplitude of active power components of currents and voltage templates. These are formulated as follows:

$$i_{sa}^{*} = I_{spt}\sin\theta_{ap}, \quad i_{sb}^{*} = I_{spt}\sin\theta_{bp}, \quad i_{sc}^{*} = I_{spt}\sin\theta_{cp} \tag{29.6}$$

Difference between sensed supply currents (i_{sa}, i_{sb}, i_{sc}) and reference supply currents (i_{sa}^{*}, i_{sb}^{*}, i_{sc}^{*}) is considered as current errors components (i_{eabc}), and these errors are used for gating pulse generation of three leg of used four-leg VSC. Sensing of neutral current is required for generating of fourth-leg gating signals. After sensing supply neutral current (i_{sn}), it is compared with reference neutral current ($i_{sn}^{*} = 0$). In the next step, estimated neutral current error component is used gating pulse generation for fouth leg of VSC.

29.4 Results and Discussion

A prototype of VSC system is developed for experimental analysis. It is also known as DTATCOM. An EPLL-based control algorithm on VSC system is analyzed using a DSP (dSPACE 1103). It is tested under linear and nonlinear loads for suppression of power quality problem. The detailed data of VSC-based shunt compensator are provided in Appendix.

29.4.1 Performance of VSC System Under Linear Loads

Figure 29.3a–f shows the recorded waveforms at end of supply currents (i_{sa}, i_{sb}, i_{sc}) and load current (i_{La}, i_{Lb}, i_{Lc}) with PCC voltage (v_{ab}) under linear loads in steady state condition. Supply power (P_s), load power (P_L), and VSC output power (P_c) are shown in Fig. 29.3g–i. The levels of supply and load power factor are 0.98 (lagging) and 0.86 (lagging), respectively. After compensation, supply neutral current is presented in Fig. 29.3j which is under limit.

Figure 29.4a, b shows the recorded CRO waveform of supply currents (i_{sa}, i_{sb}, i_{sc}) and load currents (i_{La}, i_{Lb}, i_{Lc}) with respect to supply voltages (v_{ab}) under

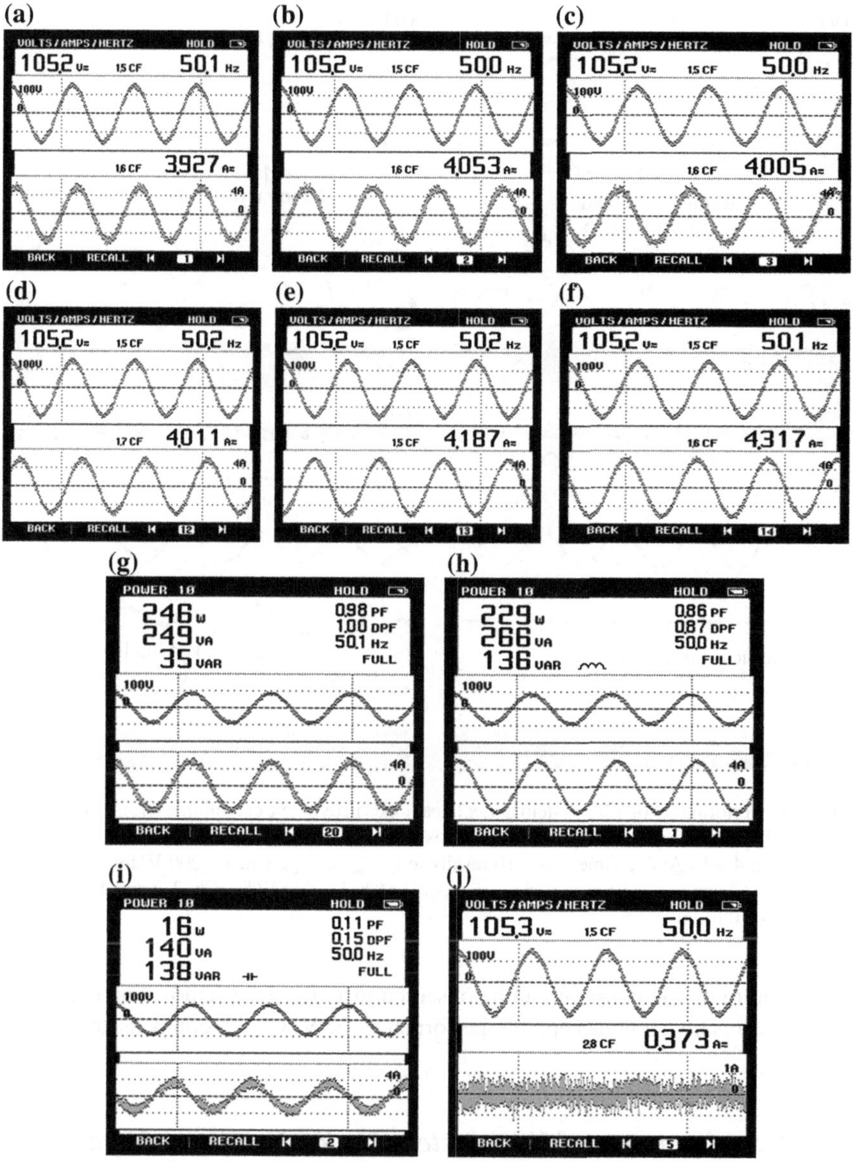

Fig. 29.3 Performance of VSC system under linear loads in steady state condition **a–c** i_{sa}, i_{sb}, i_{sc}. **d–f** i_{La}, i_{Lb}, i_{Lc}. **g–i** Per phase P_S,P_L and P_C. **j** i_{sn}

unbalanced load. Phase 'a' supply current (i_{sa}), supply neutral current (i_{sn}), and load current (i_{Ln}) are shown in Fig. 29.4c where supply neutral current is very less compared to load neutral current (i_{Ln}). Variation of dc bus voltage (v_{dc}) with supply current (i_{sa}), compensator current (i_{Ca}), and load current (i_{La}) is shown in Fig. 29.4d

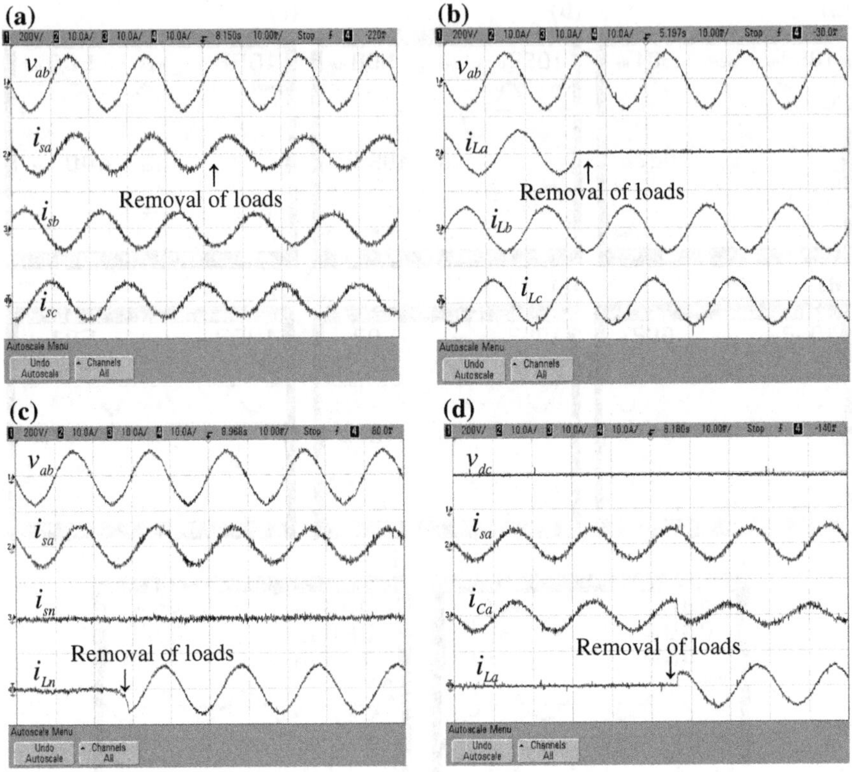

Fig. 29.4 Performance of VSC system under linear loads in dynamic condition **a** v_{ab}, i_{sa}, i_{sb}, i_{sc}. Ch. 1—200 V/div, Ch. 2, 3 and 4—10 A/div, Time axis—10 ms/div. **b** v_{ab}, i_{La}, i_{Lb}, i_{Lc}. Ch. 1—200 V/div, Ch. 2, 3 and 4—10 A/div, Time axis—10 ms/div. **c** v_{ab}, i_{sa}, i_{sn}, i_{Ln}. Ch. 1—200 V/div, Ch. 2, 3 and 4—10 A/div, Time axis—10 ms/div. **d** v_{dc}, i_{sa}, i_{Ca}, i_{La}. Ch. 1—200 V/div, Ch. 2, 3 and 4—10 A/div, Time axis—10 ms/div

where nature of compensator can be observed just after load removal and before it. These test results show acceptable performance of VSC as a compensator.

29.4.2 Performance of VSC System Under Nonlinear Loads

Figure 29.5a–f shows the waveforms of supply currents (i_{sa}, i_{sb}, i_{sc}) with respect to PCC voltages (v_{ab}) and their harmonic spectra where the total harmonic distortions (THD) of three-phase supply currents are 3.3, 2.8, and 3.9 %, respectively. Same time supply reactive power is also reduced which is shown in Fig. 29.5g. Waveform of supply neutral current (i_{sn}), harmonic spectrum of supply voltage (v_{ab}), load current (i_{La}), harmonic spectrum of load current (i_{La}), and load power is shown in Fig. 29.5i–l. Nature of three compensator currents, compensator power, and load

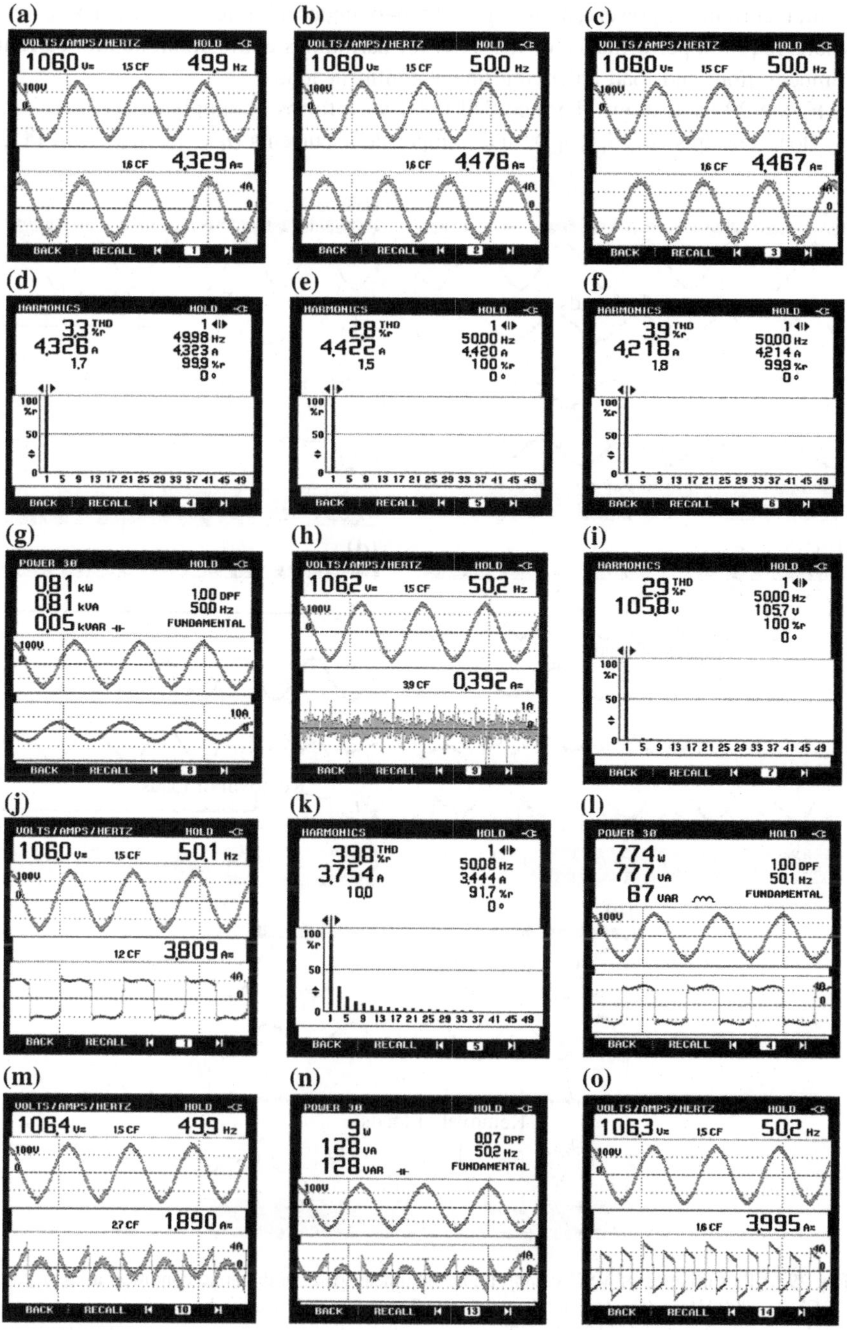

Fig. 29.5 Performance of VSC system under nonlinear loads in steady state condition **a–c** i_{sa}, i_{sb}, i_{sc}. **d–f** Harmonic spectra of $i_{sa,}$ $i_{sb,}$ i_{sc}. **g** *Ps* (h) i_{sn}. **i** Harmonic spectrum of v_{ab}. **j** i_{La}. **k** Harmonic spectra of $i_{La,}$. **l** P_L. **m** i_{Ca}. **n** P_c. **o** i_{Ln}

neutral current is presented in Fig. 29.5m–o under nonlinear loads. These test results required performance of the VSC for harmonics elimination, neutral current compensation, and load balancing under nonlinear loads.

Figure 29.6a–c shows the recorded CRO waveforms of supply currents (i_{sa}, i_{sb}, i_{sc}), load currents (i_{La}, i_{Lb}, i_{Lc}), and compensator currents (i_{Ca}, i_{Cb}, i_{Cc}) with PCC

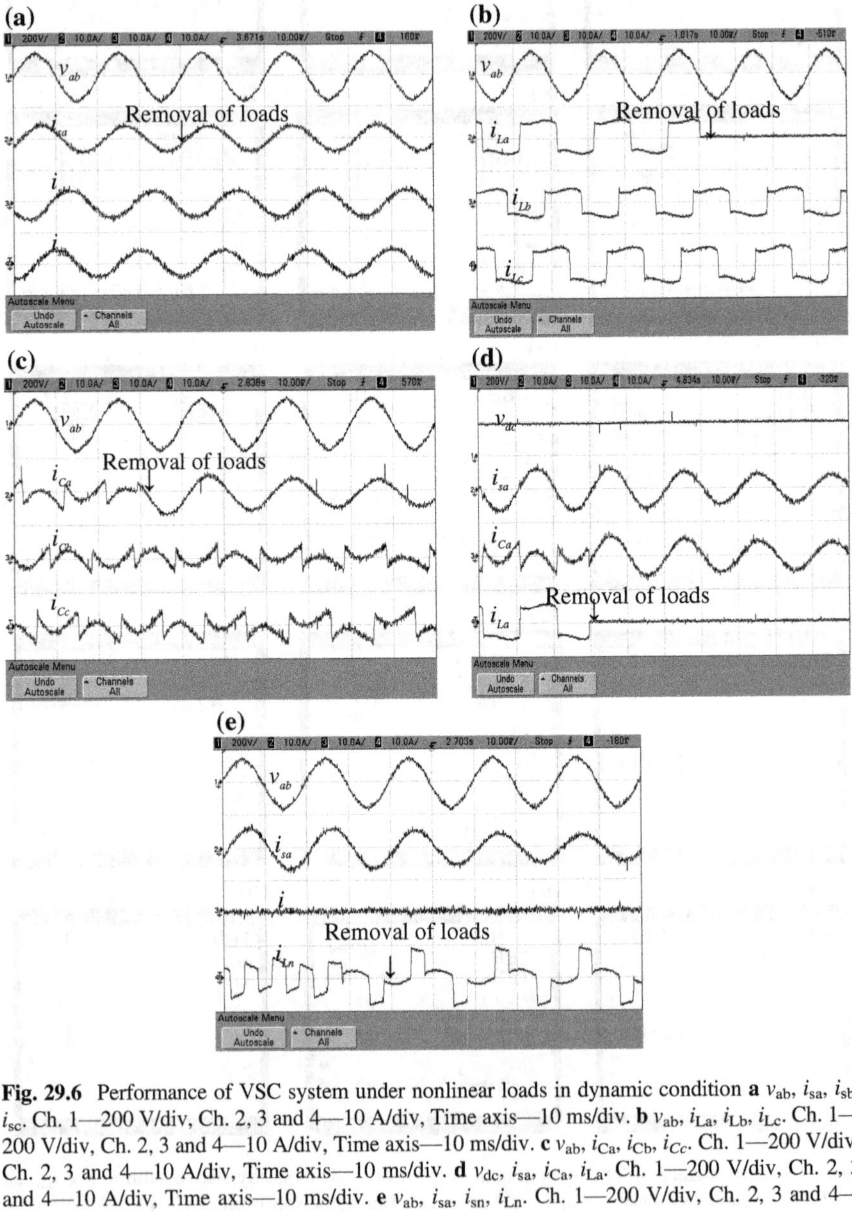

Fig. 29.6 Performance of VSC system under nonlinear loads in dynamic condition **a** v_{ab}, i_{sa}, i_{sb}, i_{sc}. Ch. 1—200 V/div, Ch. 2, 3 and 4—10 A/div, Time axis—10 ms/div. **b** v_{ab}, i_{La}, i_{Lb}, i_{Lc}. Ch. 1—200 V/div, Ch. 2, 3 and 4—10 A/div, Time axis—10 ms/div. **c** v_{ab}, i_{Ca}, i_{Cb}, i_{Cc}. Ch. 1—200 V/div, Ch. 2, 3 and 4—10 A/div, Time axis—10 ms/div. **d** v_{dc}, i_{sa}, i_{Ca}, i_{La}. Ch. 1—200 V/div, Ch. 2, 3 and 4—10 A/div, Time axis—10 ms/div. **e** v_{ab}, i_{sa}, i_{sn}, i_{Ln}. Ch. 1—200 V/div, Ch. 2, 3 and 4—10 A/div, Time axis—10 ms/div

line voltage (v_{ab}) under unbalanced nonlinear loads. Unbalanced load on three-phase system is created by removal of phase 'a' load. Effect of unbalanced load on DC bus can be observed in Fig. 29.6d where supply current (i_{sa}), compensator current (i_{Ca}), and load current (i_{La}) are shown during load removal. After observation of compensating currents, it is concluded that during unbalanced load condition supply current and compensator current of phase 'a' are in same phase and shape. Function of neutral current is seen from Fig. 29.6e where supply current (i_{sa}), supply neutral current (i_{sn}), and load neutral current (i_{Ln}) are shown during load removal. Above results show the acceptable performance of VSC system for suppression of power quality in four wire supply system.

29.5 Conclusion

The performance of VSC-based system for mitigation of power quality problems has been demonstrated under linear and nonlinear loads. Test results have shown the level of harmonics, neutral current, and nature of supply currents after compensation according to IEEE-519 standard. Dynamic performances of VSC-based system have been recorded after load removal. During recording of test results, DC bus voltage of the VSC has also been maintained to reference position without any transients.

Appendix

AC Grid Supply: 63 V (Per phase), 50 Hz; Load type (1) Linear: three parallel connected resistive and three-phase inductive element (2) Non-linear: Three single-phase full bridge diode rectifier with resistive load and filter inductance; DC bus capacitance: 2,350 μF; Interfacing inductance (L_f): 2.5 mH; DC bus voltage: 200 V; PI controller gains (DC bus voltage): $k_p = 0.25$, $k_i = 0.02$.

References

1. Vithayathil J (2010) Power electronics: principle and applications. Tata Mcgraw-Hill Education Private Ltd., New Delhi, pp 293–371
2. Sannino A, Svensson J, Larsson T (2003) Review power-electronic solutions to power quality problems. J Electr Power Syst Res 66:71–82
3. Sood VK (2004) HVDC and FACTS controllers: applications of static converters in power systems. Kluwer Academic Publishers, Boston, pp 71–191
4. Cirrincione M, Pucci M, Vitale G (2012) Power converters ac electrical drives with linear neural networks. CRC Press, Boca Raton
5. Singh B, Al-Haddad K, Chandra A (1999) A review of active filters for power quality improvement. IEEE Trans Ind Electron 46(5):960–971

6. Sankaran C (2001) Power quality. CRC Press, New York
7. Zeng Z, Yang H, Zhao R, Cheng C (2013) Topologies and control strategies of multi-functional grid-connected inverters for power quality enhancement: a comprehensive review. J Renew Sustain Energy Rev 24:223–270
8. Medina A, Segundo-Ramirez J, Ribeiro P, Xu W, Lian KL, Chang GW, Dinavahi V, Watson NR (2013) Harmonic analysis in frequency and time domain. IEEE Trans Power Delivery 28 (3):1813–1821
9. Lin BR, Wei T, Chiang HK (2004) An eight-switch three-phase VSI for power factor regulated shunt active filter. J Electr Power Syst Res 68:157–165
10. Iyer SK, Arindam Ghosh A, Joshi A (2005) Inverter topologies for DSTATCOM applications-A simulation study. J Electr Power Syst Res 75:161–170
11. Jain SK, Agarwal P, Gupta HO (2003) Design simulation and experimental investigations on a shunt active power filter for harmonics and reactive power compensation. J Electr Power Compon Syst 31:671–692
12. Crosier R, Wang S (2013) DQ-frame modeling of an active power filter integrated with a grid-connected, multifunctional electric vehicle charging station. IEEE Trans Power Electron 28 (12):5702–5712
13. Hsieh GC, Hung JC (1996) Phase-locked loop techniques—A survey. IEEE Trans Ind Electron 43(6):609–615
14. Singh B, Sharma S (2012) Design and implementation of four-leg voltage source converter based VFC for autonomous wind energy conversion system. IEEE Trans Ind Electron 59 (12):4694–4703
15. Saez V, Martin A, Rizo M, Rodriguez A, Bueno EJ, Hernandez A, Miron A (2010) FPGA implementation of grid synchronization algorithms based on DSC, DSOGI, QSG and PLL for distributed power generation systems. In: Proceedings of IEEE international symposium on industrial electronics, pp 2765–2770
16. Ghartemani MK, Hossein Mokhtari H, Iravani MR, Mohammad S (2004) A signal processing system for extraction of harmonics and reactive current of single-phase systems. IEEE Trans Power Delivery 19(3):979–986
17. Ghartemani MK, Ziarani AK (2004) Performance characterization of a non-linear system as both an adaptive notch filter and a phase-locked loop. Int J Adap Control Signal Process 18:23–53

Chapter 30
Sensor Reduction in a PFC-Based Isolated BL-SEPIC Fed BLDC Motor Drive

Bhim Singh, Vashist Bist, Ambrish Chandra and Kamal Al-Haddad

Abstract This work presents an improved power quality-based isolated bridgeless single-ended primary inductance converter (BL-SEPIC) fed brushless DC (BLDC) motor drive using a single voltage sensor. The voltage of DC bus of voltage source inverter (VSI) feeding BLDC motor is varied for controlling the speed of BLDC motor and operating the VSI in low frequency switching for electronically commutating the BLDC motor for minimal switching losses. A front-end isolated bridgeless configuration of SEPIC is used which offers reduced conduction losses. The BL-SEPIC is designed to operate in discontinuous conduction mode (DCM), thus utilizing a simple control of voltage follower. The sensorless control of BLDC motor is also used for elimination of rotor position sensors. The performance of proposed drive is evaluated over a wide range of speed control with unity power factor (PF) at AC mains. The obtained power quality indices are under the recommended limits of IEC 61000-3-2.

Keywords BL-SEPIC · BLDC motor · DCM · PFC · Power quality · VSI

B. Singh (✉) · V. Bist (✉)
Electrical Engineering Department, Indian Institute of Technology Delhi, New Delhi 110016, India
e-mail: bsingh@ee.iitd.ac.in

V. Bist
e-mail: vashist.bist@gmail.com

A. Chandra · K. Al-Haddad
Electrical Engineering Department, École de Technologie Supérieure, 1100 Notre-Dame, Montreal, QC H3C 1K3, Canada
e-mail: ambrish.chandra@etsmtl.ca

K. Al-Haddad
e-mail: kamal.al-haddad@etsmtl.ca

V. Vijay et al. (eds.), *Systems Thinking Approach for Social Problems*, Lecture Notes in Electrical Engineering 327,
DOI 10.1007/978-81-322-2141-8_30

30.1 Introduction

Power quality problems in terms of low power factor (PF) and high distortion in supply current have become a serious issue. International power quality standards such as IEC 61000-3-2 recommend limits on harmonics current drawn from AC mains. This limits the total harmonic distortion (THD) of the supply current and PF at the AC mains. A limit on supply current's THD less than 19 % and a PF higher than 0.95 has been recommended for class-A equipment (per phase current <16 A) [1]. Hence, for obtaining the performance under recommended limits, the power processing has to be done. Different configurations of single-phase PFC converters have been reported in the literature [2, 3]. The choice of best suited topology depends upon the application, cost, size, weight, and efficiency of the converter [2, 3].

Brushless DC (BLDC) motors are widely recognized for their high efficiency, high energy density, high ruggedness, and compact size [4, 5]. It has permanent magnets on the rotor and three-phase concentrated windings on the stator which are excited via a voltage source inverter (VSI). The excitation of the VSI depends on the rotor position which is sensed at a span of 60° by Hall Effect position sensors [6]. Since it has no mechanical brushes and commutator assembly, the BLDC motor is electronically commutated of which switching states depend on the rotor position [6].

In a conventional scheme, this VSI is supplied from an uncontrolled bridge rectifier with a high value of DC link capacitor. This combination of uncontrolled rectifier and DC link capacitor draws supply current from AC mains when the instantaneous value of supply voltage is higher than the DC link capacitor's voltage [7]. Hence, a harmonic-rich and peaky supply current is drawn from the AC mains which lead to a high value of supply current's THD and thus a poor PF. Such power quality indices are not recommended by IEC-61000-3-2 [1]. Hence, BLDC motor is fed by a front-end single-phase PFC converter for attaining a unity PF at AC mains [8].

A BLDC motor fed by a boost PFC with constant DC link voltage has been widely used configuration [9]. This suffers from high switching losses in VSI due to pulse width modulation (PWM) switching. The BLDC motor speed control using a variable DC link voltage at VSI has been proposed in [10]. Using this approach, a PFC-based BLDC motor drive has been proposed in [11], but suffers from high switching losses in VSI due to PWM-based switching. This work presents the use of variable DC link voltage for controlling the speed of BLDC motor and by electronically commutating the BLDC motor for minimal losses in VSI due to high frequency switching. Hence, a configuration capable of both bucking and boosting the voltage is used for achieving a wide range of DC link control. Moreover, the bridgeless PFC converter is recommended because it offers low conduction losses in the rectifier due to its complete or partial elimination [12, 13]. A bridgeless single-ended primary inductance converter (SEPIC) feeding a sensorless-based BLDC motor drive is explored in this paper.

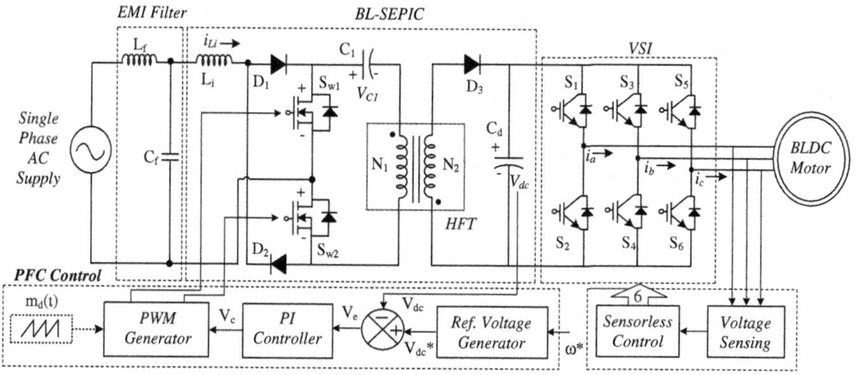

Fig. 30.1 Proposed BLDC motor drive fed by a front-end isolated BL-SEPIC

30.2 Proposed PFC Isolated BL-SEPIC Fed BLDC Motor Drive

Figure 30.1 shows the proposed isolated BL-SEPIC fed sensorless-based BLDC motor drive. The DC link voltage of VSI is controlled for controlling the speed of BLDC motor. The isolated BL-SEPIC is designed to operate in discontinuous conduction mode (DCM) of operation using a voltage follower approach and thus requiring a single voltage sensor for dual operation of PFC and voltage control. The BLDC motor is commutated electronically for reducing the switching losses in VSI. Rotor position sensors are eliminated by using the sensorless scheme of BLDC motor. The proposed drive is designed to operate for wide range of speed variation with unity PF operation at AC mains.

30.3 Operation of PFC Isolated BL-SEPIC

The current in magnetizing inductance (L_m) of high frequency transformer (HFT) becomes discontinuous in a switching period for the operation of isolated BL-SEPIC in DCM. Figure 30.2a–f shows the operation of BL-SEPIC in different modes of operation during a complete switching cycle for the two half cycles of supply voltage, respectively. Three different modes of operation for an isolated BL-SEPIC in a complete switching cycle are described below.

Mode I: As shown in Fig. 30.2a, when switch S_{w1} is turned on, the input inductor (L_i) and magnetizing inductance (L_m) of the HFT begin charging. The intermediate capacitor C_1 discharges and DC link capacitor, C_d, charges in this mode of operation as shown in Fig. 30.3.

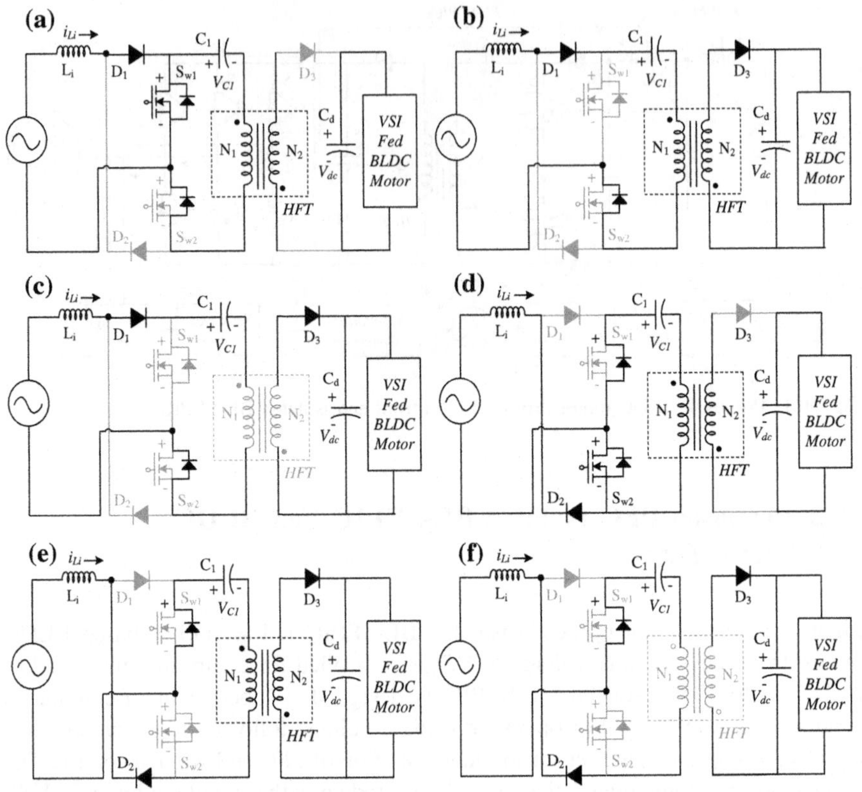

Fig. 30.2 Operation of PFC isolated SEPIC in different modes of operation for positive (a–c) and negative (d–f) half cycle of supply voltage

Fig. 30.3 Waveforms of different variables of PFC BL-SEPIC operating in DICM mode of operation

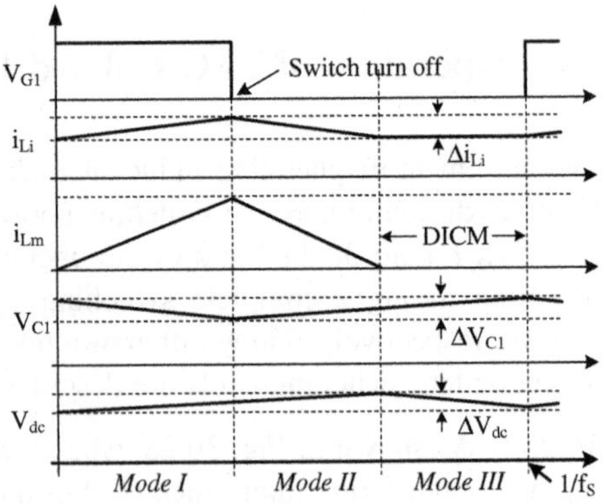

Mode II: As shown in Fig. 30.2b, when switch S_{w1} is turned off, the energy stored in input, and magnetizing starts discharging and supplies the required energy to the intermediate capacitor C_1 and the DC link capacitor, respectively. The DC link capacitor continues charging and the DC link voltage continues to increase as shown in Fig. 30.3.

Mode III: In this mode, the HFT is completely discharged and the magnetizing current (i_{Lm}) becomes zero as shown in Fig. 30.2c. The DC link capacitor supplies the required energy to the load as shown in Fig. 30.3.

In a similar way, the operation for the negative half cycle of supply voltage can be realized.

30.4 Design of PFC Isolated BL-SEPIC

The design of an isolated BL-SEPIC converter includes the estimation and proper selection of parameters such as input inductors (L_i), turns ratio and magnetizing inductance of HFT (n and L_m), intermediate capacitor (C_1), and DC link capacitor (C_d). A PFC of 300 W is designed for feeding a BLDC motor of 251 W (complete specifications of BLDC motor are given in Appendix).

The PFC converter is designed to control the DC link voltage from 40 V (V_{dcmin}) to 130 V (V_{dcmax}).

The supply voltage (v_s) is given as,

$$v_s = V_m \sin(2\pi f_L t) = 220\sqrt{2}\sin(314t) \tag{30.1}$$

where V_m is the amplitude of supply voltage, i.e., 311 V, and f_L is supply frequency, i.e., 50 Hz.

Average input voltage appearing across rectifier is as [7],

$$V_{in} = \frac{2V_m}{\pi} \approx 198\,\text{V} \tag{30.2}$$

The output voltage, V_{dc} of isolated BL-SEPIC, is as [3],

$$V_{dc} = \left(\frac{N_2}{N_1}\right)\frac{D}{(1-D)}V_{in} \tag{30.3}$$

where D is the duty ratio and the transformation ratio (N_2/N_1) is selected as 0.5 to operate at duty ratio of the order of 0.5.

From Eq. (30.3), duty ratio D is calculated for V_{dcmax} as,

$$D = \frac{V_{dc}}{V_{dc} + (N_2/N_1)V_{in}} = 0.568 \tag{30.4}$$

The equivalent load resistance (R_L) is expressed as,

$$R_L = \frac{V_{dc}^2}{P_o} = \frac{130^2}{300} = 56.33\,\Omega \tag{30.5}$$

The expression for input side inductor (L_i) is given as [3],

$$L_i = \frac{V_{in}D}{f_S \Delta i_{Li}} = \frac{198 \times 0.568}{50{,}000 \times 0.4 \times (300/198)} = 3.7\,\text{mH} \tag{30.6}$$

where Δi_{Li} is the permitted ripple current in input inductor ($I_{in} = P_o/V_{in}$) and f_S is switching frequency which is selected as 50 kHz.

The critical value of magnetizing inductance L_{mc} of the HFT to operate at boundary of CCM and DCM is given as [3],

$$L_{mc} = \frac{R_L(1-D)^2}{2Df_S(N_2/N_1)^2} = \frac{56.33 \times (1-0.568)^2}{2 \times 0.5677 \times 50{,}000 \times 0.5^2} \approx 742\,\mu\text{H} \tag{30.7}$$

The value of magnetizing inductance L_m is selected as

$$L_m \ll L_{mc} \tag{30.8}$$

Hence, from Eq. (30.8), the value of L_m is selected as 250 μH for its operation in DICM [14].

The expression for input side intermediate capacitor (C_1) given as [3],

$$C_1 = \frac{V_{dc}(N_2/N_1)D}{R_L f_S \Delta V_{C1}} = \frac{198 \times 0.5 \times 0.568}{56.33 \times 50{,}000 \times 0.1 \times 458} \approx 436\,\text{nF} \tag{30.9}$$

where ΔV_{C1} is the permitted voltage ripple across intermediate capacitor (C_1) and is given as the sum of input voltage and the reflected DC link voltage via HFT ($V_{C1} = V_{in} + 2 \times V_{dc}$).

Hence, a 440 nF capacitor is selected for this application.

The value of DC link capacitor is given and calculated as [3],

$$C_d = \frac{I_{dc}}{2\omega_L \Delta V_{dc}} = \frac{300/198}{2 \times 314 \times 0.01 \times 130} = 1{,}855.9\,\mu\text{F} \tag{30.10}$$

where ΔV_{dc} is the permitted ripple across the DC link capacitor and is selected as 1 % of V_{dc}.

Hence, a DC link capacitor of 2,200 μF is selected.

An EMI filter is designed to limit high-frequency current ripples in the supply system [15]. The maximum value of filter capacitance C_{max} is as [15],

$$C_{max} = \frac{I_{peak}}{\omega_L V_{peak}} \tan(\theta) = \frac{300\sqrt{2}/220}{314 \times 311} \tan(1°) = 344.7\,\text{nF} \tag{30.11}$$

where I_{peak} represents the input peak current, V_{peak} is the input peak voltage, and θ represents the displacement angle.

The filter capacitance C_f is selected such that C_f is lower than C_{max}, hence the value of C_f is taken as 330 nF.

The filter inductance L_f is given as [15],

$$L_f = \frac{1}{4\pi^2 f_c^2 C_f} = \frac{1}{4\pi^2 \times 5{,}000^2 \times 330 \times 10^{-9}} = 3.07\,\text{mH} \tag{30.12}$$

where f_c is the cut-off frequency such that $f_c = f_s/10$ [15].

Hence, the filter inductance of the order of 3 mH is selected.

30.5 Control of PFC-Based BL-SEPIC

The control of the front-end PFC converter generates the PWM pulses for the PFC converter switches (S_{w1} and S_{w2}) for DC link voltage control with PFC operation. The reference DC link voltage (V_{dc}*) is generated as,

$$V_{dc}^* = k_v \omega^* \tag{30.13}$$

where k_v is motor voltage constant and ω* is reference speed.

The reference DC link voltage (V_{dc}*) is compared with the sensed DC link voltage (V_{dc}) to generate the voltage error signal (V_e) given as,

$$V_e(k) = V_{dc}^*(k) - V_{dc}(k) \tag{30.14}$$

where k represents the kth sampling instant.

This error voltage signal (V_e) is given to the voltage PI controller to generate a controlled output voltage (V_{cc}) as,

$$V_{cc}(k) = V_{cc}(k-1) + k_p\{V_e(k) - V_e(k-1)\} + k_i V_e(k) \tag{30.15}$$

where k_p and k_i are gains of PI controller.

This output of voltage controller is compared with a high-frequency saw-tooth signal (m_d) to generate PWM pulses as,

$$\left\{\begin{array}{l} \text{if } m_d(t) < V_{cc}(t) \text{ then } S_{w1} = S_{w2} = \text{'ON'} \\ \text{if } m_d(t) \geq V_{cc}(t) \text{ then } S_{w1} = S_{w2} = \text{'OFF'} \end{array}\right\} \tag{30.16}$$

where S_{w1} and S_{w2} are the PWM signal to BL-SEPIC switches.

30.6 Sensorless Operation of BLDC Motor

A position sensorless scheme is used to eliminate the requirement of Hall Effect position sensors which drastically affect the overall cost of the system [16]. In this chapter, a sensorless technique using the line back-EMF detection for estimation of rotor position is used. The complete operation is classified into open loop starting, switching from starting to sensorless mode, and the sensorless operation of BLDC motor.

30.6.1 Open Loop Starting

In open loop starting technique, switching pulses of low frequency are given to VSI to produce a rotating field at the stator such as to overcome the rotor inertia and friction at standstill position. A limited DC voltage corresponding to the required speed is applied via a PFC BL-SEPIC to ensure a limited current in stator windings of BLDC motor.

30.6.2 Starting to Sensorless Mode

When the rotor rotates above a certain speed in the open loop starting mode, such that the position rotor position signals are estimated accurately, the control switches the operation of BLDC motor from open loop starting mode to sensorless mode of operation.

30.6.3 Sensorless Operation

In this approach, the line voltage is determined by the sensing of phase voltage with respect to the virtual ground. A zero crossing detector (ZCD) detects the zero crossing of the sensed terminal voltage and based on these zero crossings, the virtual position signals for achieving an electronic commutation are generated.

The switching states of the VSI corresponding to the virtual position signals are shown in Table 30.1. The line back-EMFs and the corresponding switching states of the VSI based on the zero crossings (Z_{ab}, Z_{bc}, and Z_{ca}) of the line back-EMFs are shown in Fig. 30.4.

Table 30.1 Switching states of VSI based on virtual position signals

θ (°)	Position signals			Switching states					
	S_{ga}	S_{gb}	S_{gc}	S_1	S_2	S_3	S_4	S_5	S_6
NA	0	0	0	0	0	0	0	0	0
0–60	0	0	1	1	0	0	0	0	1
60–120	0	1	0	0	1	1	0	0	0
120–180	0	1	1	0	0	1	0	0	1
180–240	1	0	0	0	0	0	1	1	0
240–300	1	0	1	1	0	0	1	0	0
300–360	1	1	0	0	1	0	0	1	0
NA	1	1	1	0	0	0	0	0	0

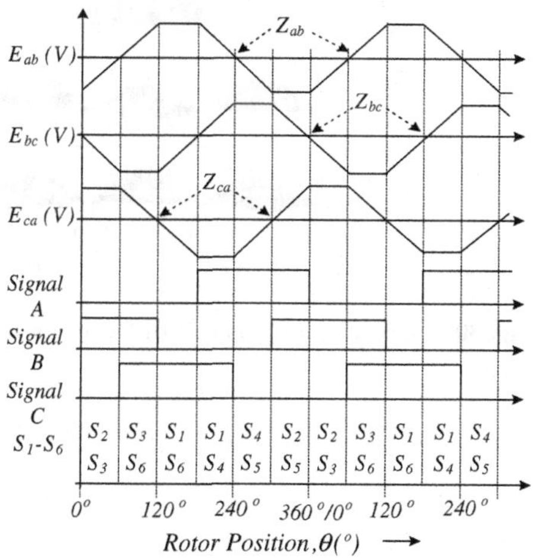

Fig. 30.4 Line back-EMF's and corresponding switching states of BLDC motor

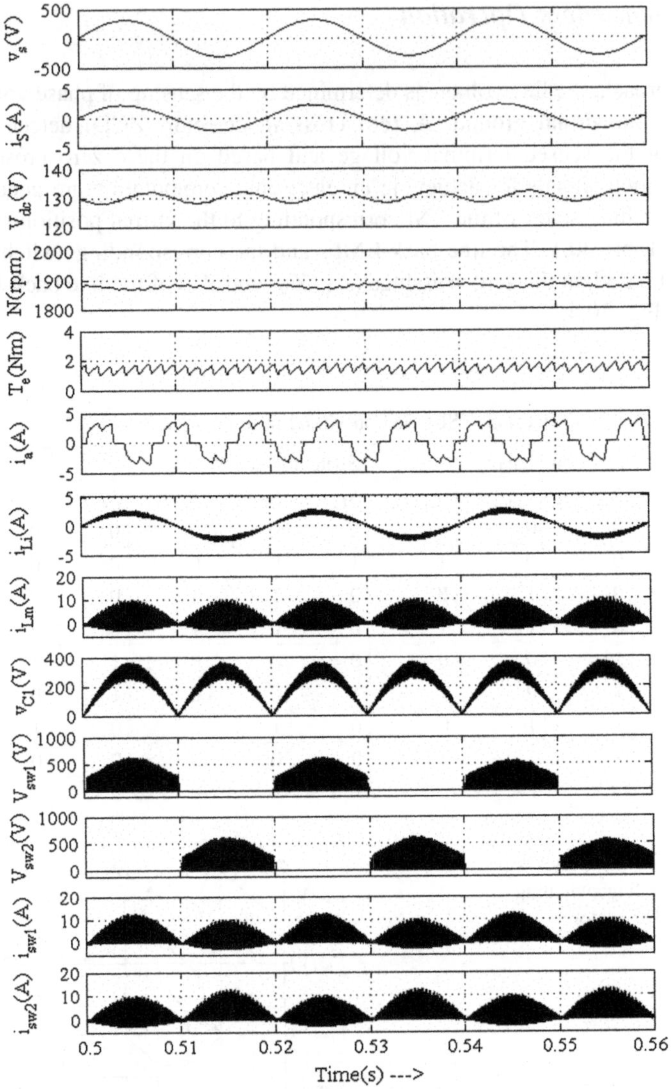

Fig. 30.5 Operation of proposed BLDC motor drive at rated condition

Table 30.2 Performance of proposed drive at speed control

S. No.	V_{dc} (V)	Speed (rpm)	DPF	THD of I_s (%)	PF	I_s (A)
1	40	450	0.9914	5.04	0.9901	0.469
2	50	590	0.9945	4.39	0.9935	0.565
3	60	750	0.9965	4.21	0.9956	0.678
4	70	910	0.9974	3.79	0.9967	0.791
5	80	1,070	0.9981	3.56	0.9975	0.908
6	90	1,240	0.9988	3.41	0.9982	1.026
7	100	1,400	0.9993	3.37	0.9987	1.149
8	110	1,560	0.9995	3.31	0.999	1.273
9	120	1,720	0.9998	3.18	0.9993	1.402
10	130	1,890	0.9999	3.06	0.9994	1.531

30.7 Results and Discussion

The proposed drive is simulated in MATLAB/Simulink environment. Power quality indices such as displacement power factor (DPF), PF, and THD of supply current at AC mains are used for evaluating the performance of proposed drive. Moreover, performance indices such as supply voltage (v_s), supply current (i_s), DC link voltage (V_{dc}), speed (ω), electromagnetic torque (T_e), stator current (i_a), inductor current (i_{Li}), intermediate capacitor's voltage (v_{C1}), and PFC converter switch's voltage and current (v_{sw1}, v_{sw2}, i_{sw1}, and i_{sw2}) are analyzed for demonstrating the overall performance of proposed BLDC motor drive.

30.7.1 Steady State Performance

Figure 30.5 shows the operation of the proposed drive at rated condition. A unity PF is achieved with a limited amount of harmonic distortion in supply current. A discontinuous current in the HFT (i_{Lm}) is obtained which confirms the DICM operation of the PFC converter. Table 30.2 shows the power quality indices for a wide variation in speed. Figure 30.6a, b shows the supply current and its harmonic spectra for the BLDC motor operating at rated load with DC link voltage as 130 and 40 V, respectively. Acceptable power quality indices within the limits of International power quality standard (IEC 61000-3-2) have been obtained [1].

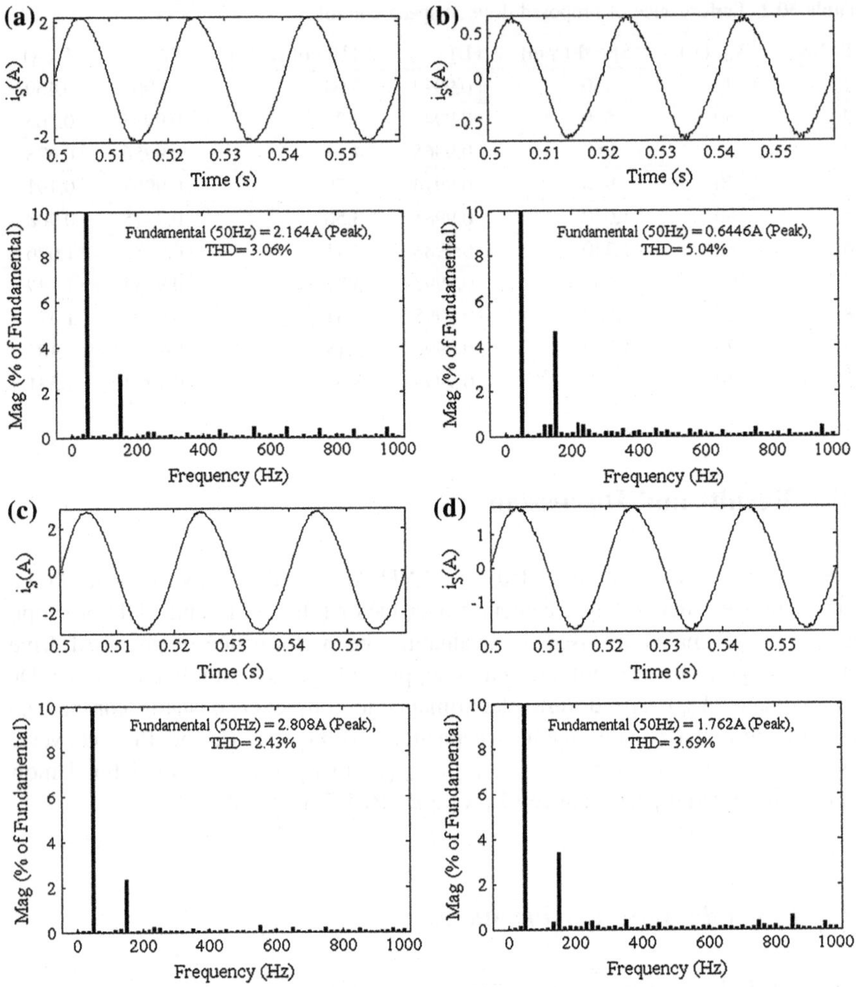

Fig. 30.6 Harmonic spectra of the supply current for BLDC motor operated at rated loading with **a** $V_s = 220$ V, $V_{dc} = 130$ V, **b** $V_s = 220$ V, $V_{dc} = 40$ V, **c** $V_s = 170$ V, $V_{dc} = 130$ V, **d** $V_s = 270$ V, $V_{dc} = 130$ V

Table 30.3 Performance of proposed drive at supply voltage fluctuations

S. No.	V_s (V)	DPF	THD of I_s (%)	PF	I_s (A)
1	170	1	2.43	0.9997	1.986
2	190	1	2.48	0.9997	1.755
3	210	0.9999	3.01	0.9994	1.602
4	230	0.9997	3.2	0.9995	1.463
5	250	0.9993	3.47	0.9987	1.345
6	270	0.9988	3.69	0.9981	1.246

Table 30.4 Stress on PFC converter switch for different loading on BLDC motor

S. No.	Load (%)	Peak voltage (V_w)	Peak current (I_w)	Rms current (I_{rms})
1	10	600	7	0.222
2	20	600	8	0.286
3	30	600	8.5	0.351
4	40	610	9	0.415
5	50	610	10	0.476
6	60	610	10.5	0.535
7	70	620	11	0.595
8	80	620	12	0.655
9	90	620	12.5	0.713
10	100	620	13.5	0.771

30.7.2 Performance During Supply Voltage Fluctuation

Figure 30.6c, d shows the supply current and its harmonic spectra for the BLDC motor operating at rated loading condition with DC link voltage as 130 V and supply voltage as 170 and 270 V, respectively. Table 30.3 tabulates the performance of the proposed drive at rated load on BLDC motor with rated DC link voltage of 130 V and supply voltage variation form 170 to 270 V. The obtained power quality indices are under the acceptable limits of IEC 61000-3-2.

30.7.3 Stress on PFC Converter Switches

Table 30.4 tabulates the stresses on the two PFC converter switches. An acceptable stress on the switches of PFC converter with a peak voltage and current stress of 620 V and 13.5 A is obtained which governs the satisfactory operation of the proposed drive.

30.7.4 Dynamic Performance of the Proposed Drive

Figure 30.7a shows the dynamic performance of the proposed drive during starting at low speed corresponding to DC link voltage of 40 V. A limited and acceptable inrush in stator and supply current is obtained. The dynamic behavior of proposed BLDC motor drive during speed change corresponding to DC link voltage change from 40 to 130 V is shown in Fig. 30.7b. Moreover, dynamic behavior of the proposed drive during supply voltage fluctuation from 270 to 170 V is shown in Fig. 30.7c. The DC link voltage is maintained constant thus demonstrating the satisfactory closed-loop performance of the proposed BLDC motor drive.

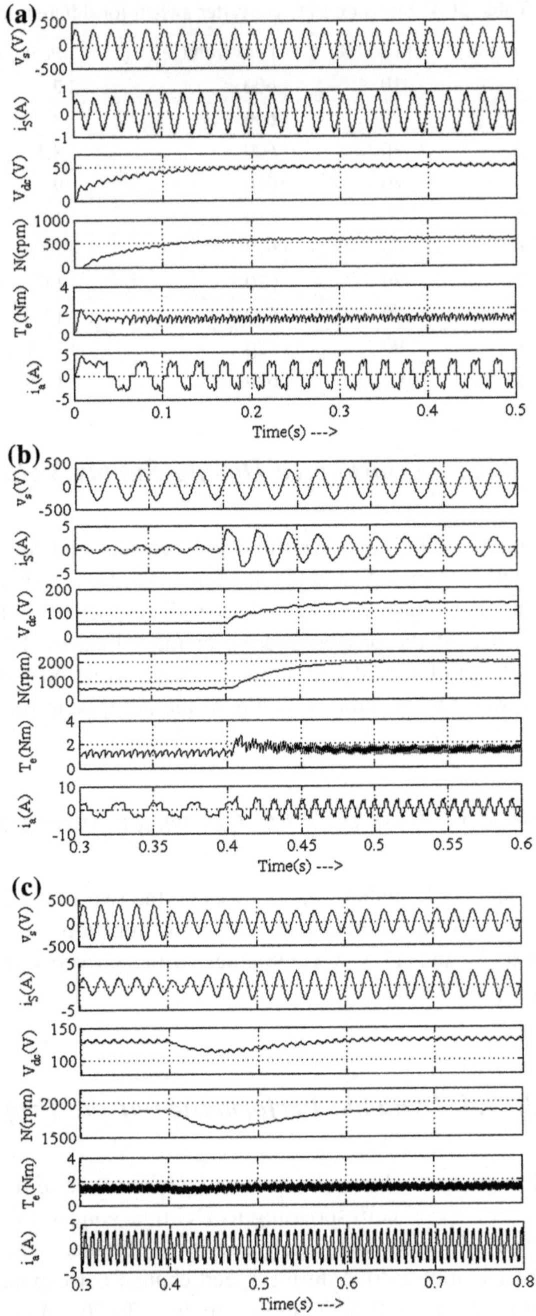

Fig. 30.7 Dynamic performance during **a** stating of BLDC motor drive at dc link voltage of 40 V, **b** speed control at change in dc link voltage from 40 to 130 V and **c** supply voltage fluctuation from 270 to 170 V

30.8 Conclusion

A PFC-based BLDC motor drive using a front-end BL-SEPIC fed has been proposed for achieving a unity PF operation at AC mains over a wide variation in speed. A PFC-based BL-SEPIC has been operated in DCM to act as an inherent PF corrector. An electronic commutation of the BLDC motor has been used for the operation of VSI in fundamental frequency switching for minimal switching losses. Moreover, a bridgeless configuration of SEPIC has been used for reduced conduction losses. Moreover, the sensorless scheme of BLDC motor has eliminated the position sensor and thus reduces the overall cost of the system. The proposed drive has shown an improved power quality in terms of high PF and low supply current distortion at AC mains for a wide range of speed control and supply voltage fluctuation.

Appendix

BLDC Motor Specifications: No. of Poles: 4 pole, Rated Power (P_{rated}) = 251.32 W, Rated DC link Voltage (V_{rated}) = 130 V, Rated Torque (T_{rated}) = 1.2 Nm, Rated Speed (ω_{rated}) = 2,000 rpm, Back-EMF Constant (K_b) = 78 V/krpm, Torque Constant (K_t) = 0.74 Nm/A, Phase Resistance (R_{ph}) = 14.56 Ω Phase Inductance (L_{ph}) = 25.71 mH, Moment of Inertia (J) = 1.8×10^{-4} Nm/s^2.

References

1. International Standard IEC 61000-3-2 (2000) Limits for harmonic current emissions (equipment input current ≤16 A per phase)
2. Singh B, Singh BN, Chandra A, Al-Haddad K, Pandey A, Kothari DP (2003) A review of single-phase improved power quality AC–DC converters. IEEE Trans Ind Electron 50 (5):962–981
3. Singh B, Singh S, Chandra A, Al-Haddad K (2011) Comprehensive study of single-phase AC–DC power factor corrected converters with high-frequency isolation. IEEE Trans Ind Inform 7(4):540–556
4. Handershot JR, Miller TJE (2010) Design of brushless permanent magnet motors. Clarendon Press, Oxford
5. Hanselman DC (2003) Brushless permanent magnet motor design. McGraw Hill, New York
6. Toliyat HA, Campbell S (2004) DSP-based electromechanical motion control. CRC Press, New York
7. Mohan N, Undeland TM, Robbins WP (2009) Power electronics: converters, applications and design. Wiley, New York
8. Singh B, Singh S (2010) Single-phase power factor controller topologies for permanent magnet brushless DC motor drives. IET Power Electron 3(2):147–175

9. Wu CH, Tzou YY (2009) Digital control strategy for efficiency optimization of a BLDC motor driver with VOPFC. In: IEEE conference on energy conversion congress and exposition (ECCE), pp 2528–2534, 20–24 Sept 2009
10. Krishnan R (2001) Electric motor drives: modeling, analysis and control. Pearson Education, India
11. Gopalarathnam T, Toliyat HA (2003) A new topology for unipolar brushless DC motor drive with high power factor. IEEE Trans Power Electron 18(6):1397–1404
12. Sabzali AJ, Ismail EH, Al-Saffar MA, Fardoun AA (2011) New bridgeless DCM sepic and Cuk PFC rectifiers with low conduction and switching losses. IEEE Trans Ind Appl 47 (2):873–881
13. Mahdavi M, Farzanehfard H (2011) Bridgeless SEPIC PFC rectifier with reduced components and conduction losses. IEEE Trans Ind Electron 58(9):4153–4160
14. Simonetti DSL, Sebastian J, Uceda J (1997) The discontinuous conduction mode Sepic and Cuk power actor preregulators: analysis and design. IEEE Trans Ind Electron 44(5):630–637
15. Vlatkovic V, Borojevic D, Lee FC (1996) Input filter design for power factor correction circuits. IEEE Trans Power Electron 11(1):199–205
16. Acarnley PP, Watson JF (2006) Review of position-sensorless operation of brushless permanent-magnet machines. IEEE Trans Ind Electron 53(2)

Chapter 31
A Frequency Shifter-Based Simple Control for Solar PV Grid-Interfaced System

Chinmay Jain and Bhim Singh

Abstract This paper deals with a grid-interfaced solar photovoltaic (SPV) energy conversion system for three-phase four-wire (3P4W) distribution system. The solar energy conversion system (SECS) is a multifunctional as it not only feeds SPV energy into the grid but also serves the purpose of grid current balancing, reactive power compensation, harmonic mitigation, and neutral current elimination. In a two-stage SPV system, the first stage is a boost converter, controlled with incremental conductance (InC) maximum power point tracking (MPPT) algorithm, and a second stage is a four-leg voltage source converter (VSC). A simple frequency shifter-based control is proposed for the control of VSC. A proportional integral (PI) controller along with feedforward term for SPV power is used for fast dynamic response. Simulations are carried out in MATLAB along with Simulink and Sim Power System toolboxes, and detailed simulation results are presented to demonstrate its required multifunctions.

Keywords Power quality · Neutral current compensation · Two-stage solar PV · Simple control

31.1 Introduction

Vanishing conventional energy sources have moved world's attention toward non-conventional energy sources such as solar photovoltaic (SPV) and wind energy. The SPV systems are of prime interest because of low maintenance and no mechanical machine involvement in the energy conversion process. The solar PV systems can be

C. Jain (✉) · B. Singh
Department of Electrical Engineering, Indian Institute of Technology Delhi, New Delhi 110016, India
e-mail: chinmay31jain@gmail.com

B. Singh
e-mail: bsingh@ee.iitd.ac.in

V. Vijay et al. (eds.), *Systems Thinking Approach for Social Problems*, Lecture Notes in Electrical Engineering 327,
DOI 10.1007/978-81-322-2141-8_31

broadly classified into two categories, stand-alone and grid-interfaced systems [1]. The stand-alone PV systems are normally used at the places which are out of reach of the grid such as electric vehicle, satellites, and remote areas. The SPV is intermittent in nature; hence, the stand-alone system requires energy storage elements such as a battery to maintain instantaneous power balance between the load and the photovoltaic (PV) source. Three-port converters are proposed by researchers for power management between load, PV source, and batteries [2]. However, the use of batteries makes the system bulky and costly.

Grid-interfaced systems are becoming popular choice where the grid is available. In case of grid-interfaced system, the grid acts as infinitely large energy storage to take care of intermittency of PV source. Grid-interfaced PV inverters can be broadly classified into single-stage and two-stage systems. The single-stage system uses variable dc link voltage for maximum power point tracking (MPPT), and they usually inject currents at unity power factor (UPF) into the grid [3].

Some researchers have proposed reactive power compensation with the single-stage converter in the night time or cloudy days [4]. However, single-stage system is not effective in partial shedding conditions [5]. Dc link voltage rating is more in the case of single-stage grid-interfaced SPV system than in the case of two-stage system. In conventional two-stage system, the first stage extracts power from PV source and performs MPPT function, and the second stage feeds extracted energy into the grid. The main area of research in the grid-interfaced SPV system includes MPPT techniques, an increase in reliability and efficiency, a decrease in size, weight, and cost [6–10]. The reduction in cost can be achieved by two means, either directly or indirectly. The direct cost reduction includes reduction in installation cost or fixed cost, and in an indirect cost reduction, installation cost may be the same or a little higher, but the same resource is used efficiently so that the effective cost is reduced. The indirect cost reduction is an interesting area of research which includes allotting several other functions to the same resource which is being underutilized. In the case of grid-interfaced SPV system, the VSC for grid connection is underutilized when the power from solar PV array is less than its peak power at full irradiance. Hence, an indirect reduced cost system is proposed, which possesses several features other than feeding extracted energy at UPF.

The use of voltage source converter (VSC) as an active power filter is well known [11]. A fuzzy logic-based d–q axis current control is proposed in [12]. The synchronously rotating reference frame theory (SRFT)-based system requires tuning of phase-locked loop (PLL) and frame transformations. An instantaneous reactive power theory (IRPT)-based system is proposed in [13].The IRPT-based system operates on power-based calculation which suffers from poor dynamic response. Many researchers have proposed several soft computing theories for the control of VSC. Complicated adaptive and neural network-based control of VSC is also proposed in the literature which requires selection of optimal learning rate, tuning of internal parameters, and selection of hidden layers, and hence, these algorithms suffer from lack of intuition [14–17].

Nonlinear loads such as switched-mode power supplies (SMPS), electronic blasts, and microwave are increasing day by day. The increase in nonlinear loads is causing severe power quality problems. In the case of nonlinear loads, the neutral conductor current is nonzero even in the case of balanced distribution of loads on three phases of the grid [18]. The neutral current increases losses in the system. In the case of unbalanced distribution of loads, the problem of neutral current becomes more severe and it may lead to bursting of neutral conductor in the case of an excessive neutral current. Several neutral current compensation techniques are reported in the literature, which include different transformer connections such as zigzag and star–delta [18, 19]. The use of a separate insulated gate bipolar transistor (IGBT) leg for neutral current compensation sharing common dc bus with 3-leg VSC is also reported in the literature [20].

In the proposed work, a three-phase, four-wire distribution system is considered for the study. A three-phase four-leg (3P4L) SECS is proposed, which not only performs the functions of grid interfacing but also possesses features such as harmonic mitigation, grid current balancing, neutral current elimination, and reactive power compensation. Adding extra features to the system helps in cost reduction of the system. A novel simple frequency shifter-based control is proposed for the control of VSC. The proposed control algorithm provides all the above-mentioned features to conventional 3P4L VSC topology. The control of the system has to perform MPPT and an active power filtering both at the same time. The proposed control algorithm is more suitable than a complicated control algorithm as it has very less number of calculations for extraction of fundamental component of load current in phase with phase voltage. The performance of the system is verified by means of MATLAB simulations. The presented system follows IEEE and IEC norms [21,22].

31.2 System Configuration

The proposed system configuration is shown in Fig. 31.1. A three-phase, four-wire distribution system is the system under consideration. The proposed system consists of a SPV array, a boost converter, a four-leg VSC, interfacing inductors, ripple filter, loads, and the grid. The system consists of SPV array, which is a series–parallel combination of small power solar panels to match the required rating. The solar array is interfaced with a dc–dc boost converter. The output of boost converter is connected to dc link of a four-leg VSC. The VSC acts as a controlled current source. The midpoints of VSC legs are connected to point of common coupling (PCC) via interfacing inductors. A ripple filter is connected in parallel at PCC to absorb switching ripple of VSC in PCC voltages. The loads to be compensated are also connected at PCC. The ratings of all the elements are given in Appendix.

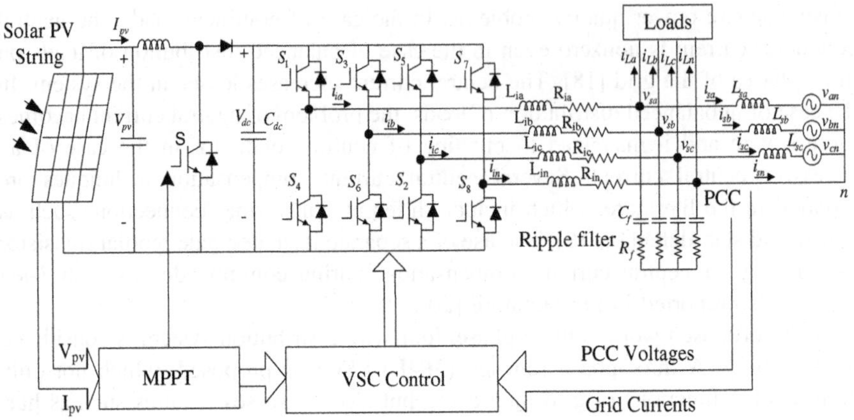

Fig. 31.1 System configuration

31.3 Control Algorithm

The control algorithm is the heart of the system. Figure 31.2 shows the proposed control algorithm. It decides the steady-state and dynamic behavior of the system. There are two main power circuits in the proposed system, i.e., the boost converter and the VSC. The output of the MPPT algorithm is the duty cycle for the boost converter. The VSC control decides the nature of injected current into the grid. Ideally, a grid-interfaced SPV system injects harmonic-free sinusoidal currents at UPF.

However, in the proposed system, the VSC provides compensation for harmonics, reactive power, neutral current, and load balancing. Hence, the currents injected by VSC are not to be sinusoidal, but the currents injected into the grid are to be sinusoidal. However, involving these extra features requires extra calculations in real time, and hence, a simple control is proposed incorporating comparatively low computations and meeting the grid norms. The two main portions of the control algorithm are as described in the following sections.

31.3.1 Maximum Power Point Tracking

An incremental conductance (InC)-based MPPT algorithm is used. The algorithm compares an InC with the conductance and takes the corrective action. The flowchart for the algorithm is given in Fig. 31.2a. For calculation of InC, ΔI_{pv} and ΔV_{pv} are estimated as follows:

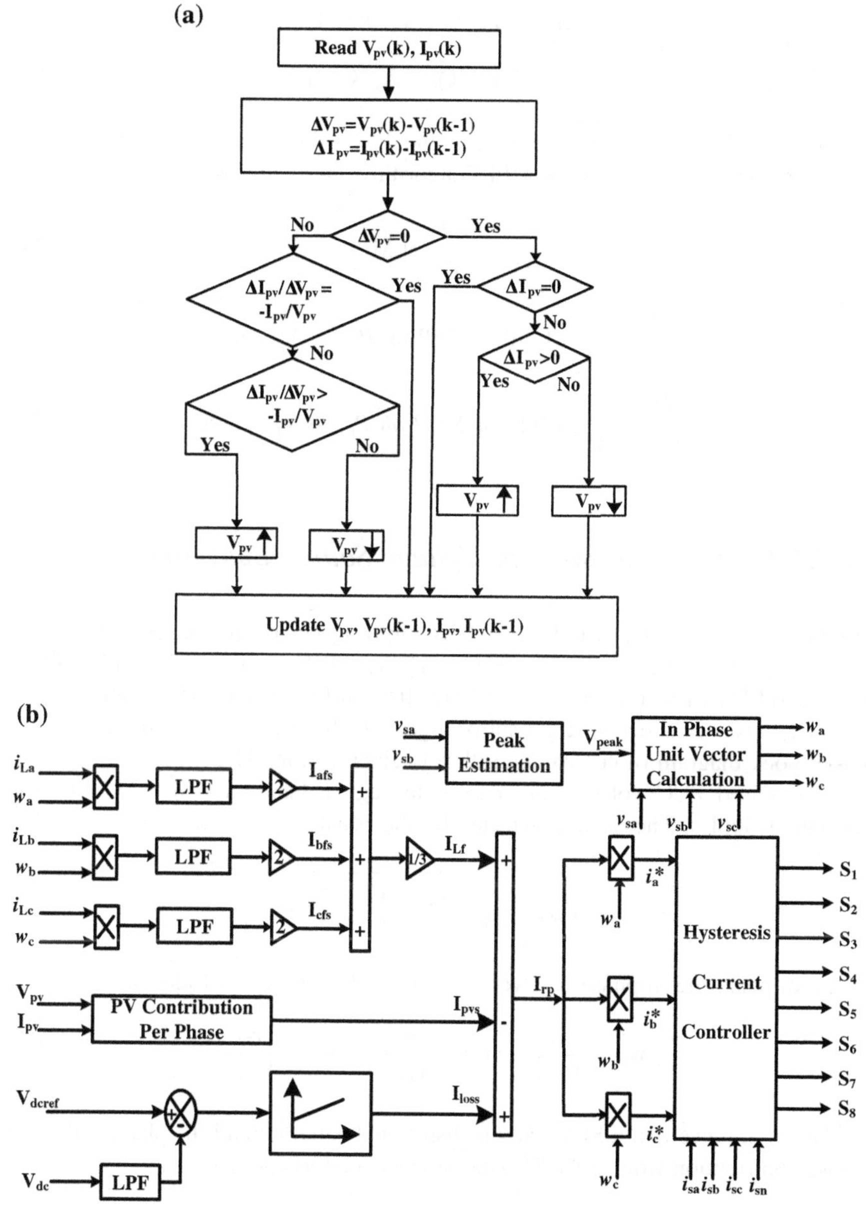

Fig. 31.2 **a** Flowchart for maximum power point tracking and **b** control strategy for VSC

$$\Delta I_{pv} = I_{pv}(k) - I_{pv}(k-1) \tag{31.1a}$$

$$\Delta V_{pv} = V_{pv}(k) - V_{pv}(k-1) \tag{31.1b}$$

where I_{pv} is the current from solar array and V_{pv} is the voltage across array.

The governing equations for MPPT algorithm are as follows:

$$\frac{\Delta I}{\Delta V} = \frac{-I}{V}, \text{ at MPP} \tag{31.2a}$$

$$\frac{\Delta I}{\Delta V} > \frac{-I}{V}, \text{ Left of MPP on } P_{pv}\text{v/s } V_{pv} \text{ curve} \tag{31.2b}$$

$$\frac{\Delta I}{\Delta V} < \frac{-I}{V}, \text{ Right of MPP on } P_{pv}\text{v/s } V_{pv} \text{ curve} \tag{31.2c}$$

31.3.2 Control Algorithm for Voltage Source Converter

For the control of VSC, two-phase PCC voltages (v_{sa}, v_{sb}), three-phase grid currents (i_{sa}, i_{sb}, i_{sc}), three-phase load currents (i_{La}, i_{Lb}, i_{Lc}), dc link voltage (V_{dc}), PV voltage (v_{pv}), and PV current (I_{pv}) are sensed. A feedforward term to accommodate changes in solar power is added in the control algorithm for fast dynamic response. The basic block diagram of control algorithm is given in Fig. 31.2b.

The sensed PCC voltages are passed through a band-pass filter to eliminate switching ripples. The peak is detected by the equation:

$$V_{peak} = \sqrt{\frac{2(v_{sa}^2 + v_{sb}^2 + v_{sc}^2)}{3}} \tag{31.3}$$

From peak, the in-phase unit vectors can be determined as follows:

$$w_a = \frac{v_{sa}}{V_{peak}}, \quad w_b = \frac{v_{sb}}{V_{peak}}, \quad w_c = \frac{v_{sc}}{V_{peak}} \tag{31.4}$$

The total power from SPV array is distributed equally to all the phases. The per phase contribution from solar PV can be calculated as follows:

$$I_{pvs} = \frac{2 \times P_{pv}}{3 \times V_{peak}} \tag{31.5}$$

The load current is multiplied with the corresponding in-phase unit templates, and the output of multiplication is then passed through a low-pass filter (LPF). The output of LPF is passed through a gain of two, and the net signal is termed as magnitude of the load current in phase with the grid voltage. The mathematical background behind this control for phase a can be described as follows:

$$i_{La} = i_{fa} + \Sigma i_{ha} \tag{31.6}$$

where i_{fa} fundamental component of phase a load current which can be rewritten as

$$i_{fa} = I_{fas} \sin \omega_0 t + I_{fac} \cos \omega_0 t \tag{31.7a}$$

$$i_{La} \times \sin \omega_0 t = (I_{fas} \sin^2 \omega_0 t + I_{fac} \cos \omega_0 t \times \sin \omega_0 t) + \Sigma i_{ha} \times \sin \omega_0 t \tag{31.7b}$$

$$i_{La} \times \sin \omega_0 t = \frac{I_{fas}}{2} + \left(\frac{I_{fac}}{2} \sin 2\omega_0 t - \frac{I_{fas}}{2} \cos 2\omega_0 t\right) + \Sigma i_{ha} \times \sin \omega_0 t \tag{31.7c}$$

From Eq. (31.7c), it can be concluded that the frequency of fundamental current is shifted to double-harmonic frequency and dc frequency, out of which in-phase fundamental component is shifted to 0 Hz with a scaling factor of half. All other harmonic components and quadrature of fundamental are shifted to second or higher order. Shifting the in-phase component to 0 Hz gives benefit of zero phase shifts in reference currents in steady-state condition. Similarly, in-phase component of fundamental current can be determined for phase b and phase c. All in-phase components of load currents are added, and the net addition gives total active component of load currents. The output of addition is divided by 3 to distribute load fundamental active power component of current equally to all the phases, which gives average per phase active power component of load current (I_{Lf}).

The output of PI controller is added to I_{Lf}. The SPV contribution per phase (I_{pvs}) is subtracted from the load current and loss component which gives net active power component for the grid (I_{rp}). The I_{rp} is then multiplied with unit templates to get the reference grid currents for three phases, and reference current for neutral leg is set zero.

The sensed and reference currents are given as the inputs to hysteresis current controller, and logic switching pulses are output of the current controller.

31.4 Result and Discussion

The proposed two-stage grid-interfaced SPV system is modeled in MATLAB along with Simulink and Sim Power System toolboxes. A SPV array rating of 25 kW is considered, and a load power rating of 5 kW per phase is considered. The performance of the system is verified via simulations. Simulation results for the system are presented and discussed under various loading conditions of the grid. Simulation results are presented for linear, nonlinear, and change in insolation levels. The harmonic analysis of the load and grid currents is also presented to demonstrate harmonic mitigation capability of the proposed system. The system parameters used for simulation study are given in Appendix.

31.4.1 Performance of System Under Linear Load

The performance of the proposed system under linear load condition is shown in Fig. 31.3a. At time $t = 0.3$ s, the system is working under power factor correction mode. At time $t = 0.35$ s, phase c load is opened, and at time $t = 0.4$ s, phase a load is opened. The load currents are unbalanced to demonstrate dynamic behavior of the system. It can be observed that load currents are unbalanced, but grid currents are balanced sinusoids at UPF. As the load is unbalanced, the neutral current can be observed in load neural (i_{Ln}). The VSC neutral current (i_{VSCn}) is equal and opposite to load neural current. The grid neutral current (i_{sn}) is compensated to zero. When the load is thrown, the load power decreases, and at the same time, if PV power is kept constant, then the power injected into the grid increases. The increment in power injected into the grid can be observed in the form of increment in grid currents. When the load is added again, the decrement in grid currents can be observed. During unbalancing, no appreciable effect is observed on dc link voltage (V_{dc}) and power from solar PV array (P_{pv}).

31.4.2 Performance of System Under Nonlinear Load

Figure 31.3b shows the performance of the proposed system under nonlinear load. A nonlinear load of 5 kW is considered on each phase. The power from solar PV array is considered to be constant. At $t = 0.3$ s, the grid currents are balanced and sinusoidal at UPF. The VSC is supplying all the harmonic currents required. For nonlinear loads, neural currents can be observed even for balanced load. The load neutral current increases in the case of unbalanced loading of the system. The load and VSC neutral currents are out of phase which results in zero current in grid neutral conductor. Phase c load is removed at $t = 0.35$ s and added at $t = 0.45$ s, respectively. Similarly, load on phase a is also varied. It can be observed that grid currents are balanced and sinusoidal even in the case of unbalanced nonlinear load on the system. The VSC currents are unbalanced to make grid currents balance sinusoids. When the load power is decreased, an increase in grid currents is observed on account of an increase in injected power into the grid. During load unbalancing, no appreciable effect is observed on dc link voltage (V_{dc}) and power from solar PV array (P_{pv}).

31.4.3 Performance of System for Step Change in Irradiance

Performance of the system for step change in irradiance is shown in Fig. 31.3c. At $t = 0.4$ s, the irradiance is 1,000 W/m^2. The load on the system is unchanged which can be observed from load currents. The power from the solar PV array is 25 kW,

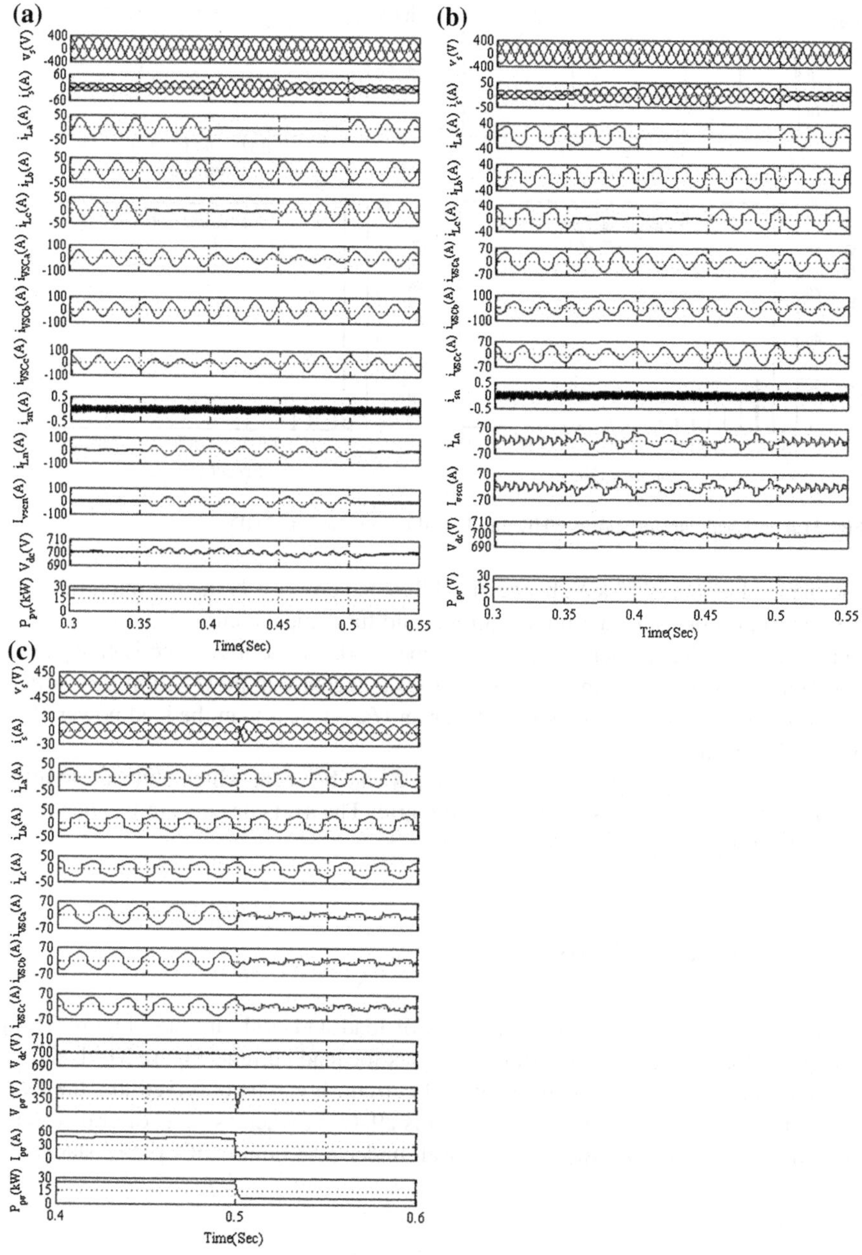

Fig. 31.3 **a** Behavior of system for linear load, **b** behavior of system for nonlinear load, and **c** behavior of system for step change in insolation

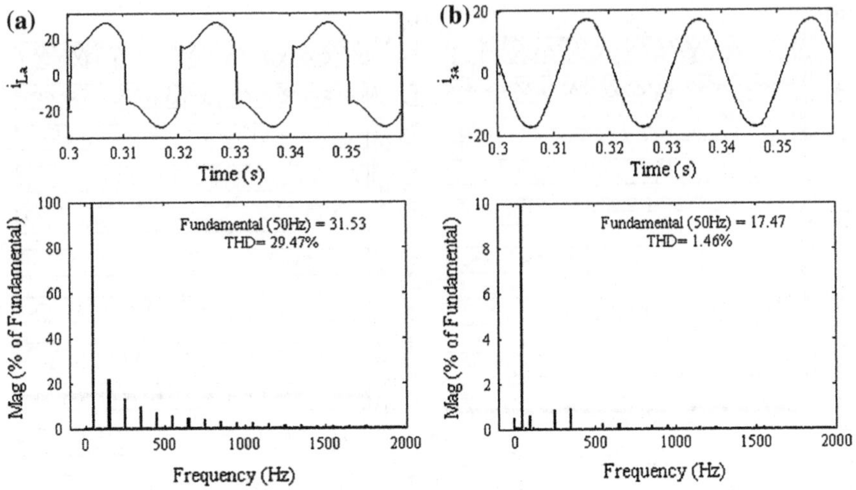

Fig. 31.4 **a** Load current and its THD and **b** grid current and its THD

and the load power is 15 kW; hence, the difference power is being fed into grid. The VSC currents consist of harmonic currents and fundamental currents corresponding to active power from solar PV array. At time $t = 0.5$ s, the irradiance is changed to 300 W/m^2. The decrease in irradiance causes a decrease in power from SPV array. It can be easily observed that solar PV power (P_{pv}) is less than the load power after a decrease in irradiance.

The reversal in the phase of grid currents can be seen, which shows that after $t = 0.5$ s, the real power is drawn from the grid. The grid currents are balanced and harmonic free.

31.4.4 Harmonic Analysis

Figure 31.4 shows the harmonic analysis of load and grid currents. Figure 31.4a shows the load current and harmonic analysis for load current. Figure 31.4b shows grid current and harmonic analysis of grid current. It can be observed that the total harmonic distortion (THD) of load current is 29.47 %, whereas THD of grid current is 1.46 %. The THD of grid current is well under 5 % limit of IEEE-519 standard.

31.5 Conclusion

A two-stage, three-phase, four-leg VSC-based system has been proposed to interface solar PV array with a 3P4W distribution system. A simple control with reduced calculations has been proposed to interface SPV to the grid. The proposed control algorithm had been found fast, accurate, and easy for real-time implementation. A simple control has selected as VSC in the proposed topology which serves not only the functions of power quality improvement at the grid but also neutral current compensation and interfacing SPV with the grid. The proposed interfacing SPV array in the distribution system reduces losses in the power system, and multi-functionality of VSC improves the utilization factor of VSC. The simplicity of frequency shifter algorithm makes feasible and easy implementation of proposed multifunctional VSC. The performance of proposed control has been demonstrated under different loading and weather conditions. Both dynamic and steady-state performances have been found satisfactory. The THD of grid current meets IEEE and IEC standards.

Acknowledgments Authors are very thankful to Department of Science and Technology (DST), Govt. of India, for funding this project under Grant Number RP02583.

Appendix

SPV Data: panel short-circuit current (I_{scn}) = 8.2 A, panel open-circuit voltage (V_{ocn}) = 32.8 V, panel current at MPP (I_{mpp} at 1,000 W/m^2) = 7.59 A, panel voltage at MPP (V_{mpp} at 1,000 W/m^2) = 27.89 V, voltage temperature coefficient (K_v) = −82e−3 V/K, current temperature coefficient (K_i) = 0.0031 A/K, number of series cell in each panel = 54, number of panels in series = 21, and number of panels in parallel = 6. Supply system parameters: supply voltage rms line to line 415 V, frequency = 50 Hz, grid source inductance = 3 mH/phase, grid source resistance = 0.312 Ω/phase, ripple filter $R = 5\ \Omega$, $C = 5\ \mu F$, $K_{ploss} = 1$, $K_{iloss} = 0.1$.

References

1. Wang Z, Fan S, Zheng Y, Cheng M (2012) Control of a six-switch inverter based single-phase grid-connected PV generation system with inverse park transform PLL. In: Proceedings of international symposium on industrial electronics (ISIE), pp 258–263
2. Chen Y, Wen G, Peng L, Kang Y, Chen J (2013) A family of cost-efficient non-isolated single-inductor three-port converters for low power stand-alone renewable power applications. In: Proceedings of twenty-eighth annual IEEE applied power electronics conference and exposition (APEC), pp 1083–1088
3. Chen Y, Smedley KM (2004) A cost-effective single-stage inverter with maximum power point tracking. IEEE Trans Power Electron 19(5):1289–1294
4. Libo W, Zhengming Z, Jianzheng L (2007) A single-stage three-phase grid-connected photovoltaic system with modified MPPT method and reactive power compensation. IEEE Trans Energy Convers 22(4):881–886
5. Kashif MF, Choi S, Park Y, Sul SK (2012) Maximum power point tracking for single stage grid-connected PV system under partial shading conditions. In: Proceedings of 7th international power electronics and motion control conference (IPEMC), vol. 2, pp 1377–1383

6. Koutroulis E, Blaabjerg F (2013) Design optimization of transformerless grid-connected PV inverters including reliability. IEEE Trans Power Electron 28(1):325–335
7. Shimizu T, Suzuki S (2011) Control of a high-efficiency PV inverter with power decoupling function. In: Proceedings of 8th international conference on power electronics and ECCE Asia (ICPE & ECCE), pp 1533–1539
8. Liang Z, Guo R, Li J, Huang A (2011) A high-efficiency PV module-integrated DC/DC converter for PV energy harvest in FREEDM systems. IEEE Trans Power Electron 26 (3):897–909
9. Gu Y, Li W, Zhao Y, Yang B, Li C, He X (2013) Transformerless inverter with virtual DC bus concept for cost-effective grid-connected PV power systems. IEEE Trans Power Electron 28 (2):793–805
10. Hendriks JW, Fransen HPW, Van Zolingen RJC (1995) Reliable cost effective photovoltaic (PV) systems system approach with one-source responsibility. In: Proceedings of 17th international conference on telecommunications energy, pp 755–757
11. Emadi A, Nasiri A, Bekiarov SB (2005) Uninterruptible power supplies and active filters. CRC Press, New York
12. Yatak MO, Bay OF (2011) Fuzzy control of a grid connected three phase two stage photovoltaic system. In: Proceedings of international conference on power engineering, energy and electrical drives (POWERENG), pp 1–6
13. Singh B, Solanki J (2009) A comparison of control algorithms for DSTATCOM. IEEE Trans Industr Electron 56(7):2738–2745
14. Kumar P, Mahajan A (2009) Soft computing techniques for the control of an active power filter. IEEE Trans Power Delivery 24(1):452–461
15. Han Y, Xu L, Khan MM, Chen C, Yao G, Zhou LD (2011) Robust deadbeat control scheme for a hybrid APF with resetting filter and ADALINE-based harmonic estimation algorithm. IEEE Trans Industr Electron 58(9):3893–3904
16. Hammoudi MY, Allag A, Mimoune SM, Ayad M-Y, Becherif M, Miraoui A (2006) Adaptive nonlinear control applied to a three phase shunt active power filter. In: Proceedings of IEEE international conference on industrial technology, pp 762–767
17. Lam CS, Choi WH, Wong MC, Han YD (2012) Adaptive DC-link voltage controlled hybrid active power filters for reactive power compensation. IEEE Trans Power Electron 27 (4):1758–1772
18. Negi A, Surendhar S, Kumar SR, Raja P (2012) Assessment and comparison of different neutral current compensation techniques in three-phase four-wire distribution system. In: Proceedings of 3rd IEEE international symposium on power electronics for distributed generation systems, pp 423–430
19. Singh B, Jayaprakash P, Kothari DP (2010) Magnetics for neutral current compensation in three-phase four-wire distribution system. In: Proceedings of joint international conference on power electronics, drives and energy systems (PEDES) and power India, pp 1–7
20. Srikanthan S, Mishra MK (2010) Modeling of a four-leg inverter based DSTATCOM for load compensation. In: Proceedings of international conference on power system technology (POWERCON) pp 1–6
21. IEEE Recommended Practices and Requirement for Harmonic Control on Electric Power System, IEEE Std. 519 (1992)
22. Limits For Harmonic Current Emissions, International Electrotechnical Commission IEC-61000-3-2 (2000)

Chapter 32
Development of Green Manufacturing System in Indian Apparel Industry

Ankur Saxena and A.K. Khare

Abstract Green manufacturing defines the reduction of hazardous substances in the design, manufacture and application of products or processes that can affect the environment and causes the concern towards global warming. It refers to a wide area including but not limited to, air, water and land pollution, energy usage and efficiency, waste generation and recycling. It is known that carbon footprint, which is a measure of production of greenhouse gases (CO_2, CH_4 and N_2O), prominently affect the global warming. Hence, it is important to reduce the carbon footprint of an industry to preserve the environment. Being a significant contributor in Indian industrial production and export, textile and apparel industry is an important engine for the nation. Due to the alarming situation of global warming, in recent years, both researchers and practitioners have devoted attention towards the impact of garment and textile industry on environment. Despite the significant relevance of the subject, a structured analysis of the problem is missing even though there is significant research work going on for reducing carbon footprint in different manufacturing industries, but not much work has been reported for garment and apparel industry. India is a significant and large emitter of greenhouse gases and most of it because of the industrial production; hence, there is a certain need to reduce these emissions, Therefore, present research work is focused to reduce the emissions of greenhouse gases in garment and apparel industry. Through this research, the researcher is investigating the level of awareness towards green manufacturing particularly of carbon footprint in garment manufacturers/decision-makers. A detailed system/instrument which will calculate CFP per garment and covers all possible aspects and concept of green manufacturing has been developed and used to attain this objective.

Keywords Textile industry · Environment · Apparel industry · CFP

A. Saxena (✉) · A.K. Khare
National Institute of Fashion Technology, Jodhpur, India
e-mail: ankur.saxena@nift.ac.in

V. Vijay et al. (eds.), *Systems Thinking Approach for Social Problems*,
Lecture Notes in Electrical Engineering 327,
DOI 10.1007/978-81-322-2141-8_32

32.1 Introduction

In recent years, public concern about climate change has grown significantly. Emissions of greenhouse gases such as carbon dioxide, methane and nitrous oxide from industrial activities have long been known to be the major contributors to global warming. This trend has led to significant interest in the increased use of energy technologies with inherently low-carbon footprints (e.g. renewable energy sources such as wind, solar or biomass) as well as in retrofitting of existing ones (e.g. via carbon capture and storage) to reduce greenhouse gas emissions. At the same time, there has been increased research on the development of new modelling techniques to analyse and simulate the effects of these technologies on carbon emissions, and furthermore to optimise the deployment of appropriate technologies in order to meet environmental goals while simultaneously considering technical and economic constraints [1, 2].

The industrial sector currently accounts for about one-half of the world's total energy consumption, and the consumption of energy by the sector has almost doubled over the last 60 years. Furthermore, industrial energy consumption is expected to increase 40 % from 175 quadrillion Btu in 2006 to 246 quadrillion Btu in 2030 [3].

There are good and significant research work happening for reducing carbon footprint in different manufacturing and service industries; for example, Cagiao et al. calculated the carbon footprint in cement industry [4], but very less research work has been reported in apparel industry.

At the present time, the most important factor to consider when deciding on alternative engineering is still the economic issue. However, the increasing social awareness in the fight against climate change and the need to achieve more sustainable commerce than what is currently being practiced will make the labelling of companies, products, public and private institutions, domestic consumption, etc., a key element in the not-too-distant future [4].

32.2 Indian Textile and Apparel Industry

The Indian textile sector has its roots going back several 1,000 years. After the industrial revolution in Europe, this sector in India also saw growth of an industrial complex. However, over the last 50 years, the textile industry in India has shown a chequered performance.

GMF plays a very important role not only in the country's economy but also in the lives of millions of people of the country. Along with textiles, the apparel industry contributes almost one-third of the country's exports. Traditionally, the industry has been divided into two sectors, exports and domestic. According to the annual report of ministry of textiles in the year 2012, the Indian textiles industry has an overwhelming presence in the economic life of the country. Apart from providing one of the basic necessities of life, the textiles industry also plays a pivotal role through its contribution to industrial output, employment generation and the export earnings of the country. Currently, it contributes about 14 % to industrial production, 4 % to the GDP and 17 % to the country's export earnings. It provides direct employment to over 35 million people, which includes a substantial number of SC/ST, and women (MOT, annual report, 2012).

Figures 32.1 and 32.2 show the Indian textile market share and Indian textile market size, respectively, for the year 2011, while Table 32.1 shows the value chain and capacities of different segments of the textile sector.

The textiles sector is the second largest provider of employment after agriculture. Thus, the growth and all-round development of this industry has a direct bearing on the improvement of the economy of the nation; India's textiles and clothing industry is one of the mainstays of the national economy. It is also one of the largest contributing sectors of India's exports worldwide. The report of the Working Group constituted by the Planning Commission on boosting India's manufacturing export during 12th Five Year Plan (2012–2017), envisages India's exports of Textiles and Clothing at US$32.35 billion by the end of 11th Five Year Plan. While the report of Working Group on Textiles for the 11th Five Year Plan envisages for US$55 billion for the same. Based on historic growth rate of 10 % (CAGR), it was expected to result in exports of US$52 billion by the end of 11th Plan (MOT, annual report, 2012).

Fig. 32.1 Indian textile market share—2011

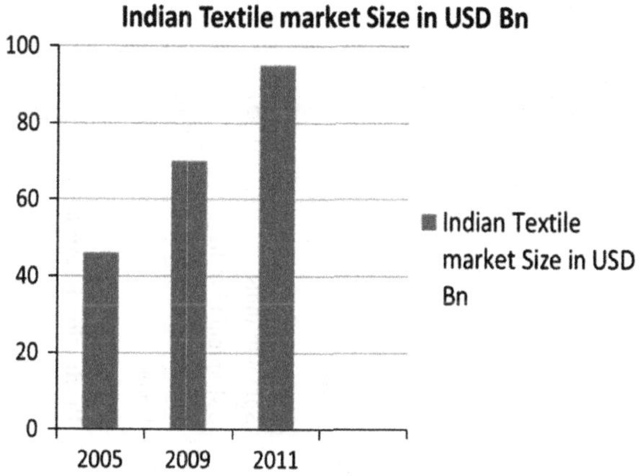

Fig. 32.2 Indian textile market size—2011

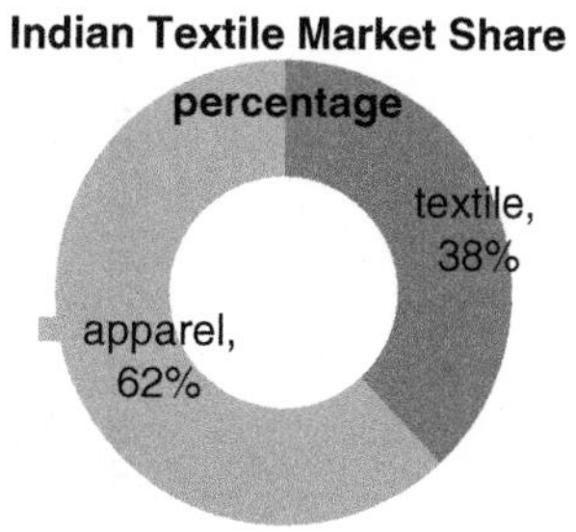

Table 32.1 Value chain and capacities

Fibre		Yarn		Fabric		Garment	
Cotton		**Capacity**		**Capacity**		**Capacity**	
Area	11 million ha	Spinning mills	1,757	Composite mills	183	Sewing machines	3 million
Production	5,780 million kg	SSI spinning mills	1,333	Weaving mills	174	Units	75,000
Man-made fibres (in mn kg)		Short staple spindles	48 mn	Handlooms	2.4 mn	**Production**	
Capacity	1,765	Rotor	0.77 mn	Power looms	2.3 mn	Pieces	8 bn
Production	1,285	Long staple spindle	1 mn	Looms	66,000		
Others (Production in mn kg)		**Production in mn kg**		**Production in mn m^2**			
Wool	45	Spun yarn	4,713	Woven fabric	47,084		
Silk	20			Knitted fabric	14,646		
Jute	1,494						

An export target of US$65 billion and creation of 25 million additional jobs has been proposed with a CAGR of 15 % during the XII Plan. At current prices, the Indian textiles industry is pegged at US$55.

32.3 Industry Perspective Towards Green

In a survey conducted by BCG and MIT in 2009, around 1,560 companies were interviewed. Executive of nearly all the companies interviewed said that sustainability-related issues have or will soon have a material impact on their business.

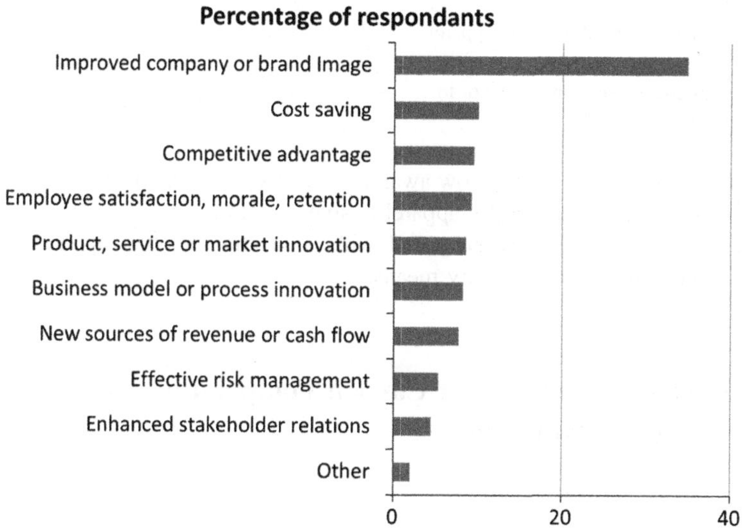

32.4 Methodology

To attain the desired objectives, methodology of this work can be broadly divided into two parts. In part one, survey has been conducted with in all the garment and apparel manufacturers in Delhi NCR and tries to find out the awareness level of green manufacturing in the industry.

In part two, carbon footprint has been calculated for a particular garment by calculating all possible energy inputs and energy is converted into carbon footprint.

32.5 Part 1: Survey for Awareness

A survey was conducted to understand the awareness level of green manufacturing from 37 professional of garment and apparel industry, and observation was as follows:

Questions asked	No. of participants answered "Yes"	No. of participants answered "No"
Are you aware of kyoto protocol or UNFCC?	7	30
Are you aware of carbon credit/CDM?	5	32
Any idea about carbon footprint/ emissions?	6	31
Can you relate any of above with apparel industry?	2	35
Can carbon credit help apparel sector to gain anything positive?	–	–

And the results proved very low awareness levels in middle management people who are associated with NCR's apparel business sector.

Everyone was clueless when posed with a question relating apparel sector's benefit from carbon credit by any means.

32.6 Part 2: Calculation of Carbon Footprint of Any Particular Garment

Calculation of total emission

Head	Electricity consumption (kW/shift)	Carbon factor in case of cole	Carbon di oxide emitted
General electricity consumption	2,116.316	2	1,923.924 kg
Electricity consumption by machinery	308.96	2	280.8727

As per the above calculation, total carbon di oxide emitted per day is 2,204.7 kg/day for 2,000 shirts, which means one shirt will be responsible for around 1.1 kg Carbon di oxide emission

General lighting, cooling and peripheral consumption of power

Department	Area	Power consumed by tubelights (watts)	Power consumed by fans (watts)	Power consumed by exhausts (watts)	Power consumed by Acs	Others (number per dept.)	Power consumed by others	Total power consumed
Reception	25	560	135	0	3,000	1	50	3,745
Meeting rooms	50	840	270	0	4,500	1	50	5,660
Fabric store	200	3,360	0	22,500	0	1	50	25,910
Trims	30	504	162	0	0	1	50	716
Testing laboratory	40	2,240	216	4,500	3,000		5,000	14,956
Cutting	150	5,880	360	16,875	0	1	50	23,165
Sewing	350	13,720	1,890	39,375	0	2	100	55,085
Finishing	150	5,880	240	16,875	0	1	50	23,045
Packing	50	1,120	270	5,625	0	2	100	7,115
Storage	300	5,040	0	33,750	0	1	50	38,840
Maintenance	40	1,120	216	0	0	0	0	1,336
Boiler	25	420	0	2,812.5	0	0	0	3,232.5
Water and sanitation	50	840	0	5,625	0	0	0	6,465
Merchandising office	20	448	1,080	0	3,000	6	300	4,828
Other offices	45	756	2,430	0	7,500	8	400	11,086
Showroom	30	3,360	1,620	0	3,000	1	50	8,030
Security	30	504	1,620	0	0	1	50	2,174
Electrical/ Generator	25	420	1,350	0	0	0	0	1,770
Canteen	70	1,176	3,780	7,875	0	0	0	12,831

(continued)

(continued)

Department	Area	Power consumed by tubelights (watts)	Power consumed by fans (watts)	Power consumed by exhausts (watts)	Power consumed by Acs	Others (number per dept.)	Power consumed by others	Total power consumed
Lobbies/Aisles	200	3,360	0	0	0	0	0	3,360
Others	100	2,240	0	0	4,500	5	250	6,990
Street	500	4,200	0	0	0	0	0	4,200
Total	2,480	57,988	15,639	155,812.5	28,500		6,600	264,539.5

Electricity consumed in lighting, cooling and other peripherals is 264,539.5 W/h or 2,116,316 W/shift

Power consumption by machines

S. no.	Machine	Wattage	Utilisation (%)	No. of machines	Power consumption/h/m/c	Power consumption/shift/m/c	Total power consumption/shift/all m/cs
Cutting							
1	Straight knife	1,200	50	2	600	4,800	9,600
2	Machine X				0	0	0
3	Machine Y				0	0	0
	Total			2	600	4,800	9,600
Sewing							
1	Single needle	650	30	68	195	1,560	106,080
2	Double needle	650	30	12	195	1,560	18,720
3	Overlock	750	30	12	225	1,800	21,600
4	Flat lock	650	30	8	195	1,560	12,480
5	Feed of the arm	650	30		195	1,560	0
6	Bartack	320	30		96	768	0
7	Button hole	300	30	2	90	720	1,440
8	Button attaching	300	30	2	90	720	1,440
9	Machine X				0	0	0
10	Machine Y				0	0	0
	Total			104	1,281	10,248	161,760
Finishing							
1	Spot cleaning	2,000	20	2	400	3,200	6,400
2	Needle detector	400	50	1	200	1,600	1,600
3	VIET	1,000	95	10	950	7,600	76,000
4	Steam iron	600	95	10	570	4,560	45,600
5	Boiler	1,000	100	1	1,000	8,000	8,000
6	Machine X				0	0	0
7	Machine Y				0	0	0
	Total			24	3,120	24,960	137,600

32.7 Conclusion

Considering only the activities involved at manufacturers end, 3.365 kg of carbon is released per T-shirt, and carbon emission of the factory is 6,729.8 kg/shift of 8 h. If CER certificates are traded at 20 Euros each then price of one CER will be INR 1,200, if exchange rate of Euro is considered as INR 60. Considering above saving of 1 kg of carbon emission can lead to earning of INR 12. Hence is can be concluded that T-shirt manufacturers have a potential of earning INR 40.38 per piece by following the green manufacturing system.

References

1. Tjan W, Tan RR, Foo DCY (2009) A graphical representation of carbon foot print reduction for chemical processes. J Cleaner Prod: 848–856
2. Díaz E, Fernández J, Ordóñeza S, Canto N, González A (2012) Carbon and ecological foot prints as tools for evaluating the environmental impact of coal mine ventilation air. Ecol Indic 18:126–130
3. Fang K, Uhan N, Zhao F, Sutherland JW (2011) A new approach to scheduling in manufacturing for power consumption and carbon foot print reduction. J Manuf Syst 30:234–240
4. Cagiao J, Gómez B, Doménech JL, Mainar SG, Lanza HG (2011) Calculation of the corporate carbon foot print of the cement industry by the application of MC3 methodology. Ecol Ind 6:1526–1540

Chapter 33
Significant Patterns Extraction to Find Most Effective Treatment for Oral Cancer Using Data Mining

Neha Sharma and Hari Om

Abstract Development of cancer in oral mucosa as classified by the World Health Organization is a two-stage process that initially shows up as a premalignant, precancerous sore and that subsequently develops into the malignant cancerous stage. Early evaluation of oral precancerous lesions has a dramatic impact on oral cancer mortality rates as the medicine is very effective in early stage diagnosis. This paper aims at extracting the patterns that help finding the most effective course of oral cancer treatment and its post-treatment management. The Apriori algorithm is used to mine a set of significant rules for prevention of oral cancer by adopting the most efficient treatment. We attempt to find the association among various treatments, histopathology, follow-up symptoms, and follow-up examination. The experimental results show that all the generated rules hold the highest confidence level, thereby making them very useful for deciding effective treatment to cure oral cancer and its follow-up.

Keywords Data mining · Association rule mining · Apriori · Oral cancer · WEKA · Effective treatment · Cancer management

N. Sharma (✉)
Padmashree Dr. D.Y. Patil Institute of Master of Computer Applications, Akurdi, Pune 411044, Maharashtra, India
e-mail: nvsharma@rediffmail.com

N. Sharma
University of Pune, Pune, Maharashtra, India

H. Om
Computer Science and Engineering Department, Indian School of Mines, Dhanbad, Jharkhand, India
e-mail: hari.om.cse@ismdhanbad.ac.in

V. Vijay et al. (eds.), *Systems Thinking Approach for Social Problems*, Lecture Notes in Electrical Engineering 327,
DOI 10.1007/978-81-322-2141-8_33

33.1 Introduction

With the advancements in healthcare industries and health technologies in last few decades, the disease type has changed to some extent such that the chronic diseases have become the conventional diseases instead of the acute infective diseases. Malignancy or cancer is a most serious chronic disease that is hard to cure. The body function is damaged gradually, and different sorts of incapacities or even death are incurred. For successfully regulating the medicine course of malignancies and enhancing the quality of life, the development of an effective and straightforward malignancy prediction system is urgently needed. Recently, utilization of computerized system in healthcare services has been expanded incredibly [1]. The medical community is conscious that the health data that is being continuously collected should be useful in extracting distinctive knowledge [2]. The extensive amount of data unequivocally surpasses the capacities of people for productive utilization without specific instruments for examination [1, 3, 4]. The situation is depicted as rich in data, but poor in information. With a specific end goal to fill this developing gap, the diverse approaches of data mining are applied. Data mining—knowledge discovery—is the science of extracting critical information from the large amount of existing raw data and deploying that information across the organization [1, 4, 5]. It is basically a process of discovering patterns and trends that go beyond straightforward examination and can answer inquiries that cannot be answered through basic question and reporting systems. But data mining does not work by itself. It uses sophisticated mathematical algorithms to segment the data and evaluate the probability of future events. Automatic discovery of patterns, prediction of likely outcomes, creation of actionable information, and focus on large datasets and databases are the key properties of data mining [6–8]. However, it does not dispense with the need to know the business, to comprehend the data or to comprehend analytical techniques. Data mining discovers hidden information in data, but cannot tell the value of the information to your organization. Important patterns might already be known as a result of acquaintance to business domain and working with data over time. Notwithstanding, data mining can confirm or qualify such empirical observations in addition to finding new patterns that may not be immediately apparent through simple observation.

Oral cancer that is of noteworthy public health importance in India has been undertaken as a study. Public health authorities, private treatment centers, and scholastic medicinal centers in India have distinguished oral disease as a grave issue [9]. Conceivable signs and manifestations of oral malignancy on examining a patient may contain the following: a bump or thickening in the oral delicate tissues, soreness or a feeling caught in the throat, difficulty in biting or swallowing, ear torment, trouble in moving the jaw or tongue, roughness, deadness of the tongue or other areas of the mouth, or swelling of the jaw that causes dentures to fit inadequately or get uncomfortable. The aforementioned data must be utilized proficiently for unanticipated identification and medication of oral cancer, as they are significant and can improve the survival rate appreciably and bring about an improved personal satisfaction for survivors. In this paper, we use the Apriori algorithm to extract a set of noteworthy rules pertaining to most effective course of treatments such as LFT, FNAC of neck node, surgery, radiotherapy, chemotherapy, and survivability of the oral cancer patients.

This paper is organized as follows. Section 33.2 reviews various related approaches in literature. Section 33.3 provides the information about oral cancer, and Sect. 33.4 gives brief description about association rule mining. Section 33.5 presents the experimental results, and Sect. 33.6 concludes the paper with future directions.

33.2 Literature Review

Singh et al. [10] apply the Apriori algorithm with transaction reduction on cancer symptoms. They consider five different types of cancer and try to find out the symptoms helping spreading the cancer and faster spreading type of cancer. Many authors in their research papers presented the applications of mining frequent item sets and association rules from the viewpoint of the user's interaction with the system [11–15]. RuthRamya et al. [16] apply association rules to improve the accuracy in diagnosis and obtain the valuable rules and information in case of chest pain. Their proposed algorithm also synchronizes the rule generation and classifier building phases, shrinking the rule mining space for building a classifier to speed up rule generation. Swami et al. [17] discuss the multidimensional association rule generation and its model of smoking habits in order to take some preventive measures to reduce the various habits of smoking in youths. Ha and Joo [18] discuss a hybrid method by combining association rules and classification trees to make fast and accurate classification of chest pain disease by a physician. Srikant et al. [19] consider the problem of integrating constraints in the form of Boolean expression that appoints the presence or absence of items in rules.

Nahar et al. [20] extract the significant prevention factors for a particular type of cancer by constructing a prevention factor dataset through an extensive literature review and using three association rule mining algorithms: Apriori, Predictive Apriori, and Tertius algorithms. They report experimentally that the Apriori algorithm is the most useful association rule mining algorithm for discovery of prevention factors. Milovic et al. [21] present the applicability of data mining in health care and explain how these patterns can be used by physicians to determine diagnoses and prognoses and apply treatments for patients. Anuradha and Sankaranarayanan [22] have done a detailed survey on various methods adopted by the researchers for oral cancer detection at an earlier stage by making a comparison among the various methods for identification and classification of cancers.

Kaladhar et al. [23] predict oral cancer survivability using classification algorithms that include CART, Random Forest, LMT, and Naïve Bayesian classification algorithms. These algorithms classify the cancer survival using tenfold cross-validation and training dataset. The Random Forest classification technique correctly classifies the cancer survival dataset and reduces the absolute relative error compared to other methods. Chuang et al. [24] consider DNA repair genes. They chose single nucleotide polymorphisms (SNPs) dataset with 238 samples of oral cancer and control patients for disease prediction. All prediction experiments were conducted using the support vector machine, and they reported that the performances of the holdout cross-validation were superior to tenfold cross-validation, and the best classification accuracy was 64.2 %. Gadewal and Zingde [25] compiled and enlarged the oral cancer gene database (DB) to include 374 genes by adding 132 gene entries to enable fast retrieval of updated information.

33.3 Oral Cancer

Oral cancer is a subtype of head and neck cancer that is any cancerous growth located in any subsites of the oral cavity [26]. The early detection of oral cancer depends upon a thorough oral cancer examination, usually by a dentist or other qualified healthcare provider, for possible signs and symptoms of this disease. Most cancers of the head and neck are squamous cell carcinomas. About one-third of oral cancers are diagnosed in the mouth cavity and a similar proportion on the tongue [27]. Lip cancer is the least frequent type of oral cancer. Oral cavity cancer includes tumors of the buccal mucosa, retromolar triangle, alveolus, hard palate, anterior two-thirds of the tongue, floor of the mouth, and mucosal surface of the lip [28].

Mouth cancer is more likely to affect people over 40 years of age, although an increasing number of young people are developing the condition—especially women [29]. Five decades ago, mouth cancer was five times more common in men than women; now, it is only twice as common, as more and more women are diagnosed [30]. In India, the incidence is about 50 % of all cancers (male incidence rates up to 6.5 per 100,000 per annum). In Europe, the incidence of oral cavity cancer is very low, less than 5 % of all cancers [30]. In France, (male incidence rates up to 8 per 100,000 per annum), it is the third most common cancer in males and the second most common cause of death from cancer [31]. Heavy smoking (or chewing tobacco products), heavy alcohol consumption, and poor dentition are the principle risk factors in Western countries [32]. The "tumor, nodes, and metastases" (TNM) staging system is used for staging head and neck cancers. T is the extent of the primary tumor; N is the involvement of regional lymph nodes; and M is the presence of metastases. Oral cancer treatment may include surgery, radiation therapy, chemotherapy, or a combination of treatments.

33.3.1 Surgery

Surgery, a common treatment for oral cancer, is used to remove the tumor in the mouth or throat. Sometimes, the surgeon also removes lymph nodes in the neck. Other tissues in the mouth and neck may be removed as well. Patients may have surgery alone or in combination with radiation therapy.

33.3.2 Radiation Therapy

Radiation therapy is a type of local therapy that affects the cells only in the treated area. It is used alone for small tumors or for the patients who cannot have surgery. It may be used before surgery to kill cancerous cells and shrink the tumor. It also may be used after surgery to destroy cancerous cells that may remain in the area.

33.3.3 Chemotherapy

Chemotherapy uses anticancer drugs to kill cancer cells. It is also called systemic therapy because it enters the bloodstream and can affect cancer cells throughout the body.

The depth of infiltration is predictive of prognosis [28]. Although there is evidence that a visual examination as part of a population-based screening program reduces the mortality rate of oral cancer in high-risk individuals, there is otherwise little evidence to support any screening program for early detection of oral cavity cancers [33].

33.4 Association Rule Mining

In data mining, association rules are useful for analyzing and predicting the future event. Association rules are if/then statements that help uncover relationships between apparently unnecessary data in a relational DB or other information repository [34–40]. An association rule has two parts: a forerunner or antecedent (if) and a subsequent or consequent (then). An antecedent is an item found in the data and a consequent is an item that is found in combination with the antecedent.

33.4.1 Apriori

The Apriori is a classic algorithm for frequent item set mining and association rule learning over the transactional databases [41]. It proceeds by identifying the frequent individual items in the DB and extending them to larger and larger item sets as long as those item sets appear sufficiently often in the DB. The frequent item sets determined by a Apriori can be used to determine association rules, which highlight general trends in the DB [42]. Association rules mining using Apriori algorithm uses a “bottom-up” approach, breadth-first search, and a hash tree structure to count the candidate item sets efficiently. A two-step Apriori algorithm is explained with the help of flowchart as shown in Fig. 33.1, and the algorithm is mentioned below.

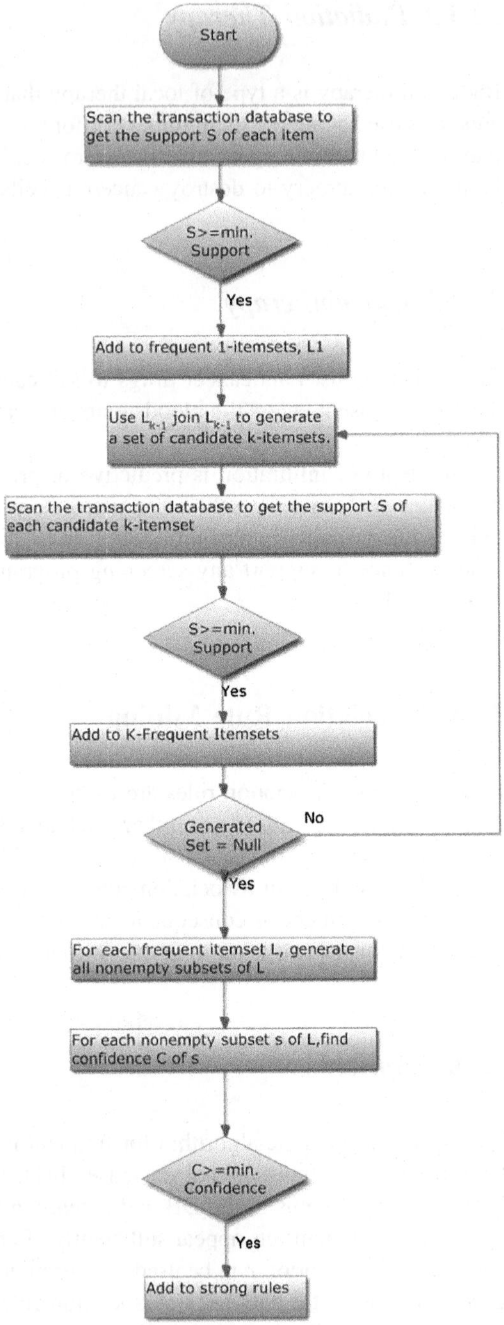

Fig. 33.1 Flowchart of Apriori algorithm

33.4.2 *Apriori Algorithm: Candidate Generation—Test Approach*

Step 1: Initially scan DB once to get frequent 1-itemset
Step 2: Generate length $(k + 1)$ candidate itemsets from length k frequent itemsets
Step 3: Test the candidates against DB
Step 4: Terminate when no frequent or candidate set can be generated

33.4.3 *Interestingness Criteria*

To select interesting rules from the set of all possible rules generated, constraints on various measures of significance and interest can be used. The best-known constraints are minimum thresholds on support and confidence.

Support
The rule holds with support supp in T (the transaction dataset) if supp % of transactions contain $X \cup Y$ [34].

$$\text{Supp}(X \to Y) = P(X \cup Y).$$

Confidence
The rule holds with confidence conf in T if conf % of transactions that contain X also contain Y [34, 43].

$$\begin{aligned}\text{Conf }(X \to Y) &= P(Y|X) = \text{Supp}(X \cup Y)/\text{Supp}(X)\\ &= P(X \text{ and } Y)/P(X)\end{aligned}$$

Lift
It is the probability of the observed support to that expected, if X and Y were independent [44].

$$\begin{aligned}\text{Lift}(X \to Y) &= \text{Supp}(X \cup Y)/\text{Supp}(X) \times \text{Supp}(Y)\\ &= P(X \text{ and } Y)/P(X)P(Y)\end{aligned}$$

Leverage
It measures the difference of X and Y appearing together in the dataset and what would be expected if X and Y were statistically dependent [45].

$$\mathrm{Lev}(X \rightarrow Y) = P(X \text{ and } Y) - (P(X)P(Y))$$

Conviction

It is the probability of the expected frequency that X occurs without Y (that is to say, the frequency that the rule makes an incorrect prediction) [46].

$$\begin{aligned}\mathrm{Conv}(X \rightarrow Y) &= 1 - \ \mathrm{Supp}(Y)/1 - \mathrm{conf}(X \rightarrow Y) \\ &= P(X)\,P(\text{not } Y)/P(X \text{ and not } Y)\end{aligned}$$

33.5 Experimental Results

The DB for this work is created by collecting data through a retrospective chart review and the entire process is presented in [47]. There are a total of 33 variables, and 1,025 records of patients were created for the analysis. A data mining tool—WEKA 3.7.9 [48]—has been used to explore the behavior of the Apriori algorithm for finding the association rules toward most effective course of treatments. The oral cancer data are initially stored in MS Excel sheet and then converted into comma-separated values (.csv file) and subsequently to attribute relation file format (.arff file), which is the acceptable format to WEKA tool. Minimum support defined by the tool for the generated rule is 0.1 (103 instances), and minimum confidence is 0.9. Association rules for most effective course of treatment and survivability of the oral cancer patients are mentioned below, and the same is presented in the graphical form in Fig. 33.2.

Rule 1 Survival = alive 449 ⇒ Surgery = Y 449
⟨conf:(1)⟩ lift:(1) lev:(0) [0] conv:(0)

Rule 2 Radiotherapy = N 449 ⇒ Survival = alive 449
⟨conf(1)⟩ lift:(2.28) lev:(0.25) [252] conv:(252.32)

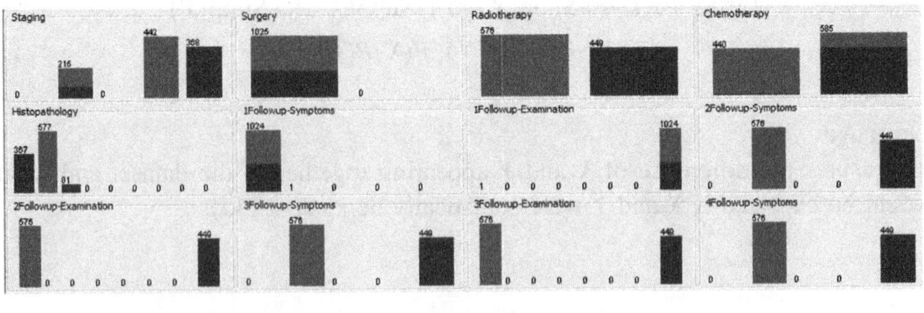

Fig. 33.2 Effective course of treatment and cancer management

Rule 3 Survival= alive 449 ⇒ Chemotherapy = N 444
⟨conf(0.99)⟩ lift:(1.73) lev:(0.18) [187] conv:(32.12)

Rule 4 Diagnosis = SCC 576 ⇒ Radiotherapy = Y 576
⟨conf(1)⟩ lift:(1.78) lev:(0.25) [252] conv:(252.32)

Rule 5 Radiotherapy = Y and Chemotherapy = Y 435 ⇒ Diagnosis = SCC 435
⟨conf(1)⟩ lift:(1.78) lev:(0.19) [190] conv:(190.55)

Rule 6 Radiotherapy = N 449 ⇒ Chemotherapy = N 444
⟨conf(0.99)⟩ lift:(1.73) lev:(0.18) [187] conv:(32.12)

Rule 7 Radiotherapy = Y and Chemotherapy = Y 435 ⇒ Staging = IV 434
⟨conf(1)⟩ lift:(2.31) lev:(0.24) [246] conv:(123.71)

Rule 8 Histopathology = SCC 577 ⇒ Survival = Dead 576
⟨conf(1)⟩ lift:(1.78) lev:(0.25) [251] conv:(126.38)

Rule 9 Survival = alive 449 ⇒ 1Followup-Symptoms = NAD. And 1Followup-Examination = NAD 449
⟨conf(1)⟩ lift:(1) lev:(0) [0] conv:(0.44)

Rule 10 2Followup-Symptoms = ulcer and 2Followup-Examination = Reccurrence 576 ⇒ Survival = Dead 576
⟨conf(1)⟩ lift:(1.78) lev:(0.25) [252] conv:(252.32)

Rule 11 3Followup-Symptoms = ulcer and 3Followup-Examination = Reccurrence 576 ⇒ Survival = Dead 576
⟨conf(1)⟩ lift:(1.78) lev:(0.25) [252] conv:(252.32)

Rule 12 4Followup-Symptoms = ulcer and 4Followup-Examination = Reccurrence 576 ⇒ Survival = Dead 576
⟨conf(1)⟩ lift:(1.78) lev:(0.25) [252] conv:(252.32)

Rule 13 5Followup-Symptoms = ulcer and 5Followup-Examination = Reccurrence 576 ⇒ Survival = Dead 576
⟨conf(1)⟩ lift:(1.78) lev:(0.25) [252] conv:(252.32)

The above rules are summarized and explained below:

A. *Rule Explanation* 1:
Radiotherapy is associated with an improvement in overall survival and loco-regional control in patients with oral cancers.
B. *Rule Explanation* 2:
Chemotherapy, in addition to radiotherapy and surgery, is associated with improved overall survival in patients with oral cancers.
C. *Rule Explanation* 3:
There are some evidences of radiotherapy and chemotherapy combined with surgery being more effective than radiotherapy alone with surgery and may benefit outcomes in patients with more advanced oral cancers.

D. *Rule Explanation* 4:
If histopathology shows SCC (squamous cell carcinoma), then it clearly indicates presence of oral cancer. This may lead to high mortality if appropriate treatment is not initiated.

E. *Rule Explanation* 5:
The disease sometimes returns because undetected cancer cells remain in the body after treatment, which may be confirmed by regular follow-up symptoms and examination. With reoccurrence of cancer in oral cavity, the chance of survival gets meager.

33.6 Conclusions

The Apriori, an association rule mining algorithm, is used to extract the most effective course of treatment depending on the stage of cancer and accordingly plan its future management. The experimental results demonstrate that all the generated rules hold the highest confidence level, thereby making them very useful for practitioners to choose the right treatment or combination of treatments. In future, we intend to use different data mining tools to extract the significant association rules for early detection and prevention of oral cancer and compare their results with that of WEKA.

Acknowledgments The authors would like to thank Dr. Vijay Sharma, MS, ENT, for his valuable contribution in understanding the occurrence and diagnosis of oral cancer. The authors devote their sincere thanks to the management and staff of Indian School of Mines for their constant support and motivation.

References

1. Han J, Kamber M, Pei J (2011) Data mining: concepts and techniques, 3rd edn. Morgan Kaufmann Publishers, Massachusetts. ISBN 978-0123814791
2. Shital CS, Andrew K, Michael A, Donnell O (2006) Patient-recognition data mining model for BCG-plus interferon immunotherapy bladder cancer treatment. Comput Biol Med 36:634–655
3. Hen LE (2008) Performance analysis of data mining tools cumulating with a proposed data mining middleware. J Comput Sci 4: 26
4. Fayyad UM, Piatetsky-Shapiro G, Smyth P (1996) From data mining to knowledge discovery in databases. Am Assoc Artif Intell (AAAI-AI Magazine), pp 37–54
5. Fayyad UM, Piatetsky-Shapiro G, Smyth P (1996) From data mining to knowledge discovery: an overview. Advances in knowledge discovery and data mining. AAAI Press/MIT Press, Cambridge, pp 1–36
6. Data Mining Curriculum. ACM SIGKDD. May 30, 2006
7. Clifton C, (2010) Encyclopedia britannica: definition of data mining
8. Hastie T, Tibshirani R, Friedman (2009) The elements of statistical learning: data mining, inference, and prediction

9. Coelho KR (2012) Challenges in oral cancer burden in India. J Cancer Epidemiol 701932:17
10. Singh S, Yadav M, Gupta H (2012) Finding the chances and prediction of cancer through Apriori algorithm with transaction reduction. Int J Adv Comput Res 2(2):23–28 ISSN (print): 2249-7277 ISSN (online): 2277-7970
11. Anh TN, Hai DV, Tin TC, Bac LH (2011) Efficient algorithms for mining frequent itemsets with constraint. In: Proceedings of the third international conference on knowledge and systems engineering
12. Bayardo RJ, Agrawal R, Gunopulos D (2000) Constraint-based rule mining in large, dense databases. Data Mining Knowl Disc 4(2–3):217–240 Kluwer Academic Publication
13. Cong G, Liu B, (2002) Speed-up iterative frequent itemset mining with constraint changes. ICDM. pp 107–114
14. Lee AJ, Lin WC, Wang CS (2006) Mining association rule with multi-dimensional constraints. J Syst Softw 79:79–92
15. Nguyen RT, Lakshman VS, Han J, Pang A (1998) Exploratory mining and pruning optimizations of constrained association rules. In: International conference on management of data, ACM-SIG-MOD pp 13–24
16. RuthRamya K et al (2012) A class based approach for medical classification of chest pain. Int J Eng Trends Technol 3(2):89–93
17. Swami S et al (2011) Multidimensional association rules extraction in smoking habit database. Int J Adv Net Appl 03(03):1176–1179
18. Ha SH, Joo SH (2010) A Hybrid data mining method for medical classification of chest pain. World Acad Sci Eng Technol 37:608–613
19. Srikant R, Vu Q Agrawal R (1997) Mining association rules with item constraints. In: Proceeding KDD97, pp 67–73
20. Nahar J, Kevin ST, Ali ABMS, Chen YP (2009) Significant cancer prevention factor extraction: an association rule discovery approach. J Med Syst. doi:10.1007/s10916-009-9372-8
21. Milovic B, Milovic M (2012) Prediction and decision making in health care using data mining. Int J Public Health Sci 01(02):69–78
22. Anuradha K, Sankaranarayanan K (2012) Identification of suspicious regions to detect Oral cancers at an earlier stage—a literature survey. Int J Adv Eng Technol 03(01):84–91
23. Kaladhar DSVGK, Chandana B, Kumar PB (2011) Predicting cancer survivability using classification algorithms. Int J Res Rev Comput Sci (IJRRCS) 02(02):340–343
24. Chuang LY, Wu KC, Chang HW, Yang CH (2011) Support vector machine-based prediction for Oral cancer using four SNPS in DNA repair genes. In: Proceedings of the international multi conference of engineers and computer scientists, 16–18 March, 2011
25. Gadewal NS, Zingde SM (2011) Database and interaction network of genes involved in oral cancer: version II. Bioinformation 06(04):169–170
26. Werning JW (2007) Oral cancer: diagnosis, management, and rehabilitation. 16 May 2007, p 1. ISBN 978-1588903099
27. Scully C, Bagan JV (2009) Recent advances in oral oncology 2008; squamous cell carcinoma imaging, treatment, prognostication and treatment outcomes. Oral Oncol 45(6):e25–e30 (Epub 26 Feb 2009)
28. SA Barbellido, Trapero JC, Sanchez CJ et al (2008) Gene therapy in the management of oral cancer: review of the literature. Med Oral Patol Oral Cir Bucal 13(1):E15–E21
29. Warnakulasuriya S (2009) Global epidemiology of oral and oropharyngeal cancer. Oral Oncol 45(4):309–316
30. www.cancerresearchuk.org/cancerinfo/cancerstats/types/oral/incidence/ukoral-cancer-incidence-statistics
31. Diagnosis and management of head and neck cancer, Scottish Intercollegiate Guidelines Network–SIGN, 2006
32. Shiboski CH, Schmidt BL, Jordan RC (2005) Tongue and tonsil carcinoma: increasing trends in the U.S. population ages 20–44 years. Cancer 103(9):1843–1849
33. Gosselin EJ, Meyers AD (eds) (2011) Malignant tumors of the mobile tongue. Medscape

34. Agrawal R, Imielinski T, Swami A (1993) Mining association rules between sets of items in large databases. In: Proceedings of the 1993 ACM SIGMOD international conference on management of data, pp 207–216
35. An J, Chen YPP, Chen H (2005) DDR: An Index method for large time series datasets. Inf Syst 30:333–348
36. Chen YPP, Chen F (2008) Targets for drug discovery using bioinformatics. Expert Opin Targets. 12(04):383–389
37. Lau RYK, Tang M, Wong O, Milliner SW, Chen YPP (2006) An evolutionary learning approach for adaptive negotiation agents. Int J Intell Syst 21(01):41–72
38. Ordonez C (2006) Association rule discovery with the train and test approach for heart disease prediction. IEEE Trans Inf Technol Biomed 10(02):334–343
39. Ordonez C, Omiecinski E (1999) Discovering association rules based on image content. In: IEEE advances in digital libraries conference (ADL'99), pp 38–49
40. Ordonez C, Santana CA, Braal L (2000) Discovering interesting association rules in medical data. ACM DMKD Workshop, pp 78–85
41. Agrawal R, Srikant R (1994) Fast algorithms for mining association rules in large databases. In: Proceedings of the 20th international conference on very large data bases, VLDB, pp 487–499
42. Zaki MJ (2004) Mining non-redundant association rules. Data Min Knowl Disc 09:223–248
43. Hipp J, Güntzer U, Nakhaeizadeh G (2000) Algorithms for association rule mining—a general survey and comparison. ACM SIGKDD Explorations Newsl 2:58. doi:10.1145/360402.360421
44. Brin S, Motwani R, Ullman JD, Tsur S (1997) Dynamic itemset counting and implication rules for market basket data. In: Proceedings of the ACM SIGMOD international conference on management of data (SIGMOD 1997), Tucson, Arizona, May 1997, 265–276
45. Piatetsky-Shapiro G (1991) Discovery, analysis, and presentation of strong rules. Knowledge Discovery in Databases, AAAI/MIT Press, Cambridge, pp 229–248
46. Brin S, Motwani R, Ullman JD, Tsur S (1997) Dynamic itemset counting and implication rules for market basket data. In: Proceedings of the ACM SIGMOD international conference on management of data (SIGMOD 1997), Tucson, Arizona, May 1997, pp 255–264
47. Sharma N, Om Hari (2012) Framework for early detection and prevention of oral cancer using data mining. Int. J Adv Eng Technol 4(2):302–310
48. Witten IH, Frank E (2005) Data Mining: practical machine learning tool and techniques, 2nd edn. Morgan Kaufmann Publishers, Elsevier, San Francisco

Chapter 34
Admission Control in Communication Networks Using Discrete Time Sliding Mode Control

Rutvij C. Joshi and Vishvjit K. Thakar

Abstract Discrete sliding mode control (DSMC) has been investigated since last three decades. It has reached to point wherein a well-established theory as well as implementation reports is available in the literature. Advances in computer communication network with wired and wireless configurations have imposed a challenging problem for control community. The paper explores the utilization of DSMC approach for the admission control (AC) in communication network. To provide better QoS, communication networks require proper admission, congestion traffic, and collision and power control mechanisms. For the past several years, many feedback control methods have been used for the control of various parameters of communication networks. The work presented here proposes a new application of sliding mode control for AC in communication network. Proper AC mechanism may avoid uncertain congestion and traffic access in the network which may reduce power wastage. As computer communication networks deal with digital data and are inherently discrete in nature, the DSMC has been applied to discrete time system. AC mechanism restricts the number of users in the network to maintain better QoS by considering available bandwidth, congestion indicator, and buffer size as feedback.

Keywords Admission control · SMC · DSMC

R.C. Joshi (✉)
SICART, Sardar Patel University, V.V. Nagar, Gujarat, India
e-mail: ec.rutvij.joshi@adit.ac.in

V.K. Thakar
ECE Department, A.D.I.T., New VV Nagar, Gujarat, India
e-mail: head.ec@adit.ac.in

V. Vijay et al. (eds.), *Systems Thinking Approach for Social Problems*,
Lecture Notes in Electrical Engineering 327,
DOI 10.1007/978-81-322-2141-8_34

34.1 Introduction

Day by day increase in network traffic demands energy-efficient algorithms to regulate resources of communication network. The ultimate goal of any communication system is to provide better QoS, and the QoS depends on flow and congestion control, admission control (AC), buffer management, fair scheduling, and power control.

34.1.1 Admission Controller

Admission controller is the first control step in controlled communication network. After admission, controller connection controller, traffic control, bandwidth controller, power controller, congestion controller, collision controller, and error controller play their roles in controlled communication network.

Main objective of any AC schemes is [1]

- Utilize all available resources of the network in optimized way
- Provide better QoS
- Maintain delay in the network (specifically for high-speed networks)
- Increase probability that a node will be admitted.

In wireless networks, lack of unavailability of enough radio resources is the major problem, and that is why, the efficient radio resource management is very essential. The AC is one of the radio resource management technique, and this plays dominant role in effectively managing the resources. The AC in the wireless networks will reduce the call-blocking probability in the wireless networks by optimizing the utilization of the available radio resources [2]. Traditionally, many AC protocols are available for homogeneous network, but heterogeneous network contains very few methods for managing admissions in the network. AC is used in many traditional telecom systems, especially in all connection-oriented communication systems, and all connections will be provided to the node after successfully passing through admission procedure. AC needs to be seen in the context of other necessary functions, especially performance measurements and control. The performance manager consists of the following functions: a performance meter that collects measurement data, AC that handles requests to join the network, and performance control that maintains the quality of service for the admitted sensor nodes. The performance meter provides feedback measurement data for AC and performance control. A request from a sensor node to join the network is handled by the AC based on feedback from the meter. The performance control function is responsible for maintaining the desired quality of service once the sensors are allowed to use the wireless channel. AC limits the traffic on the queuing system by determining if an incoming request for a new user can be met without disrupting the service guarantees to established data flows. Basically, when a new request is received, the AC or call admission control (CAC) in the case of an ATM network is executed to decide whether to accept or reject the request.

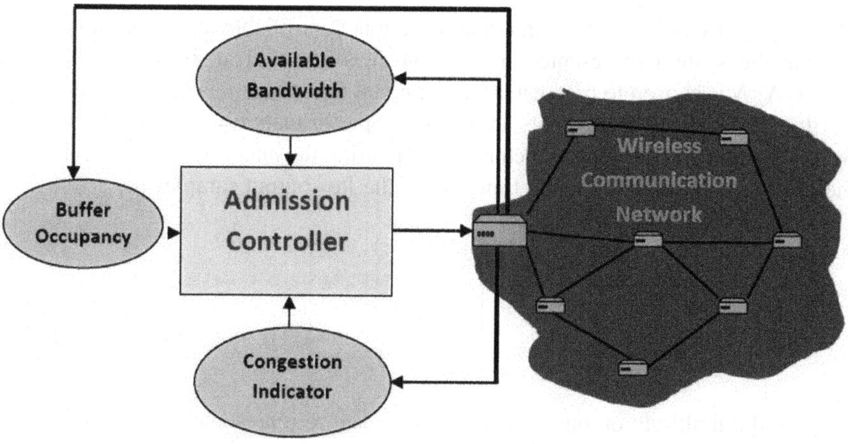

Fig. 34.1 Structure of admission controller

As shown in Fig. 34.1, AC systems are also feedback-based control system which contain bandwidth, congestion indicator, and buffer size as feedback parameters to select/reject new node in the network. In any type of communication network, better quality of service may be achieving if all available resources utilized in optimize way. AC algorithm has to monitor continuously network utilization by considering following parameters: MAC layer congestion, queue length, number of collisions, delay, and channel utilization time. The 1st three methods provide little or no information regarding network utilization if a node is not actively transmitting.

34.2 Multirate Output Feedback-Based Discrete Sliding Mode Control

In general, all control systems contain feedback, but multirate output feedback is somewhat different from them. As the name suggests, multirate output feedback is a feedback control scheme which quantizes the control input of the system at multiple sampling rates. Multirate output feedback-based control systems are more stable compare to static feedback control systems. The multirate output feedback, the control input, or sensor output are sampled at faster rate than other [3–5]. Consider the discrete system

$$\begin{aligned} x(k+1) &= \phi_\tau x(k) + \Gamma_\tau u(k) \\ y(k) &= Cx(k) \end{aligned} \tag{34.1}$$

where τ is the sampling period, x is N-dimensional vector which represents states of the variable to be control, u is a scalar input, and y is the output vector. Φ, τ, and C are the input state matrix, control input matrix, and output matrix, respectively.

For further discussion, we assume that system is controllable and observable. Now assume the system represented in Eq. (34.1) is sampled at the rate $1/\Delta$, where $\Delta = \tau/N$. N is chosen to be greater than or equal to υ.where υ is observability index of (Φ, C). Output measurements are taken at time instants $t = l\Delta$, where $l = 0, 1 \ldots N-1$, the control signal $u(t)$ which is applied during the interval $k\tau \leq t \leq (k+1)\tau$. For the discrete time system having time $t = k\tau$, the fast output samples [6]

$$y_k = \begin{bmatrix} y(k\tau - \tau) \\ y(k\tau - \tau + \Delta) \\ \vdots \\ y(k\tau - \Delta) \end{bmatrix} \quad (34.2)$$

Now the multirate output sampled system can be represented as

$$\begin{aligned} x(k+1) &= \phi_\tau x(k) + \Gamma_\tau u(k) \\ y_{k+1} &= C_0 x(k) + D_0 u(k) \end{aligned} \quad (34.3)$$

where

$$C_0 = \begin{bmatrix} C \\ C\phi \\ \vdots \\ C\phi^{N-1} \end{bmatrix} D_0 = \begin{bmatrix} 0 \\ C\Gamma \\ \vdots \\ C\sum_{j=0}^{N-2} \phi^j \Gamma \end{bmatrix} \quad (34.4)$$

In discrete sliding mode control (DSMC), the observation, analysis, and control operations are performed at regular interval of time, which is nothing but sampling time of the discrete time sliding mode control. Equation (34.3) represents discrete time systems where k is the index of sampling time. Fundamentally, in discrete time sliding mode control systems, control signal is computed at particular sampling instant. Discrete variable structure control may follow quasi-sliding mode motion. During such motion, the state of the system can approach the switching surface but cannot generally stay on it [7]. Figure 34.2 represents the generalized working of multirate output feedback-based discrete time sliding mode control.

34.2.1 Design of Switching Surface

According to the DVSC theory, all control laws must satisfy switching plane equation

$$s(k) = c^T x(k) = 0 \quad (34.5)$$

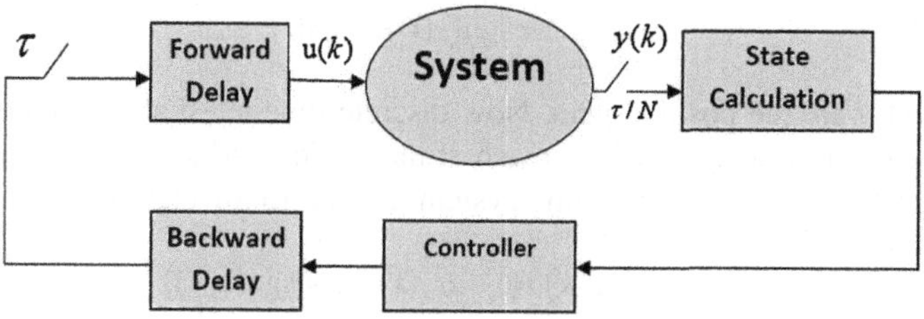

Fig. 34.2 Generalized block diagram of multirate output feedback-based discrete sliding mode controller

For the design of switching hyper plane, the system given in Eq. (34.1) is transformed into suitable form by appropriate transformation [8] as

$$\widehat{x}(k) = Mx(k)$$

The transformed system takes the form

$$\widehat{x}(k+1) = \overline{\Phi}_\tau \widehat{x}(k) + \overline{\Gamma}_\tau u(k) \tag{34.6}$$

where

$$\overline{\Phi}_\tau = \begin{bmatrix} \Phi_{11} & \Phi_{12} & \Phi_{13} \\ \Phi_{21} & \Phi_{22} & \Phi_{23} \\ \Phi_{31} & \Phi_{32} & \Phi_{33} \end{bmatrix}$$

$$\overline{\Gamma}_\tau^T = [0 \quad \Gamma_2 \quad 0]$$

$$\widehat{x}(k) = [\widehat{x}_1(k) \quad \widehat{x}_2(k) \quad \widehat{x}_3\,(k)]$$

The switching planes are

$$\begin{aligned} s(k) &= c^T M^{-1} \widehat{x}(k) = 0 \\ s(k) &= (\overline{c}^T 1)[\widehat{x}_1(k) \quad \widehat{x}_2(k) \quad \widehat{x}_3(k)]^T = 0 \end{aligned} \tag{34.7}$$

From Eqs. (34.6) and (34.7), we have

$$\widehat{x}(k+1) = \Phi_{11}\widehat{x}_1(k) + \Phi_{12}\widehat{x}_2(k) + \Phi_{13}\widehat{x}_3(k) \tag{34.8}$$

34.2.2 Reaching Law-Based Discrete Time Quasi-sliding Mode Control Law

The reaching law is a differential equation which specifies the switching function s. The reaching law for the continuous time VSC is

$$\dot{s}(t) = -\varepsilon\,\mathrm{sgn}(s(t)) - qs(t) \tag{34.9}$$

where ε and q are the positive gain. Now discrete time quasi sliding mode control employs reaching law approach to reach at the sliding surface. A reaching law for the sliding control of a discrete time system has the following form [9].

$$s(k+1) - s(k) = -q\tau s(k) - \varepsilon\tau\,\mathrm{sgn}(s(k)) \tag{34.10}$$

where $\tau > 0$ is the sampling period, $\varepsilon > 0$, $q > 0$, $1 - q\tau > 0$.

A desirable reaching mode response can be obtained by appropriate choice of parameters q and ε. The presence of signum term indicates switching across sliding surface. The trajectory will stay in specified band known as quasi-sliding mode band (QSMB). Switching function $s(k)$ is define as

$$s(k) = c^T x(k) = 0 \tag{34.12}$$

Using the reaching law in Eq. (34.10), the control law for the system represented in Eq. (34.1) may represent [6] as

$$u(k) = Fx(k) + \gamma\,\mathrm{sgn}(s(k)) \tag{34.13}$$

where

$$F = -(c^T\Gamma_\tau)^{-1}[c^T\Phi_\tau - c^T I + q\tau c^T]$$
$$\gamma = -(c^T\Gamma_\tau)^{-1}\varepsilon\tau$$

Generalized equation for switching surface and control may be represented [10] as

$$\begin{aligned} s(k) &= c^T\Phi_\tau C_0^{-1}y_k + c^T[\Gamma_\tau - \Phi_\tau C_0^{-1}D_0]u(k-1) \\ u(k) &= F\Phi_\tau C_0^{-1}y_k + F[\Gamma_\tau - \Phi_\tau C_0^{-1}D_0]u(k-1) + \gamma\,\mathrm{sgn}(s(k)) \end{aligned} \tag{34.14}$$

34.3 Proposed Admission Controller

As explain in earlier section, admission controller contains three inputs as feedback congestion indicator, available capacity (bandwidth), and buffer length. Congestion can be detected in the system by considering available link capacity and the buffer length.

34.3.1 Network Model

See Fig. 34.3.

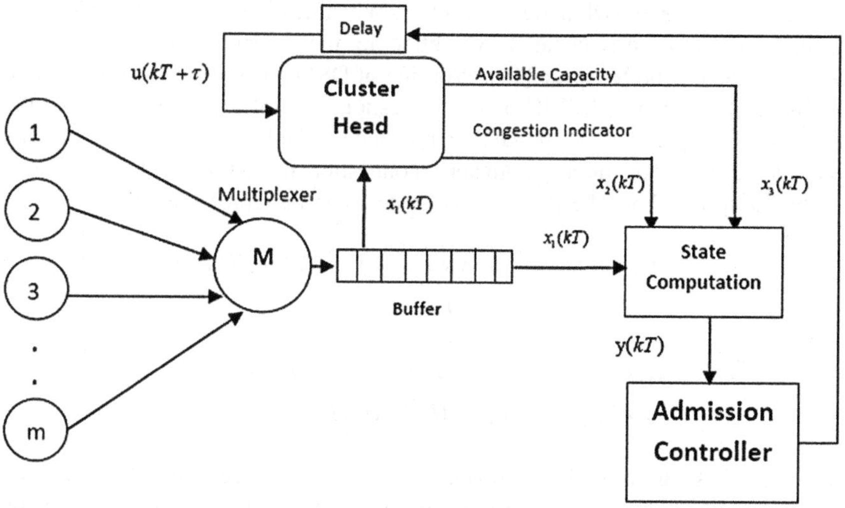

Fig. 34.3 Communication network model

where

1, 2 … m	m-number of nodes in the network
$y(kT)$	Output of the state computation
$u(kT)$	Control input
$x_1(kT)$	Current buffer occupancy
$x_2(kT)$	Current capacity of the network
$x_3(kT)$	Available buffer space

The available capacity is calculated by subtracting the current bandwidth usage from all existing sources from the maximum available bandwidth

$$\text{Available capacity} = S_{\max} - \sum_{i=1}^{m} \text{Bw}_i(kT) \tag{34.15}$$

where m is the number of existing sources in the network

$$\text{Bw}_i(k) = \frac{\text{In}_i(kT)}{\sum_{i=1}^{m} \text{In}_i(kT)} \text{Bw}(kT) \tag{34.16}$$

In Eq. (34.16), $\text{Bw}(k)$ is a bandwidth available at time instant kT, and $\text{In}_i(kT)$ is the current flow rate of source i

$$\text{Available Buffer space} = x_{\max} - \sum_{i=1}^{m} x_i(kT) \tag{34.17}$$

where $x_{\max}$ is the maximum buffer length, $x_i(kT)$ is the buffer space consumed by source i, and cluster head is responsible for continuously monitoring the network resources available at each node and routers. When new node is to be admitted, the available buffer space is updated by subtracting the ne node's buffer requirement from the available buffer space [10]. Working of DSMC-based admission controller is same as a working of DSMC-based congestion controller. Congestion controller depends on two parameters: source flow rate and available buffer length. Admission controller depends on three parameters: congestion indicator, available capacity, and buffer space. From the Eq. (34.15), system can be represented as

$$\begin{aligned} x[(k+1)T] &= \phi_\tau(x(kT)) + \Gamma_\tau u(kT) + d(kT) \\ y(kT) &= Cx(kT) \end{aligned} \tag{34.18}$$

$$\begin{aligned} x_1(kT) &= [\mathrm{sat}(x(k) - q(T_f - T_b) + I_{ni}^{(k)} - S_r^{(k)})] \\ x(kT) &= [x_1(kT), x_2(kT), x_3(kT)] \end{aligned}$$

where $x_1(kT)$ is current buffer occupancy, $x_2(kT)$ is current capacity of the network, $x_3(kT)$ is available buffer space, traffic arrival rate at destination buffer $I_{ni}^{(k)}$, bottleneck queue level $q(T_f - T_b)$ and service capacity $S_r^{(k)}$, τ is sampling time of the control signal, and system order n = (RTT/T) + 1, $x(kT) \in \Re^n$, ϕ_τ is $n \times n$ state matrix, Γ_τ and C are $n \times 1$ vectors.

$$\phi_\tau = \begin{bmatrix} \phi_{11} & \phi_{12} & \cdots & \phi_{1n} \\ \phi_{21} & \phi_{22} & \cdots & \phi_{2n} \\ \vdots & \vdots & & \vdots \\ \phi_{n1} & \phi_{n2} & \cdots & \phi_{nn} \end{bmatrix} = \begin{bmatrix} 1 & 1 & \cdots & 0 \\ 0 & 0 & \cdots & 0 \\ \vdots & \vdots & & \vdots \\ 0 & 0 & \cdots & 0 \end{bmatrix}$$

$$\Gamma_\tau = \begin{bmatrix} \Gamma_1 \\ \Gamma_2 \\ \vdots \\ \Gamma_n \end{bmatrix} = \begin{bmatrix} 0 \\ 0 \\ \vdots \\ 1 \end{bmatrix}, \quad C_\tau = \begin{bmatrix} C_1 \\ C_2 \\ \vdots \\ C_n \end{bmatrix} = \begin{bmatrix} 1 \\ 0 \\ \vdots \\ 0 \end{bmatrix}$$

34.4 Simulation Results

Sliding surface and control law can be calculated from Eq. (34.14); the simulation has been carried out with following parameters.

$$A = \begin{bmatrix} 0 & 0 & 3 \\ -3 & -8 & 0 \\ 0 & 2 & -1 \end{bmatrix}, \quad B = \begin{bmatrix} 0 \\ 2 \\ 0 \end{bmatrix}, \quad C = [1 \quad 0 \quad 0]$$

$$\tau = 0.3,\ N = 3,\ q = 2,\ \varepsilon = 0.01,\ X = \begin{bmatrix} 1 \\ 50 \\ 20 \end{bmatrix}$$

For simulation purpose, capacity of the system is set at 100, and the number of nodes in the networks is 50, so control input sends beacons to master node to channels to newly joined node. If network may have more than 100 nodes in the network, then control input generates different type of beacons to the master node, and master node would not allow newly joined node.

Figure 34.4 shows the variation in various state parameters where all states are gradually reach toward zero according to the control input, Fig. 34.5 shows small oscillation around zero after reaching into steady state which shows the quasi-sliding mode band, and Fig. 34.6 shows the variation in control law over a time index. Once system is controlled or reaches to steady state, then control signal

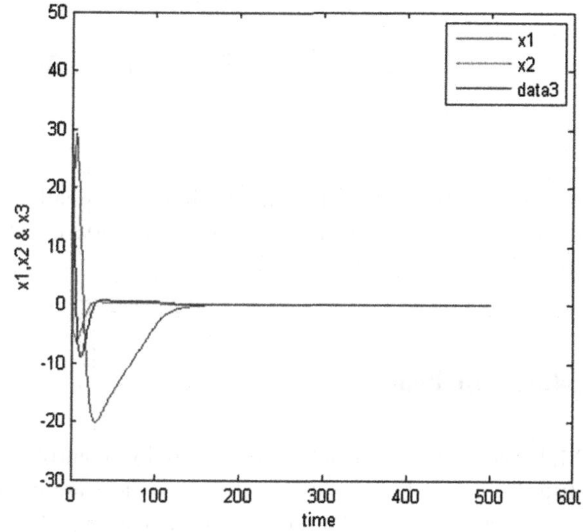

Fig. 34.4 State response of the system with reference to time index

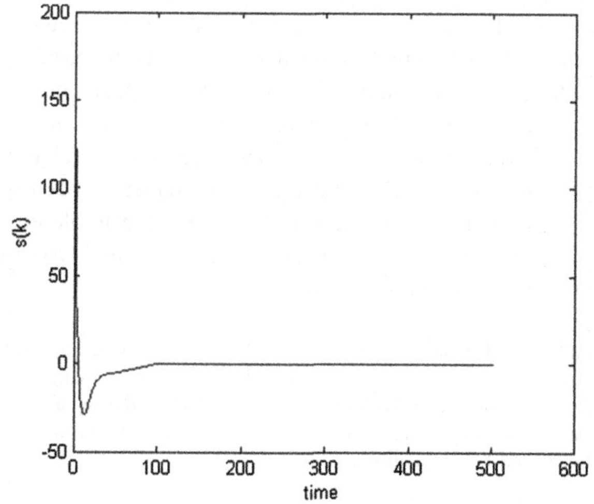

Fig. 34.5 Variation in sliding surface over a time index

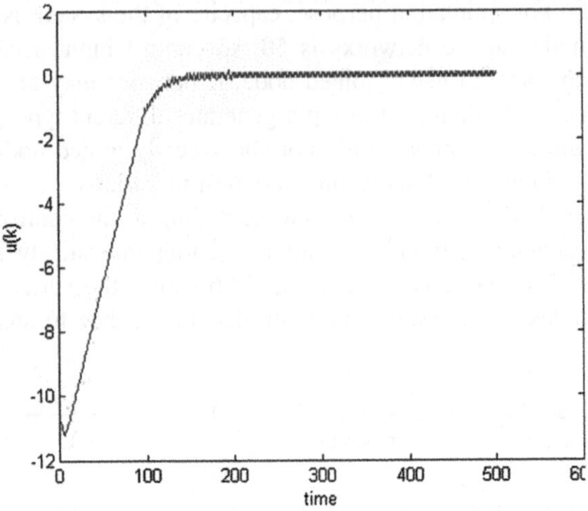

Fig. 34.6 Variation in control law with reference to time index

remains at zero level with some oscillations. Small oscillation is generated due to the switching function and may generate adverse effect if the oscillations have high frequency.

34.5 Conclusion

Multirate output feedback-based admission controller is less affected by external disturbances and traffic load conditions. There are number of methods available in the literature to restrict admission in the network. Multirate output feedback-based discrete sliding mode controller is more robust and simple. In this scheme, control law depends on congestion indicator, available bandwidth, and buffer space. Proper threshold value may be set for all these parameters. Discrete time sliding mode control-based admission controller restricts the number of node in the network in an optimal way. The admission controller proposed here allows almost all nodes those are willing to join the network if the resources are available.

The only problem in SMC-based controller is chattering effect generated because of switching law. Specifically, when chattering occurs at very high frequency, then it generated its harmonics and creates undesirable effects in control law.

As a future work, chattering elimination and QoS may be considered and robust controller can be designed.

Acknowledgments Author would like to acknowledge Sophisticated Instrumentation Centre for Applied Research and Testing (SICART), Vallabh Vidyanagar, Gujarat, and ECE Department, A. D. Patel Institute of Technology, new Vallabh Vidyanagar where this work has been carried out. Author would also like to thank reference no. 6 and 10 from which mathematical parameters have been used for simulation.

References

1. Jia W, Zhao W (1999) Efficient connection admission control algorithms for adaptive QoS real-time connections over ATM networks. Eur Trans Telecommun 10(3):135–151
2. Ramesh Babu HS, Shankar G, Satyanarayana PS (2009) Call admission control performance model for beyond 3G wireless networks. Int J Comput Sci Inf Secur 6(3):224–229
3. Chammas AB, Leondes CT (1979) Pole assignment by piecewise constant output feedback. Int J Control 29(1):31–38
4. Kabamba PT (1987) Control of linear systems using generalized samples and hold functions. IEEE Trans Autom Control 32(09):772–783
5. Hagiwara T, Arkai M (1988) Design of a stable state feedback controller based on multirate sampling of plant output. IEEE Trans Autom Control 33(9):812–819
6. Thakar VK (2006) Algorithm for discrete time sliding mode control using multirate output feedback. PhD thesis, Indian Institute of Technology, Mumbai
7. Milosavljevic C (1985) General conditions for the existence of a quasi sliding mode on the switching hyper plane in the discrete variable structure system. Autom Remote Control 46:307–314
8. Saaj CM, Bandyopadhyay B, Unbehauen H (2002) A new algorithm for discrete-time sliding mode control using fast output sampling feedback. IEEE Trans Ind Electron 49(3):518–523
9. Gao W, Wang Y, Homaifa A (1995) Discrete time variable structure control systems. IEEE Trans Ind Electron 42(02):117–122
10. Sarangapani J (2010) Wireless AdHoc and sensor networks: protocols, performances, and control. CRC Press, Taylor & Francis Group, New York (special edition, Reprint)
11. Ephremides A, Verdu S (1989) Control and optimization methods in communication network problems. IEEE Trans Autom Control 34(9):930–941
12. Utkin VI, Drakunov SV (1989) On discrete time sliding mode control. In: Preprints of IFAC conference on automatic control, vol 34, pp 1021–1022
13. Furuta K (1990) Sliding mode control of discrete systems. Syst Control Lett 14:145–152
14. Spurgeon SK (1991) Sliding mode control design for uncertain discrete time systems. In: Proceedings of conference on decision and control, UK, pp 2136–2141
15. Bartoszewicz A (1998) Discrete time sliding mode control strategies. IEEE Trans Ind Electron 45(1):633–637
16. Nilsson J, Bernhardsson B (2007) LQG control over a Markov communication network. In: Proceedings of the 36th conference on decision and control, San Diego, California, USA, pp 4586–4591
17. Tuenbaeva AN, Nazarov AA (2009) Investigation of mathematical model of communication network with unsteady flow of requests. Transp Telecommun 10(4):28–34

Chapter 35
A Comparison of NMRS with Other Market Power Indices in Deregulated Electricity Market

S. Prabhakar Karthikeyan, I. Jacob Raglend, D.P. Kothari, Sarat Kumar Sahoo and K. Sathish Kumar

Abstract Assessment of market power becomes an important issue when more than one firm/supplier and buyer enters the market. As the electricity market has moved from the conventional bundled system to the deregulated environment, the market power assessment has become vital. Most of the researchers have used Herfindahl-Hirschman Index (HHI) and must-run share (MRS) as an index in their literatures for determining the market power of a generation company. In this paper, a comparative study is made on the market power indices such as HHI, MRS with nodal must-run share (NMRS) under various system constraints such as generation and transmission line outages, and loading conditions. To illustrate and for better understanding, IEEE 24-bus reliability test system (RTS) is taken for which the indices are calculated and compared. It is found that NMRS reflects the complete information about the generation company's market power on any load at any bus. All simulations are carried out using MATLAB (R2009a) version.

Keywords Market power · Index · HHI · MRS · NMRS · Generation company

S. Prabhakar Karthikeyan (✉) · S.K. Sahoo · K. Sathish Kumar
School of Electrical Engineering, VIT University, Vellore 632014, Tamil Nadu, India
e-mail: spk25in@yahoo.co.in

I. Jacob Raglend
Department of Electrical and Electronics Engineering, NI University, Kumaracoil 629180, Tamil Nadu, India

D.P. Kothari
MVSR Engineering College, Hyderabad 501510, India

V. Vijay et al. (eds.), *Systems Thinking Approach for Social Problems*, Lecture Notes in Electrical Engineering 327,
DOI 10.1007/978-81-322-2141-8_35

35.1 Introduction

35.1.1 Market Power Indices

The following indices are made available in this paper to relate with the results simulated below. As the aim of this paper is to show the significance of nodal must-run share (NMRS) index, the comparison is restricted with the other two indices [Herfindahl-Hirschman Index (HHI) and must-run share (MRS)] which are commonly used in the literature.

35.1.1.1 Herfindahl-Hirschman Index (HHI)

It is a common measurement of market concentration that reflects the number of participants and the inequality of their market shares [1]. HHI is the weighted sum of market shares of all participants in the market. It is defined as the sum of the squares of market shares of all participants. It is given by the Eq. (35.1).

$$\mathrm{HHI} = \sum_{i=1}^{N} S_i^2 \tag{35.1}$$

where N is the number of participants and S_i is the ith participant's market share in per unit or in percentage. This index is very simple where it cannot reveal the complete information when the system is subjected to variation in load and with transmission constraints.

35.1.1.2 Must-Run Generation (MRG)

Two market power indices which include the impact of load variation, transmission constraints are MRS and NMRS [2]. Before understanding MRS, the term must-run generation (MRG) has to be defined. It is the minimum capacity which is provided by the generator to the total load considering generation and transmission constraints. It is determined by the following linear optimization problem.

$$\mathrm{Min}\ \mathrm{Pg}_k \tag{35.2}$$

Such that

$$e^T = (\mathrm{Pg} - \mathrm{Pd}) = 0 \tag{35.3}$$

$$0 \le \mathrm{Pg} \le \mathrm{Pg}_{\max} \tag{35.4}$$

$$-\mathrm{Pl}_{\max} \leq F(\mathrm{Pg} - \mathrm{Pd}) \leq \mathrm{Pl}_{\max} \quad (35.5)$$

where e is a vector with all ones, Pg is the power dispatch vector, Pd is the demand vector, $\mathrm{Pl}_{\max}$ is the line limit vector, and F is distribution factors matrix [3]. Equation (35.3) denotes power balance equation, and Eqs. (35.4) and (35.5) denote the generator output limits and transmission line limits, respectively.

35.1.1.3 Must-Run Share (MRS)

This index helps in representing the changes in market power with load level. The must-run share MRS_k of generator k in a power market is given by Eq. (35.6).

$$\mathrm{MRS}_k = \mathrm{Pg}_k^{\mathrm{must}}/\mathrm{Pd} \quad (35.6)$$

where Pd is the total demand in the power market. It is to note that when MRS is greater than zero for a generator, then it possesses market power.

35.1.1.4 Nodal Must-Run Share (NMRS)

Here, the MRS is applied to each load bus so that the effect of geographical difference of market power is included. $\mathrm{NMRS}_{k,i}$, which represents the minimum capacity that must be provided by the must-run generator k to supply a given load at node i, is given by Eq. (35.7).

$$\mathrm{NMRS}_{k,i} = \mathrm{Pg}_{k,i}^{\mathrm{must}}/\mathrm{Pd}_i \quad i = 1, 2 \ldots N \quad (35.7)$$

The NMRS is determined from the following expression

$$\mathrm{NMRS}_{k,i} = \frac{\mathrm{Pg}_{k,i}^{\mathrm{must}}}{\mathrm{Pd}_i} = \frac{[M^{-1}]_{ik}\mathrm{Pg}_k^{\mathrm{must}}}{\sum_{j \in N} [M^{-1}]_{ij} Pg_j} \quad (35.8)$$

where $\mathrm{Pg}_{k,i}^{\mathrm{must}}$ is the contribution of the must-run generator k to Pd_i, N is the number of buses, Pd_i is the load at bus i, j is the bus which is directly connected to bus i through transmission lines and $[M]$ is the distribution matrix. This matrix is used to show how the power supplied at a node is contributed from all the generators in a system [2]. This index is used to understand the impact of FACTS devices in market power both with and without FACTS device [4]. A complete survey on various indices which are used to measure the market power is available in [5].

35.2 Algorithm Used

Step 1: Form Y_{bus} matrix using line data.
Step 2: Find the generation schedules for all generators using lambda iteration method.
Step 3: Check for limits and recalculate generator schedules (for any violation in the generation limits, fixing the values equal to the limit). Go to Step 2.
Step 4: Find the slack bus. Slack bus is the bus with max generation.
Step 5: Run DC load flow and calculate line flows and bus data.
Step 6: Calculate Pg_{must} values for must-run share.
Step 7: Calculate must-run share.
Step 8: Calculate distribution matrix M_{ij}.
Step 9: Calculate Pg_{must} for NMRS
Step 10: Calculate NMRS.

35.3 Simulation and Results

35.3.1 Sample System 1

A three-bus sample system (as shown in Fig. 35.1) with two generators and one load is chosen to illustrate the measure of market power. The load is assumed to be inelastic and the analysis is done for the given transaction.

All the line reactances are taken to be 0.2j pu. The cost functions of both the generators are taken to be $F(\mathrm{Pg}_i) = 0.02 * \mathrm{Pg}_i^2 + 2 * \mathrm{Pg}_i$ (where $c_i = 0$). The maximum and minimum generations are taken as 100 and 0 MW in order to simulate zero market power as a base case. In this system, both the generators are located at an equal distance from the load; hence, both have the same fighting chance of supplying the load (Fig. 35.2).

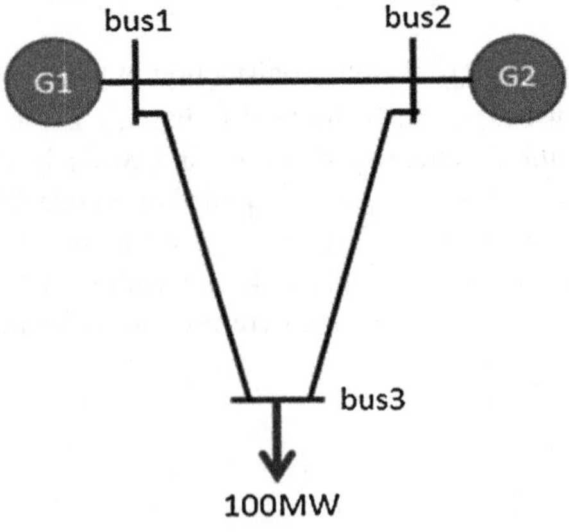

Fig. 35.1 Three bus sample system

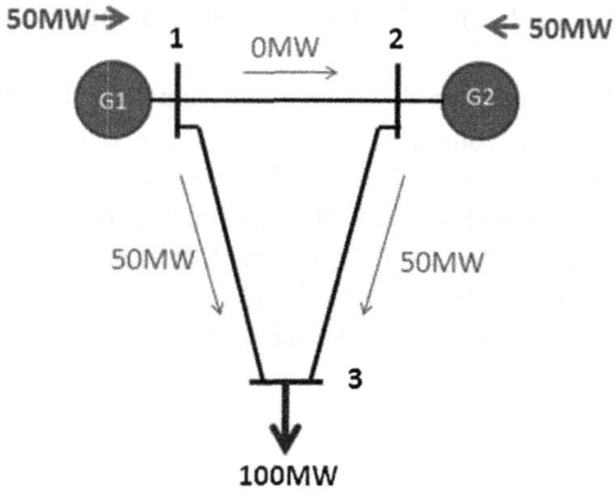

Fig. 35.2 Power flow for case 1

35.3.1.1 Case 1: Without Any Constraints

Since the maximum generation of both the generators is 100 MW, one generator can run without the other; hence, the MRG values of both the generators are zero. Therefore, the MRS and the NMRS values are also zero. No generator enjoys any market power. Hence, there is perfect competition or the market is said to be in equilibrium. HHI values indicate some market power. NMRS and MRS can reflect the exact market position (Tables 35.1 and 35.2).

When a transmission limit of 45 MW is imposed on the line connecting buses 1 and 3, G2 will have an advantage than G1. This is because most of the power from G1 will flow through the line, and hence, its generation has to be reduced.

Table 35.1 Generator data for case 1

Gen. No.	Price ($/MW)	MW prod	Total cost	Income	Profit
G1	4	50	150	200	50
G2	4	50	150	200	50

Table 35.2 Market data for case 1

Gen. No.	P_{max}	Price ($/MW)	MW prod	MS	HHI	MRG	MRS	NMRS
G1	100	4	50	0.5	0.25	0	0	0
G2	100	4	50	0.5	0.25	0	0	0

35.3.1.2 Case 2: With Line Constraints

MRG of both the generators cannot be 100 MW as this would violate the transmission constraint. We have to use linear programming to find the value of the must-run generation of all the generators. The MRG of the second generator has been found to be 65 MW and that of the first was found to be 0 MW as G2 can run freely without G1. Hence, the MRS and the NMRS values of G2 have gone up from 0 to 0.65. Since there is only one load, the MRS and NMRS values are the same (Fig. 35.3; Tables 35.3 and 35.4).

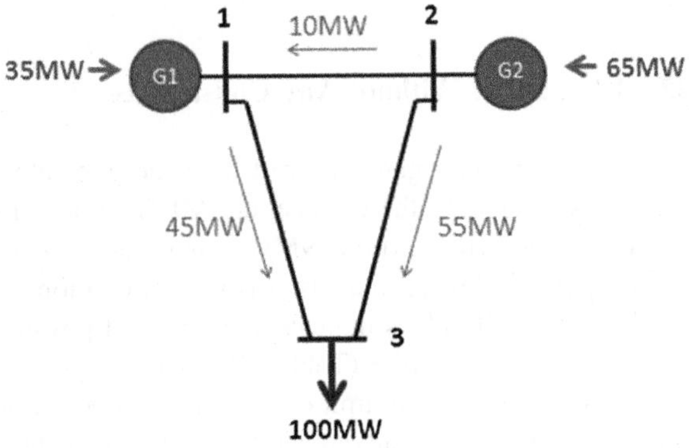

Fig. 35.3 Power flow for case 2

Table 35.3 Generator data for case 2

Gen. No.	Price ($/MW)	MW prod	Total cost	Income	Profit
G1	3.4	35	94.5	119	24.5
G2	4.6	65	214.5	299	84.5

Table 35.4 Market data for case 2

Gen. No.	Price ($/MW)	MW prod	MS	HHI	MRG	MRS	NMRS
G1	3.4	35	0.35	0.1225	0	0	0
G2	4.6	65	0.65	0.4224	65	0.65	0.65

35.3.2 Sample System 2

35.3.2.1 Case 1: Without Any Constraints

Table 35.5 shows the cost coefficients 'a' and 'b' (with $c = 0$) of the generators present at various buses in 24-bus RTS. Every generator in the system is assumed to be a generation company so that the companies' market power is calculated using the indices discussed above. Generator at bus number 14 is not included in market power calculation as it compensates only reactive power for the system.

Table 35.6 shows the active power limits (minimum and maximum) in MW. It is to note that limits are fixed in such a way that the schedules of the generators are well within the limits.

Generator schedules are determined using conventional lambda iterative approach without considering losses and the schedules are shown in Table 35.7. These schedules are calculated for the following load available at the respective buses as shown in Table 35.8, i.e., Total load = Total generation = 2850 MW.

Table 35.5 Cost coefficient of the generators

	Bus number									
	1	2	7	13	15	16	18	21	22	23
a	0.0375	0.0375	0.0325	0.0334	0.035	0.035	0.0325	0.0345	0.0325	0.0353
b	5	3.34	4.23	4.25	5.3	4.5	4.63	2.43	1.22	1.24

Table 35.6 Generator limits in MW

	Bus number									
	1	2	7	13	15	16	18	21	22	23
P_{max}	250	275	300	300	275	275	300	325	350	325
P_{min}	100	100	100	100	100	100	100	100	100	100

Table 35.7 Generator schedules for case 1

Bus No.	Pg (MW)
1	242.0478
2	264.1811
7	291.132
13	297.9578
15	255.0512
16	266.4797
18	284.9782
21	300.3418
22	337.4397
23	310.3907

Table 35.8 Load data

Bus No.	Pd (MW)
1	108
2	97
3	180
4	74
5	71
6	136
7	125
8	171
9	175
10	195
13	265
14	194
15	317
16	100
18	333
19	181
20	128

Table 35.9 gives the detail of Pg_{must} in MW of each generator to the loads. For example, generator at bus 1 exhibits its market power (in terms of Pg_{must}) on the load at bus number 1, 3, 5, 6, 8, 9, and 10, i.e., generator at bus 1 has to generate a minimum power of 54.273939, 14.04851, 35.680089, 4.03977, 0.109826, 2.104645, and 14.74322 MW to meet the loads at bus number 1, 3, 5, 6, 8, 9, and 10, respectively. In other words, generator at bus 1 has the above shares (in MW) on the said loads.

In Table 35.10, first row represents the bus number where loads are present and the first column represents the bus number where generators are present so that the Table 35.10 gives a complete matrix of NMRS of each generator to each load. From Table 35.10, it is inferred that generator 1 shows its nodal must-run share of 0.502 on load 1. At the same time, generator 1 does not have any share on load at bus number 7, 13, 14, 15, 16, 18, 19, and 20.

Table 35.11 gives the details of other indices for the same condition. These indices simply reflect the generator's share. For example, generator at bus No. 1 has its Pg_{must} of 125 MW to meet the system load of 2,850 MW i.e. MRS = 0.0439. It is to note that the Pg_{must} of a generator used in NMRS is different from Pg_{must} used in MRS. Similarly, HHI reflects only the detail such as number of participants and the difference in market shares.

Table 35.9 Pg_{must} share of each generator to each load

Generator bus No.	Load bus No.	Pg_{must} in MW
1	1	54.273939
1	3	14.04851
1	5	35.680089
1	6	4.03977
1	8	0.109826
1	9	2.104645
1	10	14.74322
2	1	1.649389
2	2	55.075856
2	3	0.426935
2	4	42.016632
2	5	1.084321
2	6	47.005046
2	8	0.113646
2	9	2.177846
2	10	0.450328
7	6	0.551258
7	7	75.137729
7	8	97.299184
7	10	2.011828
13	6	7.692864
13	8	1.296162
13	9	24.838999
13	10	28.075256
13	13	113.09672
15	3	53.862091
15	6	0.58855
15	8	0.517979
15	9	9.926297
15	10	2.147925
15	14	10.978435
15	15	112.29771
15	16	5.658987
16	6	2.513099
16	8	0.415474
16	9	7.961941
16	10	9.171603
16	14	47.068518
16	16	24.262123
18	6	0.823738

(continued)

Table 35.9 (continued)

Generator bus No.	Load bus No.	Pg_{must} in MW
18	8	0.136183
18	9	2.609748
18	10	3.006249
18	14	15.428019
18	16	7.952587
21	3	24.89255
21	6	0.416399
21	8	0.263259
21	9	5.044953
21	10	1.519657
21	14	7.778208
21	15	51.898774
21	16	4.009385
22	3	15.389082
22	6	1.955543
22	8	0.44349
22	9	8.498825
22	10	7.136791
22	14	36.613134
22	15	32.08488
22	16	18.872749
23	6	13.539006
23	8	2.32687
23	9	44.590967
23	10	49.410868
23	13	46.67647

35.3.2.2 Case 2: With Generator Constraint (P_{max})

From case 1, the active power maximum limit of generator at bus No. 7 is limited to 250 MW, and then, the generator schedules are reallocated. The schedules are shown in Table 35.12 (generator constraint of 250 MW is shown in bold).

Table 35.10 NMRS of generators (column) to the loads (row) at various buses for case 1

	1	2	3	4	5	6	7	8	9	10	13	14	15	16	18	19	20
1	0.502	0	0.078	0	0.502	0.029	0	0.00064	0.012	0.075	0	0	0	0	0	0	0
2	0.015	0.567	0.002	0.567	0.015	0.345	0	0.00066	0.012	0.002	0	0	0	0	0	0	0
7	0	0	0	0	0	0.004	0.601	0.569	0	0.010	0	0	0	0	0	0	0
13	0	0	0	0	0	0.056	0	0.0076	0.141	0.144	0.214	0.136	0.426	0	0	0	0
15	0	0	0.299	0	0	0.004	0	0.003	0.056	0.011	0.028	0	0	0.056	0.354	0.056	0
16	0	0	0	0	0	0.018	0	0.0024	0.045	0.047	0.120	0	0	0.242	0	0.242	0
18	0	0	0	0	0	0.006	0	0.00079	0.014	0.015	0.039	0	0	0.079	0	0.079	0.194
21	0	0	0.138	0	0	0.003	0	0.0015	0.028	0.007	0.019	0	0	0.040	0.163	0.040	0.034
22	0	0	0.085	0	0	0.014	0	0.0026	0.048	0.036	0.093	0	0	0.188	0.101	0.188	0.421
23	0	0	0	0	0	0.099	0	0.0136	0.254	0.253	0.088	0.495	0.176	0	0	0	0

Table 35.11 HHI and MRS of generators for case 1

Bus No.	Schedules in MW	Market share	HHI	MRS
1	242.0478	0.0849	0.0072	0.0439
2	264.1811	0.0927	0.0086	0.0526
7	291.132	0.1022	0.0104	0.0614
13	297.9578	0.1045	0.0109	0.0614
15	255.0512	0.0895	0.008	0.0526
16	266.4797	0.0935	0.0087	0.0526
18	284.9782	0.1	0.01	0.0614
21	300.3418	0.1054	0.0111	0.0702
22	337.4397	0.1184	0.014	0.0789
23	310.3907	0.1089	0.0119	0.0702

Table 35.12 Generator schedule for case 2

Bus No.	Pg (MW)
1	246.5931
2	268.7264
7	**250**
13	300
15	259.9212
16	271.3497
18	290.2228
21	305.2824
22	342.6843
23	315.2193

Similar procedure is followed in calculating Pg_{must} from which NMRS is calculated. Tables 35.13 and 35.14 show the NMRS and other indices, respectively, for case 2.

In case 1, it is inferred that the market power of the generator at bus No. 7 was on the load at bus No. 6, 7, 8, and 10. Due to the constraint on P_{max} limit of generator 7, its market power on load at bus No. 6 and 10 gets removed.

Similar to Table 35.11, Table 35.9 gives the generator share in meeting the load. This is applicable to both HHI and MRS. For example, generator at bus No. 1's Pg_{must} has increased from 125 MW (case 1) to 175 MW in meeting the system load. This is due to the reduction in maximum limit of generator at bus No. 7. So the MRS of generator at bus No. 1 has increased from 0.0439 to 0.0614.

Table 35.13 NMRS of generators (column) to the loads (row) at various buses for case 2

	1	2	3	4	5	6	7	8	9	10	13	14	15	16	18	19	20
1	0.693	0	0.102	0	0.693	0.041	0	0.013	0.017	0.11	0	0	0	0	0	0	0
2	0.017	0.744	0.002	0.744	0.017	0.468	0	0.004	0.024	0.002	0	0	0	0	0	0	0
7	0	0	0	0	0	0	0.7	0.511	0	0	0	0	0	0	0	0	0
13	0	0	0	0	0	0.067	0	0.046	0.170	0.180	0.259	0.173	0.538	0	0	0	0
15	0	0	0.394	0	0	0.005	0	0.015	0.081	0.015	0.039	0	0	0.075	0.462	0.075	0
16	0	0	0	0	0	0.024	0	0.016	0.060	0.064	0.162	0	0	0.314	0	0.314	0
18	0	0	0	0	0	0.007	0	0.005	0.019	0.021	0.053	0	0	0.103	0	0.103	0.251
21	0	0	0.171	0	0	0.003	0	0.007	0.038	0.010	0.026	0	0	0.050	0.201	0.050	0.042
22	0	0	0.103	0	0	0.017	0	0.015	0.060	0.046	0.116	0	0	0.225	0.121	0.225	0.500
23	0	0	0	0	0	0.116	0	0.0821	0.300	0.313	0.107	0.609	0.223	0	0	0	0

Table 35.14 HHI and MRS of each generator for case 2

Bus No.	Schedules in MW	Market share	HHI	MRS
1	246.5931	0.0865	0.0075	0.0614
2	268.7264	0.0943	0.0089	0.0702
7	250	0.0877	0.0077	0.0614
13	300	0.1053	0.0111	0.0789
15	259.9212	0.0912	0.0083	0.0702
16	271.3497	0.0952	0.0091	0.0702
18	290.2228	0.1018	0.0104	0.0789
21	305.2824	0.1071	0.0115	0.0877
22	342.6843	0.1202	0.0145	0.0965
23	315.2193	0.1106	0.0122	0.0877

Table 35.15 Generator schedules for case 3

Bus No.	Pg (MW)
1	226.5898
2	250.8915
7	297.5756
13	300
15	257.9139
16	269.8377
18	288.3193
21	303.4127
22	340.7659
23	314.6912

35.3.2.3 Case 3: With Line Constraint

From case 1, the line between bus No. 1 and 5 is limited with a constraint of 100 MW, where all other limits remain same as in case 1. Then, the generator schedules are as shown in Table 35.15.

In Table 35.16, the line between bus No. 1 and 5 is restricted by 100 MW (as shown in bold) from which Pg_{must} and NMRS are calculated. NMRS of each generator to the entire load is shown in Table 35.17. From Table 35.17, for case 3, it is found that the generator 1 extends its market power to the load at bus No. 4.

It is also to note that in all the above three cases, load at bus No. 7 (i.e., 125 MW) is fully dominated by the generator at bus No. 7. The change in NMRS is only due to the change in Pg_{must}, i.e. 75.137729 MW, 87.5 MW, and 73.510722 MW for case 1, case 2, and case 3, respectively.

Table 35.16 Line flow in MW for case 3

From bus	To bus	Flow in MW
1	2	−6.3252
1	3	24.9051
1	**5**	**100.0000**
2	4	70.7352
2	6	76.831
3	9	27.7336
3	24	−182.8285
4	9	−3.2648
5	10	29.0098
6	10	−59.169
7	8	172.5756
8	9	−6.7831
8	10	8.3588
9	11	−72.414
9	12	−84.9002
10	11	−102.1571
10	12	−114.6433
11	13	−86.1132
11	14	−88.4579
12	13	−64.2624
12	23	−135.2811
13	23	−115.3756
14	16	−282.4579
15	16	100.1147
15	21	−171.0146
15	21	−171.0146
15	24	182.8285
16	17	−257.4711
16	19	244.9655
17	18	−103.947
17	22	−153.5241
18	21	−74.3139
18	21	−74.3139
19	20	31.9828
19	20	31.9828
20	23	−32.0172
20	23	−32.0172
21	22	−187.2443

Table 35.17 NMRS of generators (column) to the loads (row) at various buses for case 3

	1	2	3	4	5	6	7	8	9	10	13	14	15	16	18	19	20
1	0.536	0	0.064	4.25E−04	0.536	0.026	0	3.65E−04	0.009	0.061	0	0	0	0	0	0	0
2	0.016	0.597	0.001	0.571	0.016	0.338	0	1.10E−05	2.92E−04	0.001	0	0	0	0	0	0	0
7	0	0	0	0	0	0.008	0.588	0.565	0	0.018	0	0	0	0	0	0	0
13	0	0	0	0.006	0	0.063	0	0.005	0.143	0.144	0.207	0.135	0.421	0	0	0	0
15	0	0	0.307	0.002	0	0.005	0	0.002	0.057	0.011	0.028	0	0	0.055	0.349	0.055	0
16	0	0	0	0.002	0	0.021	0	0.001	0.047	0.048	0.121	0	0	0.239	0	0.239	0
18	0	0	0	6.99E−04	0	0.007	0	6.00E−04	0.015	0.016	0.040	0	0	0.08	0	0.08	0.194
21	0	0	0.143	0.001	0	0.003	0	0.001	0.029	0.008	0.020	0	0	0.039	0.162	0.039	0.033
22	0	0	0.088	0.002	0	0.016	0	0.001	0.050	0.038	0.094	0	0	0.186	0.100	0.186	0.414
23	0	0	0	0.011	0	0.111	0	0.009	0.257	0.255	0.087	0.487	0.176	0	0	0	0

Table 35.18 Generator schedules for case 4

Bus No.	Pg (MW)
1	242.0478
2	264.1811
7	291.132
13	297.9578
15	255.0512
16	266.4797
18	284.9782
21	300.3418
22	337.4397
23	310.3907

35.3.2.4 Case 4: Line Outage

From case 1, the line between bus No. 1 and 5 is removed and the impact on NMRS is analyzed. In this case, a line to be removed is randomly selected. Generator schedules are calculated and shown in Table 35.18.

From Tables 35.17 and 35.19, the pattern of impact of NMRS of the generators to the load at bus No. 8, 9, 10, 13, 14, 15, 16, 18, 19, and 20 is more or less same, while for the other loads, the generators have considerable impact which is reflected on the NMRS.

From Tables 35.11, 35.14, 35.20, and 35.21, it is inferred that HHI can not able to reflect the transmission constraint or transmission outages and similarly MRS simply reflects the variation in market power with the system load. As MRS depends only on Pg_{max}, the value of MRS remains unchanged for case 1, case 3, and case 4.

Table 35.19 NMRS of generators (column) to the loads (row) at various buses for case 4

	1	2	3	4	5	6	7	8	9	10	13	14	15	16	18	19	20
1	0.516	0.108	0.148	0.108	3.38E −04	0.100	0	0.005	0.053	3.38E-04	0	0	0	0	0	0	0
2	0	0.449	0	0.449	5.30E −04	0.418	0	0.009	0.083	5.30E −04	0	0	0	0	0	0	0
7	0	0	0	0	0.030	0.002	0.601	0.534	0	0.030	0	0	0	0	0	0	0
13	0	0	0	0	0.159	0.010	0	0.010	0.098	0.159	0.204	0.135	0.423	0	0	0	0
15	0	0	0.253	0	0.015	0.001	0	0.007	0.065	0.015	0.032	0	0	0.062	0.355	0.062	0
16	0	0	0	0	0.055	0.003	0	0.003	0.032	0.055	0.122	0	0	0.236	0	0.236	0
18	0	0	0	0	0.018	0.001	0	0.001	0.010	0.018	0.041	0	0	0.079	0	0.079	0.196
21	0	0	0.116	0	0.010	6.87E −04	0	0.003	0.031	0.010	0.022	0	0	0.042	0.162	0.042	0.035
22	0	0	0.071	0	0.043	0.002	0	0.004	0.041	0.043	0.096	0	0	0.186	0.100	0.186	0.418
23	0	0	0	0	0.284	0.019	0	0.020	0.184	0.284	0.087	0.495	0.180	0	0	0	0

Table 35.20 HHI and MRS of each generator for case 3

Bus No.	Schedule in MW	Market share	HHI	MRS
1	226.5898	0.0795	0.0063	0.0439
2	250.8915	0.088	0.0077	0.0526
7	297.5756	0.1044	0.0109	0.0614
13	300	0.1053	0.0111	0.0614
15	257.9139	0.0905	0.0082	0.0526
16	269.8377	0.0947	0.009	0.0526
18	288.3193	0.1012	0.0102	0.0614
21	303.4127	0.1065	0.0113	0.0702
22	340.7659	0.1196	0.0143	0.0789
23	314.6912	0.1104	0.0122	0.0702

Table 35.21 HHI and MRS of each generator for case 4

Bus No.	Schedule in MW	Market share	HHI	MRS
1	242.0478	0.0849	0.0072	0.0439
2	264.1811	0.0927	0.0086	0.0526
7	291.132	0.1022	0.0104	0.0614
13	297.9578	0.1045	0.0109	0.0614
15	255.0512	0.0895	0.008	0.0526
16	266.4797	0.0935	0.0087	0.0526
18	284.9782	0.1	0.01	0.0614
21	300.3418	0.1054	0.0111	0.0702
22	337.4397	0.1184	0.014	0.0789
23	310.3907	0.1089	0.0119	0.0702

35.4 Conclusions

In this paper, a 3-bus sample is taken to illustrate the significance of nodal must-run share index over other indices. A 24-bus RTS is also taken for study to show how NMRS reflects exactly the system constraints on the generators' market power.

For 24-bus RTS, in case 1, all the three indices are calculated when the system is without any constraints. In Table 35.5, NMRS gives a clear view on how the generators are contributing its share (Pg_{must}) to each load. It also shows how other indices fail to reflect this necessary information. These indices simply reflect the market share. Similar analysis is made under different system conditions such as system under generator constraint, line constraint, and line outage. This analysis can also be extended to show the impact of change in load and generator outages.

Acknowledgments The authors thank the management of VIT University, Vellore, Tamil Nadu, India, for their continuous encouragement and support rendered throughout the period of research.

References

1. Tirole J (1988) The theory of industrial organisation. MIT, Cambridge
2. Wang P, Xiao Y, Ding Y (2004) Nodal market power assessment in electricity markets. IEEE Trans Power Syst 19(3):1373–1379
3. Wood AJ, Wollenberg BF (1996) Power generation, operation and control. Wiley, New York
4. Prabhakar Karthikeyan S, Jacob Raglend I, Kothari DP (2013) Impact of FACTS devices on exercising market power in deregulated electricity market, frontiers in energy, issue 3. Springer, New York, pp 1–8
5. Prabhakar Karthikeyan S, Jacob Raglend I, Kothari DP (2013) A review on market power in deregulated electricity market. Int J Electr Power Energy Syst 48:139–147

Author Index

V. Vijay et al. (eds.), *Systems Thinking Approach for Social Problems*,
Lecture Notes in Electrical Engineering 327,
DOI 10.1007/978-81-322-2141-8

Printed in the United States
By Bookmasters